Self-Assembly Processes
in Materials

MATERIALS RESEARCH SOCIETY
SYMPOSIUM PROCEEDINGS VOLUME 707

Self-Assembly Processes in Materials

Symposium held November 26–30, 2001, Boston, Massachusetts, U.S.A.

EDITOR:

Steven C. Moss

The Aerospace Corporation
Los Angeles, California, U.S.A.

ORGANIZERS:

Bruce M. Clemens

Stanford University
Stanford, California, U.S.A.

Jerrold A. Floro

Sandia National Laboratories
Albuquerque, New Mexico, U.S.A.

Julie A. Kornfield

California Institute of Technology
Pasadena, California, U.S.A.

Yuri Suzuki

Cornell University
Ithaca, New York, U.S.A.

Materials Research Society
Warrendale, Pennsylvania

CAMBRIDGE UNIVERSITY PRESS
Cambridge, New York, Melbourne, Madrid, Cape Town,
Singapore, São Paulo, Delhi, Mexico City

Cambridge University Press
32 Avenue of the Americas, New York NY 10013-2473, USA

Published in the United States of America by Cambridge University Press, New York

www.cambridge.org
Information on this title: www.cambridge.org/9781107411982

Materials Research Society
506 Keystone Drive, Warrendale, PA 15086
http://www.mrs.org

First published 2002
First paperback edition 2013

Single article reprints from this publication are available through
University Microfilms Inc., 300 North Zeeb Road, Ann Arbor, MI 48106

CODEN: MRSPDH

ISBN 978-1-107-41198-2 Paperback

CONTENTS

*Invited Paper

*Invited Paper

*Invited Paper

OPTICAL MATERIALS

PHOTONIC CRYSTALS

MODELING/SIMULATION

PREFACE

Symposium AA, "Self-Assembly Processes in Materials," was held November 26–30 at the 2001 MRS Fall Meeting in Boston, Massachusetts, discussing ongoing advances in self-assembly that touch essentially every area of materials science. 100 papers were presented in ten sessions including one poster session. Papers included in this volume address topics including quantum dots, nanocrystals, nanowires and chains, organic materials, magnetic materials, biological materials, photonic crystals, and optical materials.

This symposium was conducted as a "virtual symposium" unifying sessions from ten diverse symposia, including Symposia K, M, N, V, W, Y, BB, DD, and HH. One of the presentations from Symposium X was also included. This virtual symposium was organized by the Meeting Chairs with the support of the Symposium Organizers from the nine symposia listed below. The Meeting Chairs gratefully acknowledge the assistance of these Symposium Organizers:
Symposium K - "Microphotonics—Materials, Physics, and Applications" - Pierre Wiltzius, Bell Labs, Lucent Technologies; Alfons van Blaaderen, Utrecht Univ; and Claude Weisbuch, Ecole Polytechnique;
Symposium M - "Surface Science and Thin-Film Growth in Electrolytes" - Dieter M. Kolb, Ulm Univ; Benjamin M. Ocko, Brookhaven National Laboratory; Peter C. Searson, Johns Hopkins Univ; and Karl Sieradzki, Arizona State Univ;
Symposium N - "Current Issues in Heteroepitaxial Growth—Stress Relaxation and Self Assembly" - Eric H. Chason, Brown Univ; Robert Hull, Univ of Virginia; Samuel D. Bader, Argonne National Lab; and Eric A. Stach, Lawrence Berkeley National Laboratory;
Symposium V - "Nanophase and Nanocomposite Materials IV" - Sridhar Komarneni, Pennsylvania State Univ; G.Q. (Max) Lu, Univ of Queensland; Jun-Ichi Matsushita, Tokai Univ; John C. Parker, Cirqon Technologies Corp; and Richard A. Vaia, Air Force Research Lab;
Symposium W - "Nanoparticulate Materials" - Rajiv Kumar Singh, Univ of Florida; Heinrich Hofmann, EPFL-Lausanne; Mamoun Muhammed, The Royal Inst of Technology (KTH); Richard Partch, Clarkson Univ; and Mamoru Senna, Keio Univ;
Symposium Y - "Nanopatterning—From Ultralarge-Scale Integration to Biotechnology" - Kenneth E. Gonsalves, Univ of North Carolina; Lhadi Merhari, CERAMEC R&D France; Elizabeth A. Dobisz, IBM Almaden Research Ctr; Marie Angelopoulos, IBM T.J. Watson Research Ctr; and Daniel J.C. Herr, SRC;
Symposium BB - "Organic Optoelectronic Materials, Processing, and Devices" - Zhenan Bao, Lucent Technologies; Vladimir Bulovic, MIT; Susan P. Ermer, Lockheed Martin Adv Tech Ctr; Alex K-Y. Jen, Univ of Washington; George G. Malliaras, Cornell Univ; and Michael D. McGehee, Stanford Univ;
Symposium DD - "Polymer Interfaces and Thin Films" - Alamgir Karim, NIST; Thomas P. Russell, Univ of Massachusetts-Amherst; Curtis W. Frank, Stanford Univ; and Paul F. Nealey, Univ of Wisconsin-Madison;
Symposium HH - "Bio-Inspired Materials—Moving Towards Complexity" - Bradley F. Chmelka, Univ of California-Santa Barbara; Konrad Knoll, BASF AG Ludwigshafen Germany; Joachim P. Spatz, Ulm Univ; and Ulrich B. Weisner, Cornell Univ.

Because of the extraordinary commitments of the Meeting Chairs, who served as Symposium Organizers for this symposium, an editor (Steven C. Moss, The Aerospace Corporation) was recruited to organize the proceedings volume.

Numerous other papers addressing self-assembly were presented in many of the other symposia. Papers on self-assembly from Symposia A, H, and I have also been included in this proceedings volume.

In addition to the Symposium Organizers, the session chairs made significant contributions. Session chairs included: Pierre Wiltzius, Francis M. Ross, Michael J. Aziz, Masaaki Oda,

Mamoru Senna, Peter C. Searson, J. Woods Halley, Willem L. Vos, Marie Angelopoulos, Elizabeth A. Dobisz, Spiros H. Anastasiadis, Georg Krausch, Michael F. Rubner, and Paula T. Hammond. The Meeting Chairs and Symposium Organizers gratefully acknowledge their contribution.

Symposium support for Symposium A was provided by MMR Technologies, Inc. and Voltaix, Inc. Symposium support for Symposium H was provided by the Air Force Office of Scientific Research and the Office of Naval Research. Symposium support for Symposium I was provided by the Air Force Office of Scientific Research, the Fritz-Haber-Institut der Max Planck Gesellschaft, Nichia Corporation, SONY Corporation, and Xerox Palo Alto Research Center. Symposium support for Symposium K was provided by the Army Research Office. Symposium support for Symposium M was provided by the Air Force Office of Scientific Research. Symposium support for Symposium N was provided by Blake Industries, Inc. Symposium support for Symposium V was provided by the Air Force Research Laboratory and the Office of Naval Research. Symposium support for Symposium W was provided by EPFL Switzerland, KTH Sweden, and the University of Florida. Symposium support for Symposium Y was provided by CERAMEC R&D France, Motorola France, Nanoresist Technologies, Inc., the National Science Foundation, and The Whitaker Foundation. Symposium support for Symposium BB was provided by Lumera Corporation, MMR Technologies, Inc., and the Universal Display Corporation. Symposium support for Symposium DD was provided by the National Institute of Standards and Technology, the National Science Foundation, and the Stanford University NSF-MRSEC. Symposium support for Symposium HH was provided by BASF AG Ludwigshafen and Philip Morris USA. The Meeting Chairs, the Symposium Organizers, and the Materials Research Society gratefully acknowledge their support.

Steven C. Moss

September 2002

MATERIALS RESEARCH SOCIETY SYMPOSIUM PROCEEDINGS

Volume 664— Amorphous and Heterogeneous Silicon-Based Films—2001, M. Stutzmann, J.B. Boyce, J.D. Cohen, R.W. Collins, J. Hanna, 2001, ISBN: 1-55899-600-1

Volume 665— Electronic, Optical and Optoelectronic Polymers and Oligomers, G.E. Jabbour, B. Meijer, N.S. Sariciftci, T.M. Swager, 2002, ISBN: 1-55899-601-X

Volume 666— Transport and Microstructural Phenomena in Oxide Electronics, D.S. Ginley, M.E. Hawley, D.C. Paine, D.H. Blank, S.K. Streiffer, 2001, ISBN: 1-55899-602-8

Volume 667— Luminescence and Luminescent Materials, K.C. Mishra, J. McKittrick, B. DiBartolo, A. Srivastava, P.C. Schmidt, 2001, ISBN: 1-55899-603-6

Volume 668— II-VI Compound Semiconductor Photovoltaic Materials, R. Noufi, R.W. Birkmire, D. Lincot, H.W. Schock, 2001, ISBN: 1-55899-604-4

Volume 669— Si Front-End Processing—Physics and Technology of Dopant-Defect Interactions III, M.A. Foad, J. Matsuo, P. Stolk, M.D. Giles, K.S. Jones, 2001, ISBN: 1-55899-605-2

Volume 670— Gate Stack and Silicide Issues in Silicon Processing II, S.A. Campbell, C.C. Hobbs, L. Clevenger, P. Griffin, 2002, ISBN: 1-55899-606-0

Volume 671— Chemical-Mechanical Polishing 2001—Advances and Future Challenges, S.V. Babu, K.C. Cadien, J.G. Ryan, H. Yano, 2001, ISBN: 1-55899-607-9

Volume 672— Mechanisms of Surface and Microstrucure Evolution in Deposited Films and Film Structures, J. Sanchez, Jr., J.G. Amar, R. Murty, G. Gilmer, 2001, ISBN: 1-55899-608-7

Volume 673— Dislocations and Deformation Mechanisms in Thin Films and Small Structures, O. Kraft, K. Schwarz, S.P. Baker, B. Freund, R. Hull, 2001, ISBN: 1-55899-609-5

Volume 674— Applications of Ferromagnetic and Optical Materials, Storage and Magnetoelectronics, W.C. Black, H.J. Borg, K. Bussmann, L. Hesselink, S.A. Majetich, E.S. Murdock, B.J.H. Stadler, M. Vazquez, M. Wuttig, J.Q. Xiao, 2001, ISBN: 1-55899-610-9

Volume 675— Nanotubes, Fullerenes, Nanostructured and Disordered Carbon, J. Robertson, T.A. Friedmann, D.B. Geohegan, D.E. Luzzi, R.S. Ruoff, 2001, ISBN: 1-55899-611-7

Volume 676— Synthesis, Functional Properties and Applications of Nanostructures, H.W. Hahn, D.L. Feldheim, C.P. Kubiak, R. Tannenbaum, R.W. Siegel, 2002, ISBN: 1-55899-612-5

Volume 677— Advances in Materials Theory and Modeling—Bridging Over Multiple-Length and Time Scales, L. Colombo, V. Bulatov, F. Cleri, L. Lewis, N. Mousseau, 2001, ISBN: 1-55899-613-3

Volume 678— Applications of Synchrotron Radiation Techniques to Materials Science VI, P.G. Allen, S.M. Mini, D.L. Perry, S.R. Stock, 2001, ISBN: 1-55899-614-1

Volume 679E—Molecular and Biomolecular Electronics, A. Christou, E.A. Chandross, W.M. Tolles, S. Tolbert, 2001, ISBN: 1-55899-615-X

Volume 680E—Wide-Bandgap Electronics, T.E. Kazior, P. Parikh, C. Nguyen, E.T. Yu, 2001, ISBN: 1-55899-616-8

Volume 681E—Wafer Bonding and Thinning Techniques for Materials Integration, T.E. Haynes, U.M. Gösele, M. Nastasi, T. Yonehara, 2001, ISBN: 1-55899-617-6

Volume 682E—Microelectronics and Microsystems Packaging, J.C. Boudreaux, R.H. Dauskardt, H.R. Last, F.P. McCluskey, 2001, ISBN: 1-55899-618-4

Volume 683E—Material Instabilities and Patterning in Metals, H.M. Zbib, G.H. Campbell, M. Victoria, D.A. Hughes, L.E. Levine, 2001, ISBN: 1-55899-619-2

Volume 684E—Impacting Society Through Materials Science and Engineering Education, L. Broadbelt, K. Constant, S. Gleixner, 2001, ISBN: 1-55899-620-6

Volume 685E—Advanced Materials and Devices for Large-Area Electronics, J.S. Im, J.H. Werner, S. Uchikoga, T.E. Felter, T.T. Voutsas, H.J. Kim, 2001, ISBN: 1-55899-621-4

Volume 686— Materials Issues in Novel Si-Based Technology, W. En, E.C. Jones, J.C. Sturm, S. Tiwari, M. Hirose, M. Chan, 2002, ISBN: 1-55899-622-2

Volume 687— Materials Science of Microelectromechanical Systems (MEMS) Devices IV, A.A. Ayon, S.M. Spearing, T. Buchheit, H. Kahn, 2002, ISBN: 1-55899-623-0

Volume 688— Ferroelectric Thin Films X, S.R. Gilbert, Y. Miyasaka, D. Wouters, S. Trolier-McKinstry, S.K. Streiffer, 2002, ISBN: 1-55899-624-9

MATERIALS RESEARCH SOCIETY SYMPOSIUM PROCEEDINGS

Prior Materials Research Society Symposium Proceedings available by contacting Materials Research Society

Organic Materials

Mat. Res. Soc. Symp. Proc. Vol. 707 © 2002 Materials Research Society   AA9.7/Y8.7

Electric Field Induced Patterning of Polymer Films.

David G. Bucknall and G. Andrew D. Briggs
Department of Materials, Oxford University, Parks Road, Oxford, OX1 3PH, U.K.

ABSTRACT

By confining a polymer film between two electrodes one of which is solid but thin enough to be flexible, a characteristic lateral morphology is produced when a strong electric field is applied across the film. A simple model to describe the observed behaviour is presented which accounts for the length scales of the observed morphology. This model demonstrates that feature sizes ranging from microns to nanometers can be obtained through selective choice of key parameters.

INTRODUCTION

The ability to use polymers to produce patterned surfaces on a variety of length scales is of potential technological importance where cheap, large-area devices produced from non-lithographic methods are required.[1] A number of methods to achieve this have been investigated including exploitation of spinoidal dewetting of thin polymer films,[2] phase separation of polymer blends in thin films,[3-5] and diblock copolymers phase separation.[6] More recently the use of electric fields applied across unstable polymer films has produced remarkable periodic lateral structures.[7] An alternative strategy has been the use of mechanical confinement of freely standing polymer films to produce a lateral morphology which is driven by dispersion forces.[8]

This paper describes the preliminary results of the stability of a thin polymer film confined between two thin capping layers, when under the influence of an electric field. The presence of a capping layer on an unstable thin polymer film such as polystyrene (PS) is known to prevent film rupture and dewetting behaviour.[9] Freely standing trilayer thin films, consisting of a homopolymer layer which has been capped on both surfaces by thin solid layers, produces a characteristic in-plane structure (lateral morphology) when annealed at elevated temperatures.[8] Similar morphology has also been observed for thin films of blends of PS and poly(methyl methacrylate) (PMMA) capped by varying thicknesses of SiO_x.[5] In the present case, the morphology is not driven by either a phase separation process,[5] or an instability due to the attractive dispersion force between the capping layer-air surfaces,[8] but rather by the force between the electrodes caused by the applied electric field.

EXPERIMENTAL

The samples were prepared by spin coating zwitterionic telechelic polystyrene (ZT-PS) ($M_w = 6.4 \times 10^3$ g/mol, $M_w/M_n = 1.05$) from a dilute cyclohexanone solution (20 mg/ml) onto a silicon wafer previously sputter coated by an approximately 50 nm thick layer of aluminium. This gave a polymer layer of thickness (d) of 87 nm as determined using neutron reflection. The polymer film was vacuum dried overnight at 60 °C and then sputter coated with another layer of Al. The Al was sputter coated onto the ZT-PS surface using a plasma coater, with a base pressure of 10^{-6}

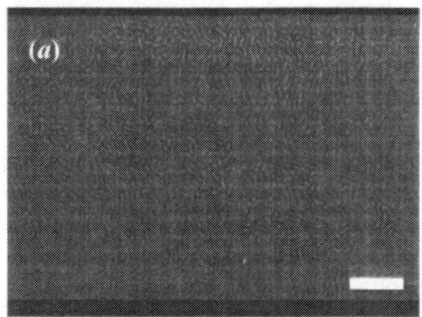 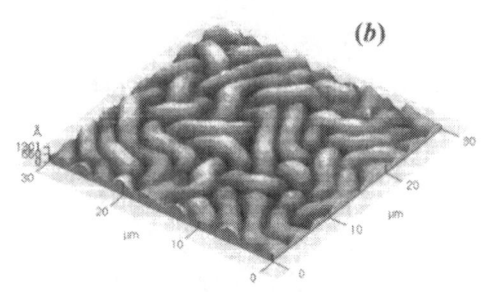

Figure 1: (*a*) Reflection mode optical micrograph of the lateral structure induced by an electric field applied across a ZT-PS film capped by an aluminium layer acting as an electrode. The white bar represents 25 mm, and the horizontal black bounding areas are not real but are simply created during file conversion. (*b*) AFM image taken using a tapping mode scan of the lateral morphology showing the 3-D nature of the film surface.

torr and an operating pressure of Ar of 2×10^{-3} torr during evaporation. The evaporation rate of 1.6 nm/s was relatively slow in order to limit heating of the ZT-PS layer. This gave an Al layer thickness l of 48 nm. These Al layers were homogeneous in thickness to within 2 nm and had a surface roughness of 1-2 nm, as determined by neutron reflectivity. Thin copper wires were attached to the base and the capping aluminium layers by initially applying silver paint (Electrolube Ltd), to make good electrical contact and subsequently coating the contacts once the paint had dried with an epoxy resin (Evo-stick epoxy, Evode Ltd), simply to ensure mechanical integrity. The wires were connected to a power supply and annealed under atmospheric conditions on a hot plate at 160 °C with a voltage of 5V. Applying such a voltage across this thin polymer film produces a large field strength of 5.7×10^{7} V/m. At this annealing temperature polymer degradation is expected to be insignificant, especially with the Al capping layer providing an oxygen barrier for the ZT-PS film. As evidenced by the change in optical quality, the lateral morphology appeared almost immediately, certainly within the first minute of annealing, and the structure did not visibly change over a subsequent period of annealing of one hour. Further experiments are required to determine whether changes are occurring during the annealing period on shorter length scales. After annealing the samples were rapidly quenched to room temperature and measured by both reflection mode optical microscopy and tapping mode AFM (Parks Scientific).

A typical example of the lateral morphology produced by applying an electric field across the ZT-PS layer is shown in Figure 1. The morphology is distinctive, with many domains which are randomly oriented but display a well-defined periodicity. From AFM imaging it can be shown that the ripples have an average peak-to-trough height of (64 ± 10) nm. The wavelength of the periodicity can be determined by performing a numerical two-dimensional fast Fourier transform (FFT) on grey scale images obtained from reflection mode optical microscopy. The resulting FFT image shown in Figure 2(*a*), displays a halo of intensity in reciprocal space. The position in the maximum of the halo in reciprocal space was determined by fitting the frequency-intensity

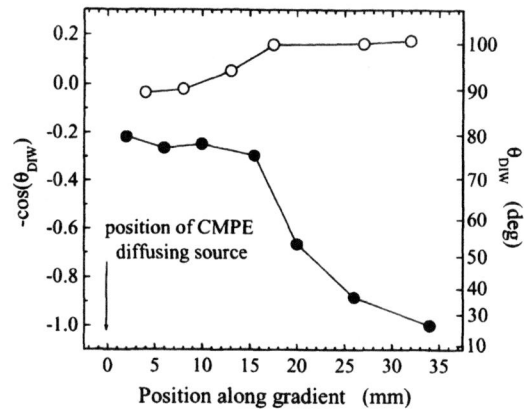

Fig. 1 Dependence of the DI water contact angle on the position along the gradient substrate measured after the CMPE-SAM formation (solid circles) and after backfilling with the OTS-SAM (open circles).

RESULTS

Figure 1 shows the variation of the contact angle of deionized water, θ_{DIW}, on CMPE-SAM covered substrate (closed circles) and the substrate that was backfilled with the OTS-SAM (open circles) as a function of the position on the substrate. The CMPE source was allowed to diffuse for 2 minutes at 88°C, the OTS was deposited for 15 minutes at room temperature. The data in Fig. 1 shows that the contact angle of CMPE decreases gradually from ≈77° down to ≈0° as one moves across the substrate starting at the CMPE side. The solid circles indicate that after the OTS deposition, the regions on the substrate far from the diffusing source are covered with a complete monolayer of OTS (contact angle ≈100°). As one traverses across the CMPE gradient, the contact angle decreases from ≈100° (OTS side) down to ≈88° (CMPE side). The minute increase of the contact angle within the CMPE-SAM is likely a result of a small interpenetration of OTS into the CMPE SAM. Ellipsometry measurements confirmed that only a single SAM monolayer was adsorbed everywhere on the substrate.

NEXAFS spectroscopy was used to provide detailed chemical and structural information about the SAMs on the substrate [9]. The NEXAFS spectra were collected in the partial electron yield (PEY) at the normal ($\theta = 90°$), grazing ($\theta = 20°$), and so-called "magic" angle ($\theta = 55°$) [9] incidence geometries, where θ is the angle between the sample normal and the polarization vector of the x-ray beam. Figure 2a shows the carbon edge K-edge NEXAFS spectra taken from CMPE-SAM (top) OTS-SAM (bottom) samples. NEXAFS spectra collected at the "magic" angle were indistinguishable from those recorded at the normal and grazing incidence geometries, revealing that the SAMs were not oriented, rather they formed a "liquid-like" structure. This observation is in accord with recent studies from Chaudhury and Allara groups that reported that the borderline between the "liquid-like" and "semi-crystalline-like" structures in hydrocarbon SAMs exists somewhere around -$(CH_2)_{12}$- [10]. The NEXAFS spectra in Fig. 2 both contain peaks at 286.0 and 288.5 eV that correspond to the 1s→σ^* transition for C-H and C-C bonds, respectively. In addition, the spectrum of CMPE also exhibits a very strong signal at 284.2 eV, which can be attributed to the 1s→π^* transition for phenyl C=C [9]. The latter signal can thus be used as an unambiguous signature of the CMPE in the sample. With the x-ray monochromator set to 284.2 eV, we collected the PEY NEXAFS signal by scanning the x-ray

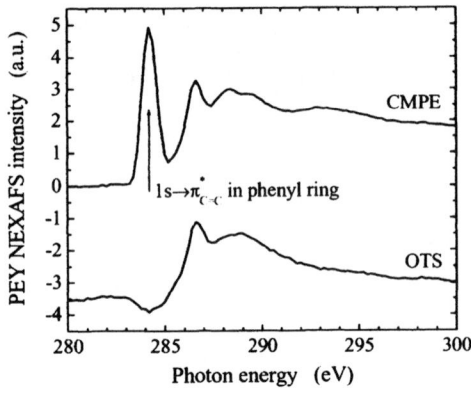

Fig. 2 Carbon K-edge PEY NEXAFS spectra collected from the CMPE-SAM and OTS-SAM. The arrow marks the position of the 1s→π* transition for phenyl C=C, present only in the CMEP-SAM sample.

beam across the gradient. Figure 3 shows the variation of the PEY NEXAFS intensity collected at 284.2 eV across the gradient sample. The functional dependence indicates that the NEXAFS intensity from the C=C phenyl bond and thus the concentration of CMPE in the sample decreases as one moves from the CMPE side of the sample towards the OTS-SAM; the functional form closely resembles that of a diffusion-like profile. Moreover, the profile width ($\approx$ 30 mm) is in accord with that obtained from the position-dependent contact angle data (cf. Fig. 1).

Fig. 3 Normalized PEY NEXAFS intensity collected at E=284.2 eV as a function of position across the substrate containing the initiator gradient.

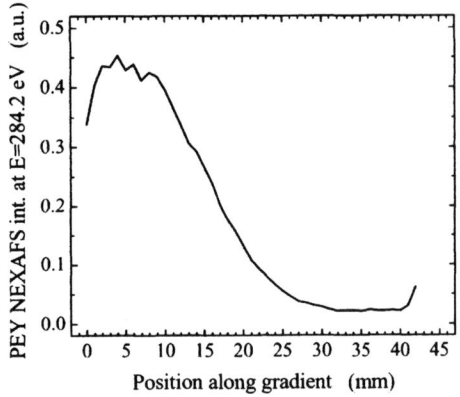

After the preparation of the CMPE-gradient substrate, the ATRP polymerization of PAAm was performed as described previously. Figure 4 shows the variation of the apparent PAAm brush thickness along the substrate (solid circles) and the DI water contact angle measured on the PAAm brushes (open circles). The CMPE initiator was deposited using the technique described earlier by diffusing CMPE for 2 minutes at 88°C. The polymerization was carried at 130°C for a) 15 hours, and b) 45 hours. In both cases the apparent thickness of the polymer brush decreases gradually as one moves across the substrate starting at the CMPE edge. Because the polymers grafted on the substrate all have roughly the same number of segments, the variation of the polymer film thickness can be attributed to the difference in the density of the CMPE grafting

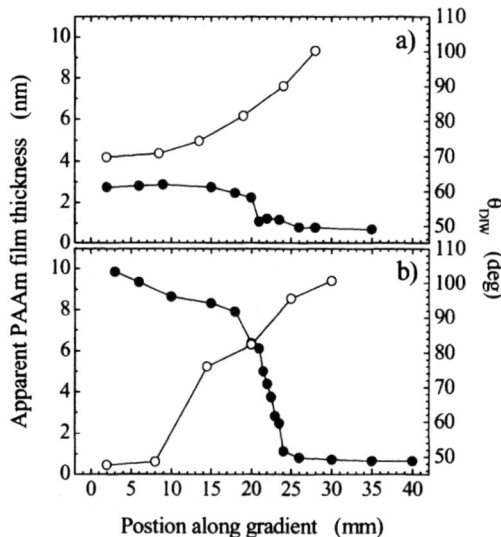

Fig. 4 Thickness (solid circles) and DI water contact angle (open circles) measured on the PAAm gradient brushes. The CMPE initiator was deposited using the technique described in the text; the deposition time and temperature were 2 minutes and 88°C, respectively, the PAAm polymerization was carried at 130°C for: a) 15 hours, and b) 45 hours.

points on the substrate. This conclusion can be further justified by the fact that the width of the polymer thickness gradient agrees with that of the CMPE gradient determined from the contact angle (*cf.* Fig. 1) and NEXAFS (*cf.* Fig. 3) measurements. This behavior thus indicates that the gradient in polymer thickness mimics that of the CMPE-SAM. The thickness of the PAAm film on the CMPE-SAM increases with increasing polymerization time. Specifically, by increasing the polymerization time from 15 hours to 45 hours, the PAAm film thickness increases from ≈2.9 nm to ≈9.2 nm. Because the grafting density of the brushes remained fixed, increasing the polymerization time resulted in the increase of the grafted polymer length. Evidence for this behavior can be gathered by examining the contact angles of water on the grafted polymer brushes. The contact angles of DI water on the dense part of the polymer brush were ≈70 and ≈47° for the 15 and 45 hours polymerizations, respectively. These results thus show that while in the former case the probing liquid penetrated pores in the PAAm brush and presumably sensed also the hydrophobic initiator, in the latter case a value expected for pure PAAm was achieved.

We are currently in the process of measuring the number of segments of the grafted polymers on the substrates. By knowing this, one can readily convert the thickness measurements into polymer grafting density. We hope that with this set up we would be able to probe both the "brush regime" (dense polymer layer) as well as the "mushroom regime" (loose polymer layer) and for the first time determine the nature of the "mushroom-to-brush" transition on a single sample. We note that this polymer gradient technology as described here can be further enhanced by combining with "mechanically assisted polymer assembly" (MAPA), a technique recently developed in our group [8]. MAPA is based on carrying out surface polymerization on elastomeric substrates, such as poly(dimethyl siloxane) (PDMS) network that have been previously mechanically stretched. After the polymerization, the strain is removed from the PDMS substrate and the grafted macromolecules stretch away form the substrate forming a dense polymer brush whose grafting density can be smoothly adjusted by varying the extension of the flexible substrate.

CONCLUSIONS

We have presented a method for preparing polymer brushes with gradient variation of their grafting densities on solid substrates. The technique consists of: i) a generation of a molecular gradient of polymerization initiator on a solid substrate, and ii) polymerization from the substrate bound initiator centers ("grafting from"). In this publication we described the preparation and properties of gradient polymer brushes of poly(acryl amide) on silica-covered substrates. Specifically, we showed that the polymer density within the gradient polymer brush can be varied by controlling the polymerization time.

ACKNOWLEDGMENTS

This research was supported by The Camille Dreyfus Teacher-Scholar award, The 3M Non-Tenured Faculty award, and NACE International. NEXAFS experiments were carried out at the National Synchrotron Light Source, Brookhaven National Laboratory, which is supported by the U.S. Department of Energy, Division of Materials Sciences and Division of Chemical Sciences. The authors thank Drs. Daniel Fischer and William Wallace (both NIST) for their assistance during the course of the NEXAFS experiments.

REFERENCES

1. Y. Xia, G. M. Whitesides, Angew. Chem. Int. Ed. Engl. **37**, 550 (1998); Y. Xia, et al., Chem. Rev. **99**, 1823 (1999).
2. A. Ulman, An Introduction to Ultrathin Organic Films from Langmuir-Blodgett to Self Assembly, (Academic Press: New York, 1991).
3. M. Husseman, M. et al., Angew. Chem. Int. Ed. Engl. **38**, 647 (1999); R. Shah, et al., Macromolecules **33**, 597 (2000); N. L. Jeon, Appl. Phys. Lett. **75**, 4201 (1999); N. Kim et al, Macromolecules **33**, 3793 (2000); B. de Boer et al., Macromolecules **33**, 349 (2000); P. Ghosh, et al., Macromolecules **34**, 1230 (2001); D. M. Jones and W. T. S. Huck, Adv. Mater. **13**, 1256 (2001); J. Hyun and A. Chilkoti, Macromolecules **34**, 5644 (2001).
4. M. K. Chaudhury, G. M. Whitesides, Science **256**, 1539 (1992).
5. K. Efimenko, J. Genzer, Adv. Mater. **13**, 1560 (2001).
6. R. R. Fuierer et al, Adv. Mater. **14**, 154 (2002).
7. X. Huang, L. J. Doneski, M. J. Wirth, Chemtech **19**, (Dec 1998), 19; Anal. Chem. **70**, 4023 (1998); X. Huang, M. J. Wirth, Macromolecules **32**, 1694 (1999).
8. T. Wu, K. Efimenko, J. Genzer, Macromolecules **34**, 684 (2001).
9. J. Stöhr, NEXAFS Spectroscopy (Springer-Verlag, Berlin, 1992).
10. M. K. Chaudhury, M. J. Owen, J. Phys. Chem. **97**, 5722 (1993); D. L. Allara, A. N. Parikh, E. Judge, J. Chem. Phys. **100**, 1761 (1994).

Mat. Res. Soc. Symp. Proc. Vol. 707 © 2002 Materials Research Society

Charge Transport in Mesoscopic Carbon Network Structures

V. Ksenevich[1], J. Galibert[2], V. Samuilov[1,2,3], Y.-S. Seo[3], J. Sokolov[3], M. Rafailovich[3]
[1] Department of Physics, State University of Belarus, 220080, Minsk, Belarus
[2] Laboratoire National des Champs Magnetiques Pulses, F-31432 Toulouse CEDEX 4, France
[3] Department of Materials Science, SUNY at Stony Brook, Stony Brook, NY 11794, USA

ABSTRACT

The charge transport and quantum interference effects in low-dimensional mesoscopic carbon networks prepared using self-assembling were investigated.
The mechanism of conduction in low-dimensional carbon networks was found to depend on the annealing temperature of the nitrocellulose precursor. The charge transport mechanism for carbon networks obtained at $T_{ann}=750$ ^{0}C was found to be the hopping conductivity in the entire investigated temperature range. The Coulomb gap near the Fermi level in the density of states was observed in the investigated carbon networks. The width of the Coulomb gap was found to be decreased with the annealing temperature of the carbon structures. The crossover from the strong localization to the weak localization regime of the charge transport in the carbon structures, obtained at $T_{ann}=950$ ^{0}C and $T_{ann}=1150$ ^{0}C, was observed in the temperature range $T>100$ K and $T>20$ K, respectively.

INTRODUCTION

High temperature treatment of polymer precursors in vacuum results in different carbon structures synthesis, like carbon fibers [1-3], carbon-black polymer composites [4,5], porous aerogels [1,6] and low dimensional carbon networks [7,8]. These disordered carbon matrixes have been used as model materials for electronic transport phenomena investigation dependently on the annealing temperature, which was found to determine the degree of disorder. As a result different quantum phenomena can be observed: strong localization, hopping conductivity, quantum interference, weak localization, metal-insulator (MI) transition.

These carbon materials could be considered as nanostructured materials with different internal morphology. We have developed a simple approach of self-organized fabrication of two dimensional mesoscopic networks with the feature size down to 50 nm and the size of the cells of the order of 500 nm. Two dimensional mesoscopic network-like carbon structures were produced by high temperature annealing of nitrocellulose precursors [8].

EXPERIMENTAL DETAILS

The technique is based on the self-organized patterning in a thin layer of complex liquid - diluted nitrocellulose solution (1% nitrocellulose in amyl acetate with ethyl alcohol as a solvent) in the presence of humid atmosphere. The sub-micron sized water droplets were trapped at the surface and self assemble in regular aggregates. The polymer was found to precipitate at the water-polymer solution interface, forming a layer encapsulating the droplets and preventing their coalescence. After the evaporation of the solvent the resulting networks were transferred onto insulating substrates. The final carbon structures (Fig.1) were produced by heat treatment of the nitrocellulose precursors at relatively low temperatures T=750-1150 ^{0}C.

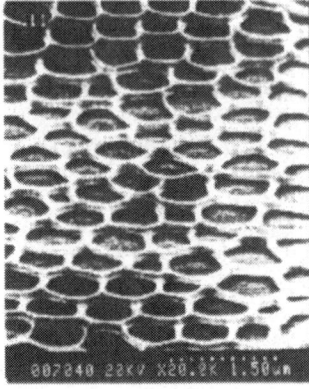

Figure1. The scanning electron microscopy image of the carbon mesoscopic network.

Four-point probe contacts in van-der-Pauw and linear geometry were produced by thermal sputtering of Al through metal masks on top of the carbon networks. Magnetoresistance measurements were carried out in pulsed magnetic fields, at the Laboratoire National des Champs Magnetiques Pulses de Toulouse (LNCMP) in the temperature range 1.6-300 K, and up to 34 T. Zero field resistance and MR was measured using the standard four-probe lock-in technique in the linear range of current-voltage characteristics.

The resistance versus temperature and magnetic field was found to show a difference in the mechanism of charge transport in mesoscopic network carbon structures dependently on the temperature of heat treatment of the initial nitrocellulose precursors. We have considered three temperature ranges of heat treatment: low, intermediate and high.

The mechanism of charge transport in carbon network structures annealed at 750 ^{0}C was found to be variable range hopping conduction with Coulomb interaction. The resistance versus temperature followed the law typical for the strong localization [9]:

$$R = R_0 \exp(T_0/T)^n, \tag{1}$$

with n=1/2 at T<125 K, and n=1/4 at T>125 K (Fig.2).
It is well-known that Eq. 1 with n=1/4 is typical for the variable range hopping mechanism [10] with the parameter T_0 :

$$T_0^M = \frac{\beta}{k_B \xi^3 g(\mu)}, \tag{2}$$

where the dimensionless β= 21.2, k_B – Boltzmann constant, ξ -the localization length, $g(\mu)$ – density of the states near the Fermi level.

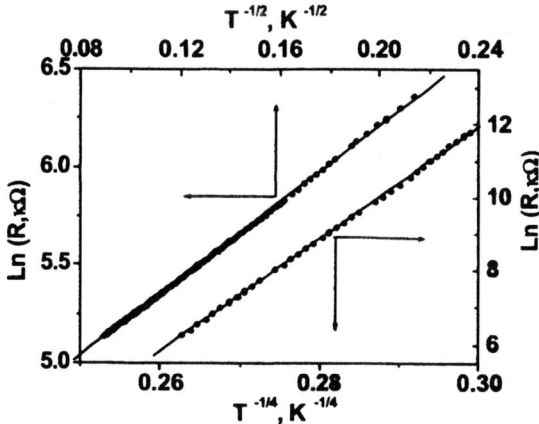

Figure 2. Resistance versus temperature of carbon network structures annealed at 750°C. The upper curve is for the temperature range T=21.86-123.8K. The lower curve is for T=125.19-209.7K.

For the case of the variable range hopping with Coulomb gap near the Fermi level (n=1/2 for the mechanism of Shklovskii-Efros (SE) [9]), the parameter T_0 is the following:

$$T_0^{SE} = \frac{\beta_2 e^2}{k_B \varepsilon \varepsilon_0 \xi},$$ (3)

with the dimensionless $\beta_2=2.8$, e – the charge of the electron, $\varepsilon \varepsilon_0$ – the dielectric constant. We have to mention that the conductivity of different disordered carbon materials, like fibers and aerogels [1,2] was described in the framework of this kind of transport mechanism, that is valid for 3-D materials.

The resistance versus magnetic field is plotted in the temperature range 30-300K as shown on the Fig.3. Almost in the whole range of magnetic fields, the resistance could be described with the typical equation for the hopping conduction [9,11]:

$$R(B)=R(0)\exp K_S B^2,$$ (4)

with Ks:

$$K_, = \frac{e^2 \xi^4}{C_1 \hbar^2} \left[\frac{T_{SE}}{T} \right]^p,$$ (5)

where e is the electron charge, ξ – the localization length, C_1 – the dimensionless constant of the order of 10^3 [9] ($C_1=660$ [12]), $\hbar$ - the Plank constant. The value of p depends on the influence of e-e interaction on the hopping transport. For the three-dimensional Mott-type hopping $p= -3/4$. If the Coulomb gap is in the density of states, $p= -3/2$. Equation 4 is used at the low field limit, when $\lambda \gg \xi$ ($\lambda=(\hbar /eB)^{1/2}$ is the magnetic length) [9]. The value of λ at B=30 Tesla is approximately 4.5 nm. The value of ξ is believed to be less than λ in the available range of magnetic fields. We have determined ξ from the dependence of R(B). Note, that this method for deducing ξ assumes that the values T_{SE} and ξ do not vary too much with the magnetic field. We have ploted $\log K_s$ versus $\log T$ on the Fig. 4.

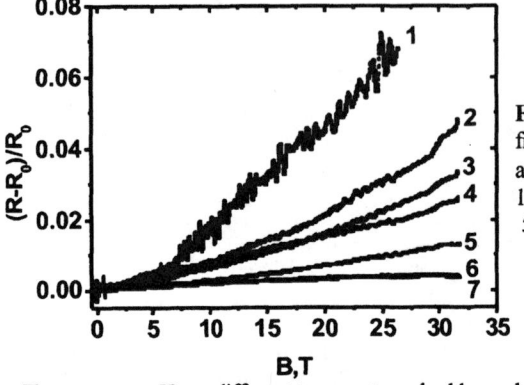

Figure 3. Resistance versus magnetic field for carbon network structures annealed at 750 ^{0}C at different T:
1) 31.4K, 2) 50.0 K, 3) 60.5 K, 4) 70.0 K, 5) 100.2 K, 6) 202.0 K, 7) 301.8 K

The parameter K_S at different temperatures had been determined in advance R(B) as $\ln[R(B)/R(0)] = K_s B^2$. The determined value of the localization length was found to be 2.7 nm which is less than λ at the highest available magnetic field, and the high magnetic fields region is unreachable. Using Eq. 4 and the experimental value of $T_o^{SE} = 2726$K, experimentally determined from the lower plot on the Fig.2 at B=0, and the determined value of the localization length ξ =2.7 nm, we have estimated the dielectric constant ε. The determined value of ε ~ 80 is believed to be too high for disordered carbon far away from the MI transition with the typical value of ε~10.

The dielectric constant could be considered as the sum of two components: the normal host dielectric response of the lattice (ε_h), and an anomalous contribution by hopping electrons [13]:

$$\varepsilon = \varepsilon_h + 4\pi e^2 g(\mu) \xi^2 . \qquad (6)$$

The second component in Eq. 6 could be anomalously higher than the first one near the MI transition.

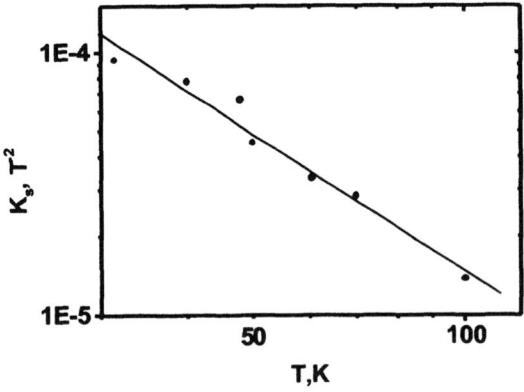

Figure 4. Ks versus T (calculated using $\ln[R(B)/R(0)] = K_s B^2$ in the low field limit) for carbon network structures heat treated at 750 ^{0}C. The value of p (Eq.5), determined from the slope of the plot, was found to be -1,72∓0,14, which is close to the value −3/2 , typical for the SE model [9].

The curve of the resistance versus temperature of the carbon network structures was well fitted by the Eq. 1 for the variable-range hopping mechanism. The transition temperature from Mott- type behaviour (n=1/4) to that of Shklovski-Efros (n=1/2) was determined with the Zabrodski method to be $T_c \approx 13K$. Decreasing of the transition temperature from Mott to Shklovskii-Efros regime at raising annealing temperature implies that the width of the Coulomb gap in the density of states near Fermi level decreased due to increasing dielectric constant. This confirms, that our mesoscopic carbon networks get closer to the MI transition as the annealing temperature increases. At temperatures higher than approximately 100K, R(T) was found to be described by a power law

$$R \sim T^n, \tag{7}$$

which is typical for weak localization mechanism of the transport [14,15]. This was due to the hopping length which decreased with increasing temperature, and which was believed to compare with the localization length. In this case the transition from hopping conduction to the diffusive transport took place. The transition from positive magnetoresistance typical for hopping conduction to the negative one for diffusive transport was observed [8]. At T=100 K the small (<0.5 %) negative magnetoresistance was observed due to magnetic field suppression of quantum interference effects in the weak localization regime [4]. As soon as the hopping distance R_h is related to the localization length ξ [9]: $R_h \sim \xi \sqrt{T_{SE}/T}$, at higher temperatures they may be comparable, and the observed small negative MR could be consistent with the weak-localization picture [4]. At low temperatures and considering the positive magnetoresistance three regions were defined: low, intermediate and high magnetic fields. At low fields, ln(R-R_0/R_0) is proportional to B^2. In the intermediate range, the MR is nearly linear with B. The MR is much less dependent on B in the high field region (sub-linear relationship) and tends to saturation at the highest fields and the lowest temperatures [8].

The localization length and the dielectric constant tend to increase at the system gets closer to the MI transition. In the carbon networks annealed at 950 ^{0}C the values of the localization length and the dielectric constant were found to be ξ=8.75 nm (much higher than the value for amorphous carbon $\xi \sim 1.2$ nm [16]) and $\varepsilon \sim 640$. The value of the density of the states near the Fermi level was found to be $7.32 \cdot 10^{19}$ eV^{-1}cm^{-3}.

The transition from variable range hopping to the diffusion transport was also observed for the carbon networks heat-treated at 1150 ^{0}C. Figure 5 shows that Ln R varies linearly with Ln T and follows the Eq.7, which is typical for a weak localization mechanism.

The conductance increases with raising temperature in the case of 3-D weak localization due to the decreasing of the time of inelastic scattering ($\tau_i \sim T^{-p}$) [17]:

$$\sigma = \sigma_0 + \frac{e^2}{\hbar\pi^3}\frac{1}{a}T^{p/2}, \tag{8}$$

where a~k_F^{-1} is the mean free path, and p depends on the scattering mechanism and the dimension of the system.

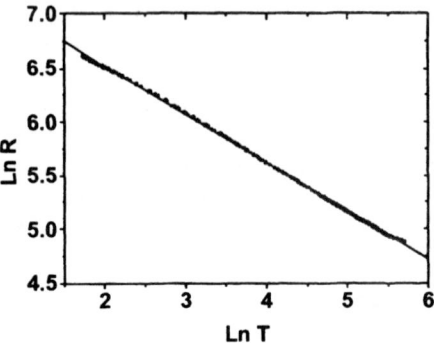

Figure 5. Resistance versus temperature for carbon network structures annealed at 1150 °C.

As it is known, in order to observe weak localization, elastic scattering should dominate over inelastic scattering, i.e. $\tau_i \gg \tau_e$ [14]. As a result weak localization effects are observed at low temperatures, mostly. But, for carbon graphite materials with turbostrate structure, in comparison with normal metals, the weak localization effects are observed at relatively high temperatures [18]. This is due to the significant difference in parameters, like Debye temperature, Fermi energy, etc. [19].

The resistance versus magnetic field for carbon network structures annealed at 1150 °C is shown on Fig.6. The exponential dependence of the Eq. 4 was observed in the low temperature range (T=2-20 K) only. At higher temperatures, the transition to the negative magnetoresistance was observed. The large value of the localization length (16.3 nm) is believed to confirm that the carbon network structures get closer to the MI transition as the annealing temperature increases.

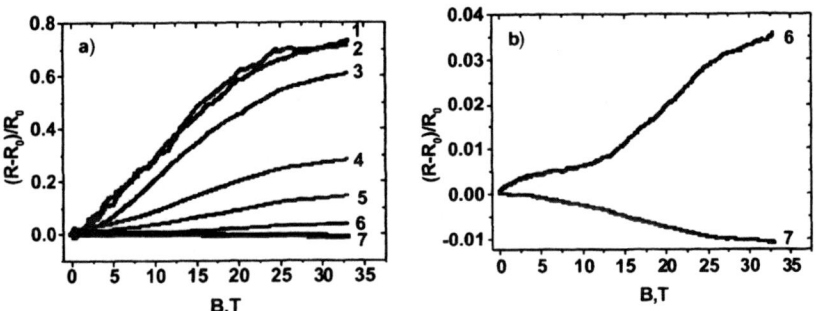

Figure 6 Resistance versus magnetic field for carbon network structures annealed at 1150 °C at different temperatures: 1) 2.98 K, 2) 3.27 K, 3) 4.2 K, 4) 8.07 K, 5) 12.0K, 6) 21.05 K, 7) 49.9 K

CONCLUSION

The dependence of the mechanism of transport on the temperature of heat treatment of the initial nitrocellulose network precursors was found to be related to the formation and growth of the carbon clusters in amorphous matrix. The transport mechanism in low temperature annealed carbon networks is variable range hopping, whereas in high temperature annealed networks it is diffusion transport.

REFERENCES

[1] A.W.P. Fung, Z.H. Wang, M.S. Dresselhaus et al. *Phys. Rev. B.* **49**, 17325 (1994).
[2] A.W.P. Fung, M.S. Dresselhaus and M. Endo., *Phys. Rev. B.* **48**, 14953 (1993).
[3] V. Bayot, L. Piraux, J.-P. Michenaud and J.-P. Issi., *Phys. Rev. B.* **40**, 3514 (1989).
[4] P. Mandal, A. Neumann, A.G.M. Jansen et al. *Phys. Rev. B.* **55**, 452 (1997).
[5] D.van der Putten, J.T. Moonen, H.B. Brom et al. *Phys. Rev .Lett.* **69**, 494 (1992).
[6] A.W.P. Fung, G.A.M. Reynolds, Z.H. Wang et al. *J. Non-Cryst. Solids.* **186**, 200 (1995).
[7] K. Yoshino, H. Kajii, Y. Kawagishi et al. *Jap. J. Appl. Phys.* **38**, 4926 (1999).
[8] V.A. Samuilov, J. Galibert, V.K. Ksenevich, et al. *Physica B*, **294-295**, 319 (2001).
[9] B.I. Shklovskii, A.L. Efros. *Electronic Properties of Doped Semiconductors* (Springer Series of Solid State Science, Springer, Berlin,1984).
[10] N.F. Mott *J. Non-Cryst. Solids.* **1**, 1 (1968).
[11] N. Mikoshiba. *Phys. Rev.* **127**, 1961 (1962).
[12] A.G. Zabrodski. *Sov. Phys. Semicond.* **14**, 670 (1980).
[13] R. Rosenbaum. *Phys.Rev. B* **44**, 3599 (1991).
[14] P.A. Lee, T.V.Ramakrishnan. *Rev. Mod. Phys.* **57**, 287 (1985).
[15] G. Bergman. *Phys. Rep.* **107**, 1 (1984).
[16] J.J. Hauser. *Solid St. Comm.* **17** 1577 (1975).
[17] T.V. Ramakrishnan. *Phys. Rev. Lett.* **42**, 673 (1979).
[18] V. Bayot, L. Piraux, J.-P. Michenaud, and J.-P. Issi. *Phys. Rev. B.* **40**, 3514 (1989).
[19] L. Langer, V. Bayot, E .Grivei et al. *Phys. Rev. Lett.* **76**, 479 (1996).

Mat. Res. Soc. Symp. Proc. Vol. 707 © 2002 Materials Research Society   AA10.4/DD11.4

Application of Self Assembled Monolayer Approach to Probe Fiber Matrix-Adhesion

E. Feresenbet[1], D. Raghavan[1], G. A. Holmes[2]
[1]Howard University
Chemistry Department
Washington, D.C. 20059
[2]Polymer Division
National Institute of Standards & Technology
100 Bureau Drive Stop 8543
Gaithersburg, Maryland 20899-8543

ABSTRACT

Adhesion at the fiber-matrix interface of composites is often related to a combination of factors such as mechanical interlocking, physico-chemical interactions, and chemical bonding of the fiber-matrix interphase region. We demonstrate the use of self-assembled monolayer (SAMs) approach for depositing silane coupling agent on glass fiber and studying the impact of the individual interactions on the adhesion process. Through some unique chemistry, functionalized and non-functionalized C11 chlorosilanes were deposited on to E-glass fiber and modified. The adhesion of diglycidyl ether of bisphenol-A (DGEBA) cured with meta-phenylene diamine (m-PDA) to SAM layer on E-glass fibers was measured by performing single fiber fragmentation tests (SFFT). The extent of adhesion between the fiber and matrix was found to be dependent on carbon chain length of coupling agents, and the functional group at the end of the SAMs layer. Furthermore, the contributions to adhesion by physico-chemical interaction and covalent bonding has been individually assessed.

INTRODUCTION

One of the major technical challenges to a composite's use in high performance applications is the reliable prediction of long-term performance (e.g., failure behavior, fatigue behavior, durability, and stiffness). When composites are manufactured, a small region (< 1 μm), known as the fiber-matrix interphase, forms between the fiber and the matrix [1]. This region exhibits properties similar to, but distinguishably different from the properties of the bulk matrix [2]. Since stress is transferred between the fiber and matrix at the fiber-matrix interphase, the efficiency of this stress-transfer process at the interphase is critical to a composite's ultimate strength and durability. Therefore, the role of the interphase structure, fiber topography, and fiber-matrix chemical bonding on ultimate composite performance needs to be understood fundamentally.

Traditionally, a mole fraction of 50 % ω-aminopropyl trimethoxysilane (APS) coupling agent is applied on glass fiber and this coupling agent promotes adhesion between the glass fiber and epoxy matrix. However, the orientation of bonding sites in coupling agents (i.e., amino group in ω-amino propyl trimethoxysilane on the glass fiber) has been called into question and its impact on covalent bonding to the host matrix (epoxy) has not been fully understood.

In recent years, molecular systems with well defined structural patterns have received growing attention for studying interfaces. A variety of experimental techniques have emerged which allow one to endow a solid surface with rich, well defined chemical patterns while keeping the surface flat on a molecular scale. SAMs offer one of the highest quality routes for the preparation of chemically and structurally well defined organic surfaces [3-6]. The solvent deposition of the trichlorosilane layer may provide an approach for developing organized self-assembled monolayers [5]. Using this approach, the impact of chemical bonding on interfacial adhesion has been investigated in a more controlled and fundamental manner [4-6]. These studies use the single-fiber fragmentation test (SFFT) to monitor both the fiber strength and the fiber-matrix interface/interphase strength. However there have been difficulties associated with controlling the fiber-matrix interface structure so as to fully understand the impact of interface structure on composites performance.

EXPERIMENTAL

The SFFT specimens were prepared with a DGEBA epoxy, Epon 828 (Shell), cured using m-PDA (Fluka Chemical). One hundred grams of DGEBA and 14.5 grams of m-PDA were weighed out in separate beakers. To lower the viscosity of the resin and melt the m-PDA crystals, both beakers were placed in a vacuum oven set at 65 °C. After the m-PDA crystals were completely melted, the silicone molds containing the coated fibers with silane agent were placed into another vacuum oven that was preheated to 75 °C at –20 kPa, for 20 min. Details of the aqueous coating procedure and SAMs coating procedure can be found elsewhere [7, 8]. The preheating of mold dries the mold and minimizes the formation of air bubbles during the curing process. At approximately 9 min before the preheated molds were removed from the oven, the m-PDA is poured into the DGEBA and mixed thoroughly. The mixture was placed into the vacuum oven and degassed for approximately 7 min. After 20 min, the preheated molds were removed from the oven and filled with the DGEBA/m-PDA resin mixture using 10 ml disposable syringes. The filled molds were then placed into a programmable oven. A cure cycle of 2 h at 75 °C followed by 2 h post-curing at 125 °C was performed.

The SFFT is an indirect micromechanics method used to calculate the degree of adhesion between a rigid fiber coated with silane agent and a more ductile polymeric matrix in fiber reinforced polymer composite. In a single fiber fragmentation test, typically, the fiber is embedded in a matrix material with a higher strain-to-failure than the fiber and the fiber breaks when longitudinal strain is applied. The test is performed by the sequential application of strain increments. The breaks occur at flaws along the fiber length, in a progressive way from the most critical to least critical flaw. A saturation limit is eventually attained, when the fragmented fiber is made up of a large number of very short fragments. Upon reaching the saturation limit, any additional strain does not cause further failure of the fiber. The resulting distribution of fiber fragment lengths represents the raw data from the single fiber fragmentation test. The details of the testing procedure, and the standard error in the testing method can be found elsewhere [9, 10]. All specimens in this research were tested with a 10 min delay between strain increments.

RESULTS AND DISCUSSION

Adhesion at the fiber-matrix interphase of composites is often ascribed to the following factors: (1) mechanical interlocking, (2) physico-chemical interactions, (3) chemical interaction, and (4) mechanical deformation of the fiber-matrix interphase region. Nardin and

Ward [11] appear to be the first to establish that fiber-matrix interface strength as a sum of the first three factors.

$$\tau_{interphase} = \tau_M + \tau_{PCL} + \tau_{CB}$$

where

$\tau_{interphase}$ denotes the total fiber-matrix adhesion as measured by the interphase strength.

τ_M denotes the adhesion at the fiber-matrix interphase due to mechanical interlocking.

τ_{PCL} denotes the adhesion at the fiber-matrix interphase due to physico-chemical interactions.

τ_{CB} denotes the adhesion at the fiber-matrix interphase due to covalent bonding.

By solvent deposition of coupling agent we prepare pure and mixed monolayers fuctionalized and non fuctionalized silane on glass fiber. The monolayer formations provide an approach for studying the impact of the three factors on interfacial adhesion in a more controlled and fundamental manner.

Figure 1 compares the number of fiber break results of aqueous deposited silane and solvent deposited trichlorosilane. The smooth lines drown on each data points in Figure 1 are just to guide the reader's eyes. The number of fiber breaks is directly proportional to the adhesion at the fiber-matrix interface. Hence, an increase in the number of breaks in the test specimen reflects an increase in the adhesion at the fiber-matrix interface. Because of the non-bonding nature of propyl trimethoxysilane (PTMS) and undecyl trichlorosilane (UTCS) towards epoxy, according to the Nardin and Ward model [11], for glass fibers coated with PTMS and UTCS system the adhesion contribution must come from mechanical interlocking and/or physicochemical interaction. A close examination of SFFT result (fiber breaks) for 100 % mole fraction PTMS coated and 100 % mole fraction UTCS coated glass fiber composite shows higher breaks for PTMS than UTCS. For the 100 % mole fraction PTMS coated glass fiber epoxy composite, the average number of breaks was (44 ± 6), where the number after ± is one standard deviation from the mean and is taken as the standard uncertainty, while for 100 % mol fraction UTCS coated glass fiber epoxy composite, the average number of breaks was (6 ± 2). The comparison of non-bonding coupling agents (PTMS and UTCS) suggests that PTMS coated fiber is more strongly adhered to the epoxy matrix than UTCS. Using atomic force microscopy (AFM), the PTMS coupling agent deposited film on fiber was observed and the film is rougher and non-uniform than the UTCS coupling agent deposited film. Based on AFM images, one can infer that the epoxy resin is likely to penetrate the porous PTMS coupling agent layer and reach the bare glass fiber surface. We expect, mechanical interlocking mechanisms that involves penetration of the epoxy resin through the porous silane layer to the glass surface to have largely contributed to increased adhesion, as measured by single fiber fragmentation test. Solvent deposited UTCS on glass fiber, the long back bone chain promote better packing of coupling agent on glass surface and there is less likelihood for epoxy to penetrate glass surface. As expected, the number of breaks for the UTCS coated glass fibers epoxy composites is small. The adhesion contribution may come largely from physico-chemical interactions.

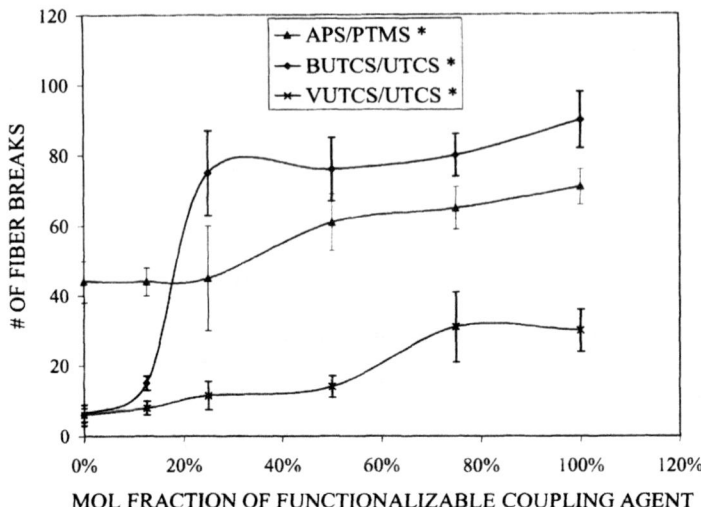

Figure 1. Average number of fiber breaks for E-glass fiber solvent deposited with short chain and long chain coupling agents as the mole fraction of functionalizable terminal groups in solution. The percentage is based on the mixing ratio of the silylating agents in solution during the deposition process [* For example, 0 % in the X-axis represents the mixture of 0 % mole fraction of bonding APS and 100 % mole fraction of non-bonding PTMS.] (Error bar represents one standard deviation from the mean and is taken as the standard uncertainty)

Since "mechanical interlocking" is minimal for long chain non-bonding silane coupling agent (UTCS), we can study the impact of purely covalent bonding on overall adhesion by considering long chain bonding silane coupling agent i.e, amino undecyl trichlorosilane (AUTCS).

Figure 1 shows the result for number of fiber breaks as a function of mole fraction of functionizable coupling agent in the solution. The deposited BUTCS was in-situ modified to AUTCS. This approach allows all amine groups of the long chain silane coupling agent on glass fiber to be oriented away from the fiber surface, and will be available for reaction with the epoxy resin. It can be seen that the magnitude of fiber breaks increases initially (shown by the line corresponding to ◆) and reaches 85 % of the saturation value at 25 % mole fraction of functionalizable coupling agent in solution and marginally increases with further increase in functional coupling agent component above this concentration.

A plausible explanation for saturation at a mole fraction of 25 % of functionalizable coupling agent is that DGEBA is a large monomeric molecule and may sterically hinder the accessibility of neighboring reactive groups (amine) to react with the epoxy functional group of another DEGBA molecule [12]. Consequently, a large fraction of the groups registered on the fiber may not have participated in the epoxy/amine reaction. Therefore, the amount of adhesion between glass fiber and epoxy is dependent not only on the bonding/non-bonding type of coupling agents, the composition of bonding/non-bonding coupling agent mixture, and the packing/registering of coupling agent molecules, but also on the structure of the reacting matrix.

A large difference in the extent of debonding for UTCS and BUTCS/ in-situ modified glass epoxy composites at the fiber matrix interface was observed during the fragmentation tests. The debond area is indirectly proportional to adhesion at the glass/fiber interface. Figure 2, shows the unstressed debond region associated with the fracture of glass specimen coated with 100 % mole fraction of in-situ modified BUTCS. The debond region is approximately 20 μm. When the fiber fractures, a matrix crack perpendicular to the fiber axis is formed in addition to the matrix debond region. Figure 3, shows the unstressed debond region associated with the fracture of a glass specimen coated with 100 % mole fraction of non-bonding UTCS. The total length of the debonded region is greater than 200 μm. These results further confirm that the covalent bonding in in-situ modified BUTCS largely contributes to adhesion at E-glass fiber/epoxy composite interface. This contribution was minimal in UTCS systems. To establish that indeed covalent bond contributes to enhanced adhesion, we compared the number of fiber breaks for PTMS with APS coated glass fiber epoxy composites. As expected, the number of fiber breaks for glass fibers treated with mixed APS and PTMS increased with increase in the APS concentration in the mixture. We performed statistical analysis of the results and detailed analysis of the results can be found elsewhere [13].

We also studied the role of terminal hydroxyl functional group on the overall adhesion of fiber/epoxy composite by solvent depositing vinyl undecyl trichlorosilane (VUTCS) (bonding silane coupling agent) and UTCS (non-bonding silane coupling agent) on glass fiber. The deposited VUTCS was in-situ modified to create hydroxyl terminal group so that hydroxyl groups can hydrogen bond to the epoxy resin matrix. The plot (Figure 1) shows a steady increase in the number of fiber breaks (shown by line correspond to X) as the percent of functionalizable group on the fiber surface is increased and then plateau at higher concentration of functionalizable groups on the E-glass fiber surface. The difference in the number of fiber breaks between the modified BUTCS interface and modified VUTCS interface is due to the type of chemical bonds formed with epoxy resin. In the case of amine terminated glass (modified BUTCS) surface, covalent bonding of terminal amine with the epoxy resin can result in strong adhesion, while hydroxyl terminal glass surface, it is speculated that hydrogen bonding of the hydroxyl group with the cured epoxy resin explain weak adhesion. Our comparison is based on the assumption that the extent of in situ modification of BUTCS to amine and VUTCS to hydroxyl-functionalized E-glass fiber is very similar. Further studies are needed to quantify the extent of in-situ modification in both systems and confirm our observations.

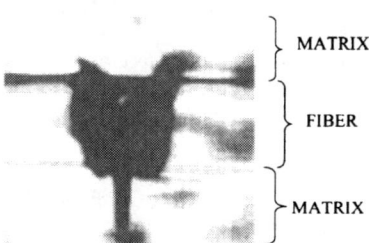

MATRIX

FIBER

MATRIX

Figure 2. Debond regions associated with the fiber breaks in in-situ modified 100 % BUTCS coated glass fiber epoxy composite.

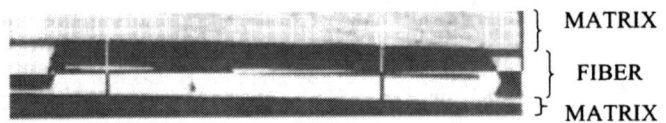

Figure 3. Debond regions associated with the fiber breaks in 100 % UTCS coated glass fiber epoxy composite.

CONCLUSION

1. The results from aqueous phase deposited silanes indicates that ~85% of the fiber breaks is obtained with ~50% of the surface covered with amino groups.
2. Solvent deposition and in-situ modification of silane was used to prepare pure and mixed functionalized silane monolayers on E-glass fiber surfaces.
3. The fragmentation data for SAMs modified glass fiber/epoxy composite indicate that ~85% of fiber breaks is obtained with ~25% of the surface covered with amino groups.
4. Hydroxyl terminated SAM was used to probe the strength of hydroxyl-hydroxyl bonding at the fiber matrix interphase. The adhesion for hydroxyl terminated SAM was found to be much lower than the adhesion results exhibited by amine functionalized SAMs surface.

REFERENCES

1) H. A. Moussawi, E. K. Drown and L. T. Drzal, Polymer. Composite, 1993, 14, 195.
2) J. L. Koenig, in Silanes Surface and Interfaces, D. E. Leyden, Ed. (Gordon and Breach Science Publishers, New York), Vol. 1, p. 43.
3) W. C. Bigelow, D. L. Pickett, and W. A. Zesman, Journal of Colloid Science 1946, 101, 201.
4) R. Maoz, and J. Sagiv, Journal of Colloid Interface Science 1984, 100, 465.
5) J. B. Brzoska, I. A. Ben, and F. Rondelez, Langmuir 1994 10, 4367.
6) N. Parikh, D. L. Allara, I. A. Ben, and F. Rondelez, Journal of Physical.Chemistry 1994, 98, 7577.
7) G. A. Holmes, R. C. Peterson, D. L. Hunston, and W. G. McDonough, Polymer Composites 2000.
8) A. Heise, H. Menzel, H. Yim, M. D. Foster, R. H. Wieringa, A. J. Schouten, V. Erb, and M. Stamm, Langmuir 1997, 13, 723-728.
9) G. A. Holmes, R. C. Peterson, D. L. Hunston, and W. G. McDonough, Polymer Composites 2000, 21, 450-465.
10) G. A. Holmes, R. C. Peterson, D. L. Hunston, and W. G. McDonough, C. L. Schutte, the Effect of Nonlinear Viscoelasticity on Interfacial Shear Strength Measurements; In Time Dependent and nonlinear effects in polymers and composites; R. A. Schapery, ed. ASTM: 2000; PP 98-117.
11) M. Nardin, and I. M. Ward, Materials Science and Technology 1987, 3, 814.
12) G.Holmes, E. Feresenbet, and D. Raghavan, Composite Interfaces, 2002, submitted.
13) E. Feresenbet, D. Raghavan, and G. Holmes, J. Adhesion, 2002, submitted.

Mat. Res. Soc. Symp. Proc. Vol. 707 © 2002 Materials Research Society　　　　　　AA10.5/DD11.5

Surface and interfacial properties in thin poly(*p*-phenylene vinylene) multilayers prepared by self-assembly method

Célio A. M. Borges, Alexandre Marletta, Roberto M. Faria and Francisco E. G. Guimarães
Instituto de Física de São Carlos, Universidade de São Paulo, CP 369, 13560-970, São Carlos-SP, Brazil.

ABSTRACT

In this work we report on the influence of interfaces (metal/film, ITO/film) on the optical properties of ultrathin layers of poly(*p*-phenylene vinylene), PPV, adsorbed by self-assembly. PPV films were prepared exploiting the self-assembly method (SA) in addition to a low temperature conversion process for PPV. Using atomic force microscopy we show that 0.5 nm thick PPV layers can be homogeneously deposited on different metals (50% transmission Au, Cu, Pt, Ag, Cr films), glass and ITO surfaces. Absorption measurements reveal that the transitions involving π-π^* conjugated states are blue shifted to different extents depending on the interface and on the film thickness. This indicates different extensions of interfacial disorder in the PPV layers. The ITO/PPV interface has the highest extension for disorder (~16 nm) and highest surface roughness (RMS~2.5 nm) while the lowest values (~5 nm and 0.6 nm, respectively) were found for the quartz/PPV interface. The metal interface induces significant luminescence quenching and spectral shifts, which depend on the type of metal used. The effect from incorporation of carbonyl and from chemical modifications of the ITO/organic interface during PPV thermal conversion is discussed.

INTRODUCTION

Organic light emitting diodes (OLEDs) have become a major focus of research due to the commercial viability of organic luminescent displays. OLEDs usually consist of organic semiconductor thin layers sandwiched between a metal cathode and transparent anode. The efficiency and stability of devices depend strongly on the electronic and structural properties of the electrode/organic interfaces [1-4]. Yet some of the fundamental aspects of charge injection processes and interface properties are not understood in detail [5-6]. Indium tin oxide (ITO), after modification with an oxidative surface treatment, has been used as hole-injecting contact, while low work function metals (Mg, Ca or Li) or metal alloys (Mg:Ag) are used as electron injecting contacts [2]. These metals are not suitable in a production environment because of their extremely high chemical reactivity. The possibility of using other cathodes thus represents an advance in OLED applications.

In this work we use ultra thin self-assembled (SA) PPV films to study the interface between metals and the organic semiconductor. We show that thin conjugated layers are suited for interface studies since any conformational change of the chain can be readily seen through optical measurements. We use a new SA method [7], which allows control of deposition of PPV layers at the monolayer level on different surfaces such as glass/ITO and metals (Au, Ag, Cu, Cr, Al). Another advantage of this procedure is that thermal conversion of PPV may be performed at considerably lower temperatures (80-100 °C) in a few minutes, producing highly ordered PPV films with very low defect incorporation [8].

EXPERIMENTAL DETAILS

Poly(xylylidene tetrahydrothiophenium chloride) (PTHT) was synthesized following a precursor route described elsewhere [9]. The SA films were produced by alternate immersion in the polycation solution (PTHT) and polyanion solution of dodecylbenzene-sulfonate (DBS), and finally dried with a flow of N_2. The PPV-precursor solution had 0.18 mg/mL and pH=5.0. The DBS solutions were prepared with concentration varying from 10^{-3} to 1 M and pH = 5.0, diluted in Milli-Q water (resistivity of 18 MΩcm). The films were adsorbed on hydrophilic quartz and substrates treated in H_2SO_4/H_2O_2 (7:3) bath for 30 min and $H_2O/H_2O_2/NH_3$ (5:1:1) bath for 30 min. The glass/ITO substrates where rinsed in $H_2O/H_2O_2/NH_3$ (5:1:1) bath for 30 min and treated for 5 min in $HNO_3:HCl:H_2O$ (1:3:20). The immersion times in PTHT and DBS solutions were fixed at 15 s and 10 s, respectively. SA-PPV films were obtained by converting the precursor film at low temperatures (80 to 120 °C) during 30 min under vacuum [7-8]. Metal films of Au, Ag, Al, Cr and Cu with 50% transmittance where deposited side by side on a glass substrate in a Balzer BAK 600 evaporation system under 8 μTorr. The UV-Vis measurements were performed with a Hitachi U-2001 spectrometer. The emission properties of the films were investigated by photoluminescence (PL) spectroscopy. Details of the PL set-up were published elsewhere [8].

RESULTS AND DISCUSSION

Figure 1a shows the UV-visible absorbance spectra for films with different numbers of bilayers of PTHT/DBS deposited on quartz substrates (continuous lines). The immersion times in PTHT and DBS solutions were fixed at 10 s and 2 s, respectively. The spectral intensity increases linearly with the number of PTHT/DBS layers (see insert of Fig. 1a). Therefore, both

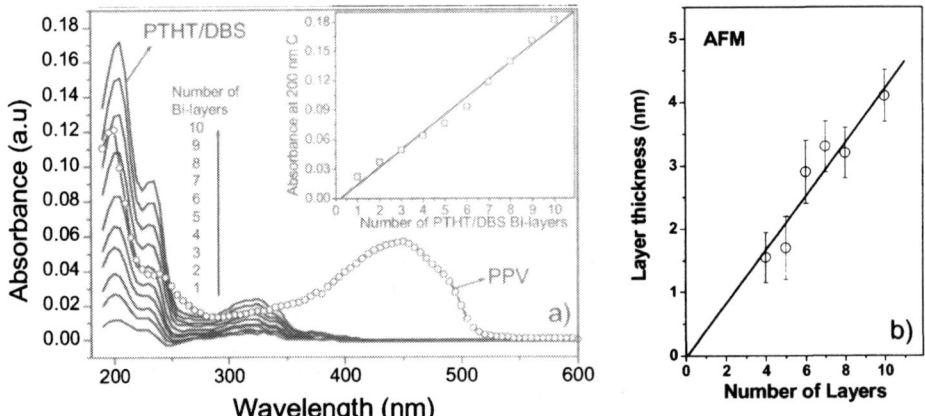

Figure 1. (a) Absorbance spectra measured after deposition of each PTHT/DBS bi-layer (continuous lines). The spectrum of the sample with 10 bi-layers after conversion at 110 °C for 30 min is shown for comparison (dotted line). The insert shows the dependence of absorbance intensity at 200 nm on number of bi-layers. b) Layer thickness measured by AFM as a function of each PTHT and DBS layers after the conversion to PPV at 110 °C. The error bars correspond to the layer roughness (RMS).

PTHT and DBS are adsorbed alternately at a constant amount in each immersion cycle. Figure 1a also displays the spectrum of a 10-bilayer sample after conversion to PPV at 110 °C during 30 min in vacuum (dotted line). Detailed data on the properties of such SA PPV can be found elsewhere [7-8]. Figure 1b shows that film thickness, obtained by AFM [10], increases linearly with the number of bilayers for films after conversion to PPV at 110 °C. This is consistent with the linear increase in Fig. 1a and demonstrates that the SA technique using PTHT and DBS can produce PPV ultra thin layers with controled thickness of approximately 0.5 nm.

The ability of the SA methodology to prepare thin layers has been exploited to study the influence of contact/organic interfaces on the optical properties of PPV. Figure 2a shows the absorption spectra of a PPV multilayer converted at 110 °C during 30 min under vacuum for different numbers of layers of PTHT and DBS, varying from 1 to 10. The absorption spectrum shifts to higher energies as the number of layers decreases. Also, in contrast to the results in figure 1, the absorption intensity increases sub-linearly with the number of layers. This indicates that the PPV effective conjugation length changes as the layers approach the interface. The blue shift has been attributed to either conformational changes of the polymer chains within the individual layers [11] or to quantum confinement of coherent electrons within the entire multilayer [12]. The wavelength of the absorbance onset asymptotes to a limiting value of 505 nm as the number of layers increases, suggesting that the influence from the quartz interface on the PPV properties lies inside 10 layers (~5 nm). The same behavior is observed for SA PPV on ITO/Glass substrates (Figure 2b). However, the substrate effects are still present within 32 layers (~16 nm) of PPV. In addition, the absorbance intensity has a significant increase only up to 16 layers. These results show that the interface plays a dominant role in the final structure and optical properties of the self-assembled ultra-thin conjugated polymers.

Further insight into the interfacial properties can be gained by examination of luminescence spectra of thin PPV multilayers prepared on different metal surfaces. Figure 3a shows the

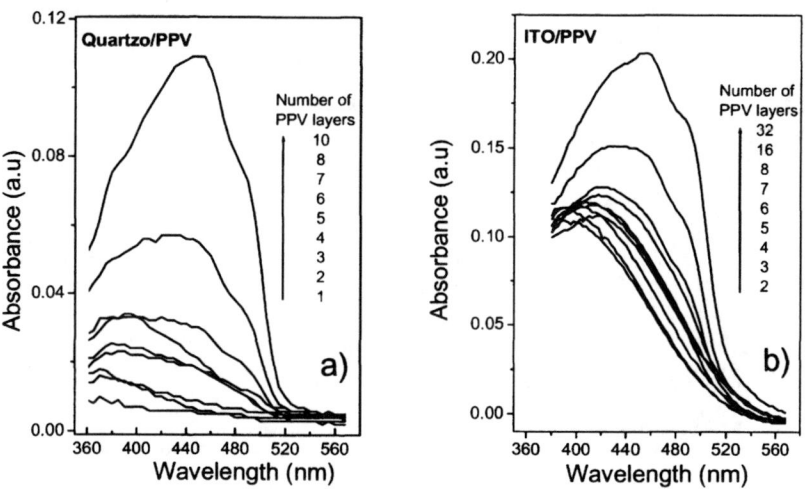

Figure 2: Absorbance spectra (300 K) of SA-PPV layers deposited on a) quartzo and b) glass/ITO substrates.

normalized luminescence, obtained at low temperature (40 K), for 32-bilayer PTHT/DBS films deposited on a glass substrate covered with various metal layers and then converted to PPV at 110 °C. The metal layers used in the experiment were 50% transmittance Cu, Ag, Au, Cr, Al layers placed side-by-side on the substrate. Note that approximately the same layer thickness was observed by AFM on all metal and glass surfaces and that all variables were held constant for PL measurements in figure 3a. Comparing the absolute PL intensities of the PPV layer prepared on glass, it is clear that the metal interface induces significant luminescence quenching. Al and Cu interfaces show the smallest decrease in luminescence while Cr and Au interfaces yield the highest quenching. In addition, figures 3b e 3c indicate that the position of the zero-phonon peak depends on the interface material and shifts to higher energies as the number of layers decreases. The glass interface shows lower blue shifts for the zero-phonon line in comparison to the metal ones (Figure 3b). Cu and Ag, which have shown the highest chemical reactivity, induce the largest change in line shape with respect to the spectral shift to the blue (Fig. 3.c), line broadening and electron-phonon coupling. The relative intensity of the vibrational progression is higher for Cu and Ag interfaces. This result is consistent with the findings that the higher the disorder of a system, the larger is the value of the electron-phonon coupling. The strong influence of Cu and Ag surfaces on PPV layers suggests that metal oxidation can induce a great amount of structural changes into the organic film near the interface. However, there is no correlation between PL quenching and disorder in our films.

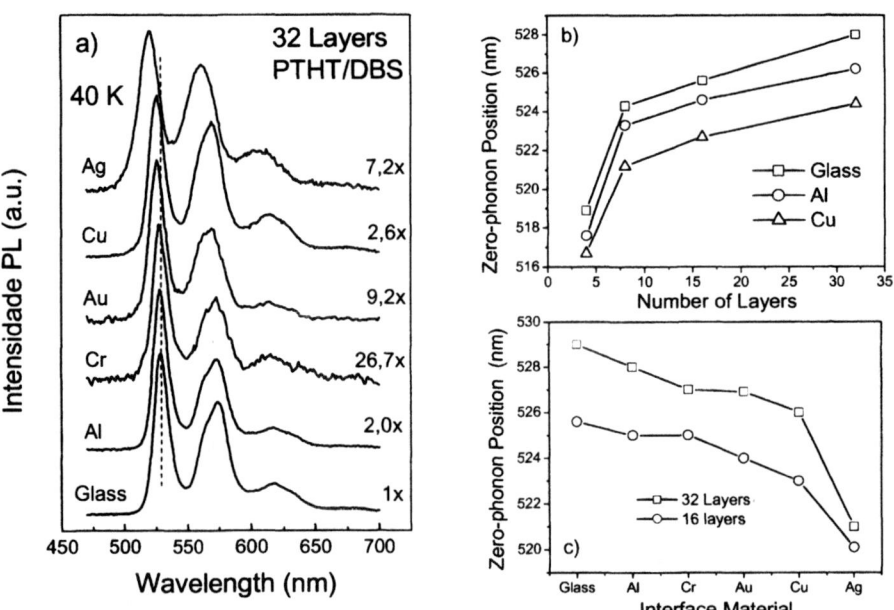

Figura 3: a) Normalized low temperature luminescence of 32 layers of SA PPV deposited on glass and on 50% transmittance layers of Cu, Ag, Au, Cr, Al on glass. b) Dependence of the zero-phonon position on the number SA PPV layers prepared on glass, Al and Ag surfaces. c) Dependence of the zero-phonon position on the interface material for films containing 32 and 16 SA PPV layers.

The thermal conversion of 50-bilayer SA PPV films on ITO/Glass substrate at different temperatures (100-230 °C), under vacuum for 30 min, indicated that a large amount of structural changes occur when high conversion temperatures are used. This can be readily seen in Figure 4 by the change in absorbance near the band edge and by the decrease of PL intensity measured at 40 K. Two identical samples were used for this experiment. Infrared (IR) measurements revealed that carbonyl groups are incorporated during the conversion procedure at high temperatures and

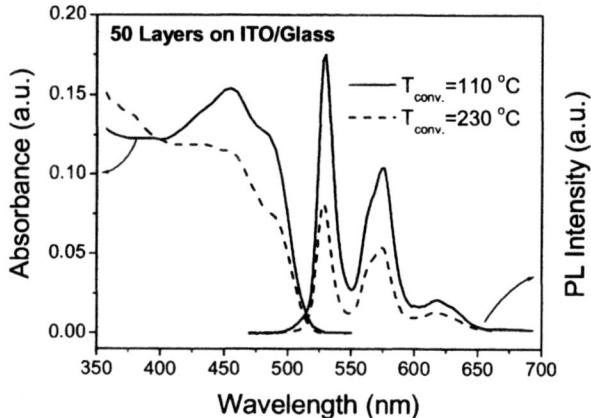

Figure 4: Absorbance (300 K) and luminescence (40 K) spectra of a SA PPV of 50 layers on ITO/Glass converted at 110 oC and 230 oC under vacuum during 30 min.

are responsible for the shortening of the conjugation length and PL quenching. When the conversion temperature is lowered (around 100 °C), no defect incorporation was detected by IR measurements. However, the conversion procedure of thin SA PPV layers under inert atmosphere also produces absorption modification when converted under standard temperatures, indicating that changes of the ITO/organic interfaces may be also occurring. Recently, ultraviolet photon electron spectroscopy (UPS) measurements have indicated that the polymer precursor for PPV interacts strongly with substrates during thermal conversion of ultra-thin films in UHV [3,6]. Some authors suggested that leaving groups, such as Cl, react with ITO during the conventional high temperature conversion procedure. We are performing experiments in order to distinguish between the effects of carbonyl incorporation at PPV surface or of ITO modifications. Anyway, the results in Figure 4 show that the low conversion temperatures used in this work (110 °C) are necessary to minimize the surface contamination and chemical modifications at the ITO/organic interface.

CONCLUSIONS

We have shown the influence of different interfaces on the optical properties of PPV ultra-thin layers prepared by a new self-assembly methodology. The advantages of this SA procedure lie in the control of mono-layer deposition, with thermal conversion being performed at considerably lower temperatures (80-100 °C) and in a few minutes. Highly ordered PPV films are produced which possess very low defect incorporation. We show that PPV layers are strongly influenced by the ITO and metal interface, as well as by the temperature of conversion. The interfacial electronic structure of such systems can be complex, and there remain many fundamental questions about the origin of contact effects on organic active layers.

AKNOWLEDMENTS

The financial assistance from the Brazilian agencies FAPESP and CNPq is acknowledged.

REFERENCES

1. I. D. Parker, J. Appl. Phys. **75**, 1656 (1994).
2. M.G. Mason, C.W. Tang, L.-S. Hung, P. Raychaudhuri, J. Madathil, L. Yan, Q. T. Le, Y. Gao, W.R. Salaneck, D.A. dos Santos, J.L. Bredas, J. Appl. Phys. **89**, 2756 (2001).
3. W.R. Salaneck, Th. Kugler, A. Andersson, P. Bröms, J. Birgerson, M. Lögdlund, "Conjugated Oligomers, Polymers, and Dendrimers: from Polyacetylene to DNA", Proceedings of the Fourth Francqui Colloquium, ed. J.L. Brédas (De Boeck Université, 1998) pp. 41-59.
4. K. Seki, E. Ito, H. Oji, D. Yoshimura, N. Hayashi, Y. Sakurai, Y. Hosoi, T. Yokoyama, T. Imai, Y. Ouchi, Synth. Met. **119**, 19 (2001).
5. J. C. Scott, G. G. Malliaras, Chem. Phys. Lett. **229**, 115 (1999).
6. R. Schlaf, C.D. Merritt, L.A. Crisafulli, Z.H. Kafafi, J. Appl. Phys. **86**, 5678 (1999).
7. A. Marletta, F.A. Castro, D. Gonçalves, O.N. Oliveira Jr., R.M. Faria, F.E.G. Guimarães, Synth. Met. **121**, 1447 (2001).
8. A. Marletta, D. Gonçalves, O. N. Oliveira Jr., R. M. Faria, F. E. G. Guimarães, Adv. Mater. **12**, 69 (2000).
9. D. D. C. Bradley, J.Phys. D: Appl. Phys. **20**, 1389 (1987).
10. R.F.M. Lobo, M. A. Pereira-da-Silva, M. Raposo, R. M. Faria and O. N. Oliveira Jr, Nanotechnology **10**, 389 (1999).
11. A.C. Fou, O Onitsuka, M.F. Rubner, J. Apll. Phys. **79**, 7501 (1996).
12. H. Hong, D. Davidov, H. Chayet, E.Z. Faraggi, M. Tarabia, Y. Avny, R. Neumann, S. Kirstein, Supramol. Sci. **4**, 67 (1997).

Mat. Res. Soc. Symp. Proc. Vol. 707 © 2002 Materials Research Society   AA10.6/DD11.6

Effect of Solvent Quality on the Friction Forces Between Polymer Brushes

Aaron M. Forster and S. Michael Kilbey II
Department of Chemical Engineering
Clemson University
Clemson, SC 29634

ABSTRACT

We have used the surface forces apparatus to measure the structural and frictional force profiles between opposing, solvated brush layers as a function of temperature. Two different polyvinylpyridine-polystyrene [PVP-PS] diblock copolymers were used to make PS brushes. The molecular weights (in thousands) of these PVP-PS materials were [114/103]k, [30/70]k, respectively. Structural and frictional force profiles in toluene and cyclohexane were measured, and the cyclohexane experiments were conducted at temperatures ranging from the theta-point to 50°C. In toluene the PS brushes needed to be compressed to ~1/5th of their equilibrium height before frictional forces were measured, but this onset of frictional forces was detected at a much lower level of compression in near-theta cyclohexane. In cyclohexane the structural force profiles were basically insensitive to the temperature change, but the frictional forces depended strongly on the solvent temperature. When the cyclohexane temperature was raised, the onset of frictional forces decreased toward the good-solvent onset. We also discuss the dependence of frictional force on shearing parameters.

INTRODUCTION

A polymer brush is formed when polymer chains are end-tethered to surfaces such that the distance between tethering points is less than the radius-of-gyration, R_g, of the free chain in solution. The untethered chains extend away from the surface to relieve the osmotic pressure created by the crowding of the chains. This stretched, upright structure is the distinguishing characteristic of a polymer brush. Because brushes straddle the solid-fluid interface, they can alter the wetting, wear, or lubricative properties of a surface.

The structural aspects of polymer brushes have been thoroughly reported on through both theory [1-3] and experiments [4-6], so few details are necessary here. In terms of the behavior of brushes subjected to shearing, previous research has shown that brush layers in a good solvent must be compressed to ~15% their equilibrium height before shear forces are measured [7-10]; however, when opposing brushes in a near-theta solvent are sheared past one another, frictional forces are measured when the brushes are only mildly compressed [8-10,13,14]. Along these lines, Grest has performed molecular dynamics simulations to explore how the shear forces between polymer brushes depend on solvent quality. Those simulations show the expected decrease in the range of the normal force profiles when the solvent condition is changed from good to near-theta; however the simulations do not show that the onset of frictional forces decreases (to smaller distances) as the solvent quality is improved [15]. We continue the study of how solvent quality affects the frictional behavior of brushes by focusing on the role of solvent temperature.

EXPERIMENTAL

A surface forces apparatus (SFA) was used to directly measure forces of interaction between two opposing polymer brush layers. The use of this device to measure static and dynamic forces between brushes has been thoroughly described [7-14]. Brushes were formed by preferential adsorption of polyvinylpyridine-polystyrene (PVP-PS) diblock copolymers from toluene. Because toluene is a poor solvent for PVP, it is driven to the mica surfaces of the SFA, leaving a layer of PS chains above the surface. The crowded PS chains stretch away from the surface to minimize the free energy of the layer [1,2].

We report on two different molecular weight PVP-PS brushes that are identified as [114/103]k and [30/70]k, where the numbers indicate the molecular weight, in thousands, of the PVP and PS blocks, respectively. It is important to assemble the brush from toluene (even for eventual work in cyclohexane) in order to prevent the PS blocks from adsorbing. Structural and frictional forces were measured in toluene at 32°C and in cyclohexane at 32°C, 40°C, and 50°C, and the system temperature was maintained within ± 0.1°C. The shearing experiments were conducted using a triangular ramp signal to drive one of the brush-covered surfaces past the other. The frictional response was measured at the opposing surface by monitoring the deflection of the spring on which this surface is mounted. The frictional forces we report are proportional to the deflection of that spring, and thus this total force is a combination of the viscous and elastic forces transmitted across the brush layers. The excitation amplitude was varied between 0.146 and 2.02 μm and the shearing frequency was varied between 0.01 and 0.2 Hz at different degrees of compression. After completing each set of shearing experiments, the normal forces were measured again to verify that the brushes remained intact.

RESULTS AND DISCUSSION

Figure 1 shows the structural force profiles – the forces of interaction as the brushes are compressed against one another – for brushes made from the [114/103]k and [30/70]k PVP-PS diblock copolymers. In all cases, the forces of interaction between the opposing, brush-covered surfaces are repulsive and long-ranged, and one-half of the distance at which the onset of repulsive forces was measured is equal to the equilibrium height of one brush layer. As the chains of the brush are concentrated by squeezing, it becomes increasingly difficult to compress the layer due to the increasing osmotic free energy. After each experiment, both the distance between tethering points, which was measured from the dry layer thickness, and universal scaling profiles confirmed that these polymer layers were stretched, i.e. brush-like.

When the solvent is changed from toluene (good for PS) to cyclohexane at 32 °C (near-theta for PS), as expected, the height of the brush shrinks dramatically. Manipulating the solvent quality by changing the temperature of the cyclohexane produces only a small change in the measured onset of repulsion between the layers.

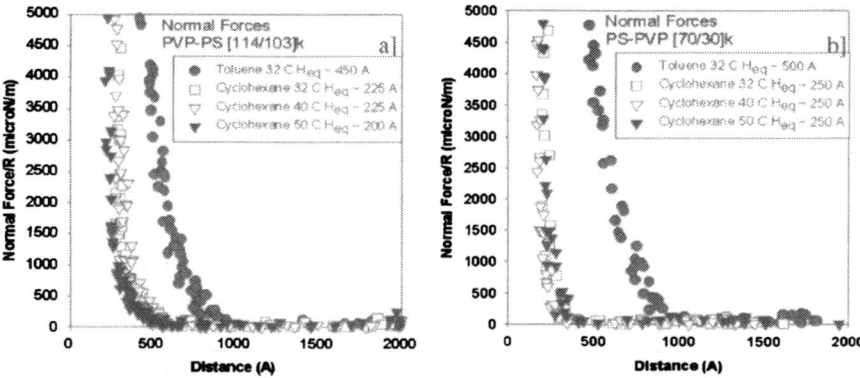

Figure 1: Structural force profiles for brushes made from the a] [114/103]k and b] [30/70]k PVP-PS diblock copolymers. The equilibrium height of the layer is one-half the distance of the onset of repulsive forces. The height of the layers decreases as the solvent quality is changed from good (toluene) to near-theta (cyclohexane at 32°C).

Although neutron reflectivity work has shown that the segment density profile extends with increasing temperature [16], it is possible that the changes in the segment density profile, which is diffuse at the outer edge of brush, do not significantly contribute to the force profiles measured with the SFA. In the case of the [30/70]k brush there is a small shift in the force profiles to larger distances as the solvent temperature is increased, and this is also seen with the [114/103]k brush when the temperature is raised from 32 to 40 °C. However, for this brush, it appears that the onset of structural forces, and therefore the brush height, shifts to smaller distances when the temperature is raised from 40 to 50°C. This is contrary to the expected behavior and may be due to the fact that this experimental data was gathered at different locations along the brush layer, and PS brushes in cyclohexane exhibit some surface heterogeneity [17]. Furthermore, after this brush was returned to toluene (from cyclohexane) the normal force profiles agreed with the previous data for this brush in toluene, suggesting that desorption was not the origin of this apparent contradiction.

The frictional forces are also affected by solvent quality. As shown in Figure 2, when bathed in toluene at 32°C, the brushes must be compressed well below their equilibrium height before frictional forces are measured with the SFA. In fact, frictional forces are not observed for the PVP-PS [30/70]k brush even when compressed to slightly less than $1/6^{th}$ of its equilibrium height. When the solvent is changed to cyclohexane at 32°C, frictional forces are measured when the brush layer is mildly compressed. This pattern of behavior agrees with previous results [8-10].

As the temperature of the cyclohexane is increased, thereby increasing the solvent quality, we observe a decrease in the distance at which we first detect frictional interactions. At 32°C, the onset of frictional forces occurred at ~288 Å and ~350 Å for the [114/103]k and [30/70]k brushes, respectively. At 40°C, frictional forces are observed at ~226 Å and ~279 Å for the two brushes, and ~209 Å and ~220 Å at 50°C. For the [114/103]k brush, the absolute distance at which we observe frictional forces in

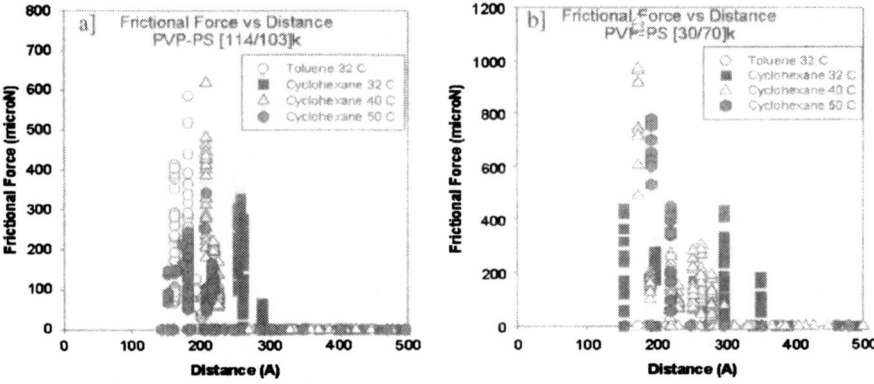

Figure 2: Frictional force profiles obtained by sliding one brush layer past the other. The force profiles correspond to the a] [114/103]k PVP-PS and the b] [30/70]k PVP-PS brushes. The onset of frictional forces decreases as the cyclohexane temperature is increased from 32 °C to 50 °C.

cyclohexane at 50°C is comparable to the distance at which we observe frictional forces in toluene (~200 Å). However, the corresponding changes in normal forces do not suggest that the solvent quality is approaching "good" because the layer height did not change significantly when the temperature was raised. A similar pattern of behavior has been reported for a single brush [9].

The data displayed in Figure 2 comprise many different runs at various shearing amplitudes and frequencies. This data can be dissected to examine the effect of excitation amplitude or frequency on the dynamic response of the brush layers. Here we will focus on the shearing results from the [114/103]k brush, although similar results were obtained from the [30/70]k PVP-PS.

In toluene at 32ºC and at a constant sliding frequency, the frictional forces appear to increase linearly with increasing shear velocity, as shown in Figure 3a. As the separation

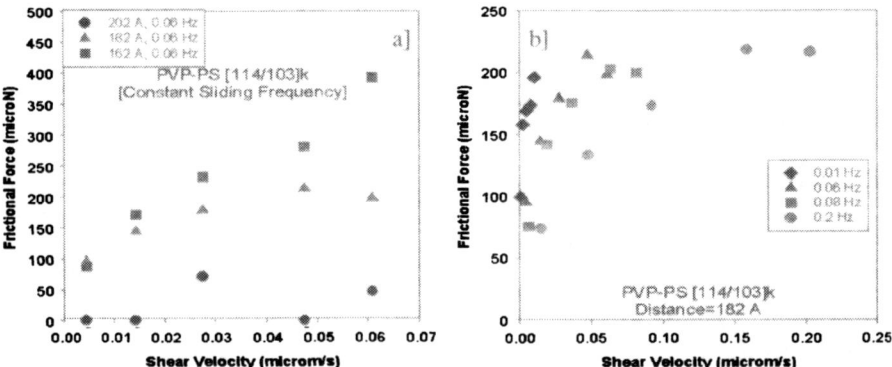

Figure 3: Frictional force dependence on shear velocity (equal to frequency*amplitude) for the PVP-PS [114/103]k brush in toluene at 32ºC. a] At constant frequency, decreasing the distance between the brush layers increases the frictional force. b] At constant separation distance, increasing the sliding frequency decreases the rise in frictional forces.

distance is decreased the frictional force increases, and the dependence on sliding velocity remains intact. These trends are similar to those reported by Schorr and coworkers [10] and the simulation results of Grest [15]. This behavior is expected because increasing the sliding amplitude exposes more of the brush layers to each other, thereby requiring more force to slide one brush layer past the other. Squeezing the layers together concentrates the polymer segments in the gap between the surfaces, so it is reasonable that as we increase the confinement, the frictional forces increase. Figure 3b shows the behavior when the brush layers are held at a constant surface separation and sheared against one another at different frequencies. The data indicate that as the sliding frequency is increased, the rate at which the frictional forces rise with sliding velocity decreases. This response indicates the mechanism of stress transmittance across the layer is also sensitive to frequency of oscillation.

The patterns of behavior observed in cyclohexane at 32, 40, and 50°C were similar to those in toluene; the frictional forces increased as the shearing velocity was increased, and the slope of that line increased with decreasing surface separation distance. Also, at a constant surface separation, the rise in the frictional forces with shearing velocity was more abrupt at lower frequencies. These behaviors are captured in Figures 4a and 4b, which are analogous to Figures 3a and 3b. The main difference between these patterns of behavior in cyclohexane and toluene is the reproducibility of the data. In cyclohexane at 32° C, the amount of scatter in the data increased and the trends were not as sharp. If one were to compare Figures 3a and 4a, it is apparent that the frictional forces in cyclohexane show a less severe dependence on shearing velocity at mild compression (larger distances). In comparing Figure 3b with 4b, it seems that at a given shearing velocity, the decrease in frictional force with increasing sliding frequency is not as pronounced in cyclohexane (compared to toluene). The exact reasons for these differences in the patterns of behavior are not known at this point. It is likely that the surface heterogeneity of the brush layers in cyclohexane contributes to the difficulty in measuring the frictional response of the layers.

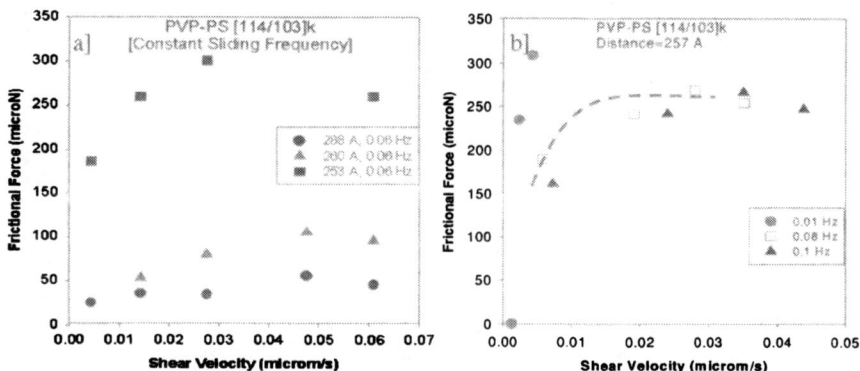

Figure 4: Frictional force dependence on shear velocity (equal to frequency*amplitude) for the PVP-PS [114/103]k brush in cyclohexane at 32°C. a] At constant sliding frequency, decreasing the distance between the brush layers increases the frictional force. The dependence is not as strong at weaker compressions. b] At constant separation distance, increasing the sliding frequency decreases the rise in frictional forces. The line for 0.08 Hz is a guide for the eye.

CONCLUSIONS

When the solvent was changed from toluene to cyclohexane, the range of the structural force profiles decreased, reflecting the contraction of the brushes. When the brushes were in cyclohexane and the temperature was raised, the SFA detected only a slight increase in the height of the layer as the temperature was increased. Because the SFA measures the force profile, but not the segment density profile, this small height change detected may not be a true indication of temperature-induced swelling of the layers. Our results show that the frictional behavior of the brushes in cyclohexane was strongly impacted by the increase in temperature. As the temperature was raised from near-theta to 50°C, the distance at which frictional forces are first measured decreased toward the good solvent limit. The dependence of frictional forces on shear velocity was similar for both solvents during constant frequency and constant surface separation experiments. However, it appears that the heterogeneity of the brush layer in cyclohexane complicates these frictional measurements.

ACKNOWLEDGEMENTS

Support for this work from the National Science Foundation (CTS-9816147) and 3M, through the Untenured Faculty Award program, is gratefully acknowledged.

REFERENCES:

1. S. Alexander, J. Phys. (France) **38**, 983 (1977).
2. P. G. de Gennes, Macromolecules **13**, 1069 (1980).
3. S. T. Milner, J. Chem. Soc. Faraday Trans. **86**, 1349 (1990).
4. S. M. Kilbey, H. Watanabe, M. Tirrell, Macromolecules **34**, 5249 (2001).
5. P. Auroy, Y. Mir, L. Auvray, Phys. Rev. Lett. **69**, 93 (1992).
6. H. Watanabe, M. Tirrell, Macromolecules, **26**, 6455 (1993).
7. J. Klein, E. Kumacheva, D. Mahalu, D. Perahia, L. J. Fetters, Acta. Polym. **49**, 617 (1998).
8. S. M. Kilbey II, P. Schorr, M. Tirrell, in *Dynamics of Small Confining Systems IV*, edited by J.M. Drake, G.S. Grest, J. Klafter, R. Kopelman, (Mater. Res. Soc. Proc., Pittsburgh, PA 1999).
9. P. A. Schorr, T. C. B. Kwan, S. M. Kilbey II, E. S. G. Shaqfeh, M. Tirrell, submitted to *Macromolecules*.
10. Schorr, P. A., Ph.D. thesis, University of Minnesota (2000).
11. J. Klein, D. Perahia, S. Warburg, Nature **353**, 143 (1991).
12. J. Klein, E. Kumacheva, D. Perahia, D. Mahalu, S. Warburg, Faraday Discuss. **98**, 173 (1994).
13. S. Granick, A. L. Demirel, L. L. Cai, J. Peanasky, J. Chem. (Israel) **35**, 75 (1995).
14. A. Dhinojwala, L. Cai, S. Granick, Langmuir **12**, 4537 (1996).
15. G. S. Grest, Adv. Poly. Sci. **138**, 149 (1999).
16. A. Karim, S., Satija, J. G. Douglas, J. F. Ankner, L. J. Fetters, Phys. Rev. **73**, (1994)
17. T. W. Kelley, P. A. Schorr, K. D. Johnson, M. Tirrell, C. D. Frisbie, Macromolecules, **31**, (1998), 4297

Mat. Res. Soc. Symp. Proc. Vol. 707 © 2002 Materials Research Society   AA10.8/DD11.8

Characterization of Thin Polymer Blend Films using ESEM
– No Charging, No Staining.

Ian C. Bache, Catherine M. Ramsdale, D. Steve Thomas, Ana-Claudia Arias,
J. Devin MacKenzie, Richard H. Friend, Neil C. Greenham and Athene M. Donald
Cavendish Laboratory, Madingley Road, Cambridge CB3 0HE, UK

ABSTRACT

Characterising the morphology of thin films for use in device applications requires the ability to study both the structure within the plane of the film, and also through its thickness. Environmental scanning electron microscopy has proved to be a fruitful technique for the study of such films both because contrast can be seen within the film without the need for staining (as is conventionally done for electron microscopy), and because cross-sectional images can be obtained without charging artefacts. The application of ESEM to a particular blend of relevance to photovoltaics is described.

INTRODUCTION

It has recently become apparent that quantum efficiencies of photovoltaic devices can be significantly enhanced in conjugated polymer blends in which phase separation occurs on a lengthscale comparable to the exciton diffusion distance[1]. In such blends, exciton dissociation is promoted at the many polymer-polymer interfaces. This leads to enhanced generation of charge carriers which can then migrate towards the appropriate electrodes, as long as there is a suitable continuous path for this to occur. One can therefore imagine an idealised structure for a photovoltaic device, consisting of an interdigitated bicontinuous polymer network, in which there is a concentration of the electron acceptor in the vicinity of the cathode and the hole acceptor polymer at the anode.

It is therefore necessary, when polymer blends are being prepared as thin films for photovoltaic device application, to examine the morphology to test how far such an idealised structure is being realised. Most commonly atomic force microscopy (AFM) is used to explore the morphology, and for many purposes this is excellent, especially where there is significant topography present[2]. However, it does have limitations, particularly if the polymers are very similar and there is little topographical detail. Also AFM provides information about the surface and, except in certain circumstances, information about the phase- and composition-distribution through the cross section of the film can only be inferred from this technique with additional data from other techniques. However, it is precisely this structure which is directly relevant to device performance in photovoltaics and LED's. This paper describes the application of environmental scanning electron microscopy (ESEM) to the study of polymer blends relevant to photovoltaic devices. It will be shown that the mode of operation of the ESEM has particular advantages for the study of phase-separated polymers, enabling composition contrast and cross-sectional information in conjugated polymer films to be obtained.

BACKGROUND TO ESEM

The ESEM is a commercial instrument, which crucially differs from a conventional SEM in that it permits the presence of a gas in the sample chamber (see e.g. [3]). Typically (but not necessarily) the gas is water vapour. For semiconductors, such as the polymer blend under study here, or insulators the presence of the gas has the effect of eliminating the standard problem of charging without the need for a conductive coating to be applied. The reason for this is that the gas molecules are ionised by the secondary electrons emitted from the sample, producing positive gaseous ions (as well as daughter electrons which lead to an amplification of the signal from the sample). These ions migrate back towards the sample surface where they can modify the normal charge build-up. Thus, ESEM offers a very real advantage over conventional SEM for the imaging of semiconductors such as polymers: the absence of a coating implies fine surface detail is not obscured. In many cases, the buildup of charge and heat in low conductivity polymer films during conventional SEM leads to poor imaging and sometimes sample degradation.

However, there is an additional advantage conferred by the lack of coating, which is that inherent spatial variations in electron emission are also not obscured. This has previously been demonstrated for the case of liquid emulsions, and a simple model proposed in terms of differences in band gap between different materials [4] – in that specific case an unsaturated oil and a water phase. Electrons generated within the material can only escape the surface and contribute to the detected signal if they can traverse the material without losing so much energy they cannot overcome the work function. This indicates that there will be a relationship between the local electronic structure (such as variations in electron affinity, density of states and band gap) and the observed emission.

Figure 1 shows a spun-cast film of a 1:1 blend of polystyrene (which has an unsaturated benzene ring) and polymethylmethacrylate. Strong contrast is seen between the two phases, and since this is a 50:50 blend, the structure is approaching a bicontinuous structure, although the lighter PMMA phase is tending to form the matrix. Thus it can be seen that, upon spin casting, no further specimen preparation is required to yield strong observable contrast, and such images provide a ready means of studying the detailed morphology of these blends.

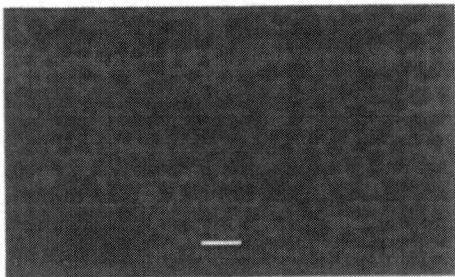

Figure 1. An ESEM image of a 50:50 blend of PS and PMMA prepared by spin casting. The darker phase is the PS phase. The bar represents 1μm.

APPLICATION TO MORPHOLOGICAL STUDY OF POLYMER FILMS FOR DEVICE APPLICATIONS

As indicated overleaf, for photovoltaic devices a bicontinuous interdigitated morphology is wanted, but also one with a continuous path for the dissociated charges to move to the appropriate electrode. Thus, phase separation on an appropriate lengthscale is required, but ideally the phase separation should also consist of phases segregated across the film thickness. ESEM has been applied to the study of this problem. The instrument used was a Philips XL30 ESEM-FEG. This was used to study films of F8BT (poly(9,9'-dioctylfluorene-co-benzothiadiazole), the electron acceptor polymer and PFB (poly (9,9, - dioctylfluorene-co-bis-N,N(4-butylphenyl)-bis-N,N-phenyl-1,4 phenylenediamine), the hole acceptor polymer.

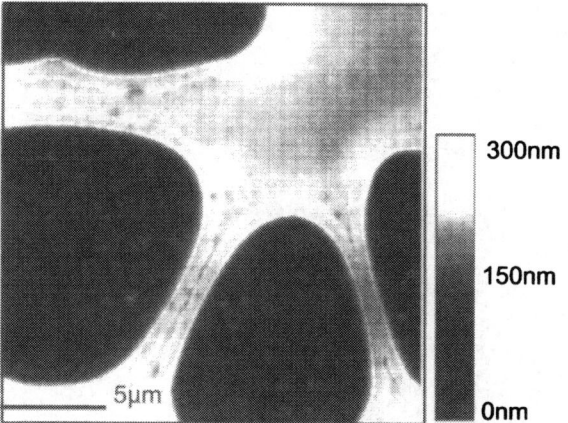

Films were spun-cast onto silicon wafers from *p*-xylene solutions at room temperature, using a mixture of 1:1 polymer composition at a concentration of 15g/l, and a spinning rate of 2500 rpm for 60s in air.

Previous AFM studies have shown that phase separation occurs on several lengthscales in such systems [1]. There is a coarse phase separation into droplets of PFB-rich material in a continuous matrix of F8BT-rich polymer. However, within both the droplets and the matrix there is a finer scale of phase separation occurring. This can be seen clearly in the AFM image shown in figure 2. Using this technique, and in this mode, the primary mode of contrast is topography,

Figure 2. AFM image of a 50:50 blend of PFB and F8BT showing droplets of PFB – rich material in a continuous matrix of F8BT. Further phase separation within the droplets is clearly visible, and to a lesser extent within the matrix.

showing that the regions of the droplets are thinner compared with the matrix. Fluorescence microscopy has shown that the high phase on the AFM image corresponds to F8BT-rich phases, and the lower phase corresponds to PFB-rich phases [2].

ESEM can also be used to image these same systems. Figure 3a shows the corresponding morphology, and the similarities are striking although the source of the contrast is not identical. Nevertheless the same hierarchy of lengthscales on which phase separation is occurring can be seen, and in the ESEM the subsidiary phase separation within the matrix can be more clearly seen than in the AFM. In this image it is presumed the contrast arises primarily from electronic structure, as discussed above, but the effects are complex and not yet completely understood, with the arguments given above simply being a first crude attempt at rationalising the many effects that are likely to contribute to the secondary electron yield.

Furthermore, if the accelerating voltage of the primary beam is now increased from 3 keV to 5 keV, an apparent inversion of the contrast is seen (figure 3b). The reason for this lies in the fact that at the higher voltages significant penetration of the electron beam can occur into the underlying silicon substrate where the sample is sufficiently thin. As the AFM image shows, the PFB droplets are thinner than the surrounding matrix, and thus penetration of the beam into the substrate will occur more readily here. The silicon is a source of a significant number of backscattered electrons (as opposed to the secondary electrons which form the bulk of the signal from the polymer film itself), which will markedly enhance the overall yield of electrons in these regions. Thus, the signal from the PFB-rich, but thinner regions, is now greater than from the F8BT-rich (and thicker) matrix, even though the intrinsic yield from the F8BT polymer may be higher. The pair of images shown in figure 3 demonstrate that, although ESEM images can show significant contrast between polymers without staining, coating or other specimen preparation, nevertheless care must be taken in interpreting the images.

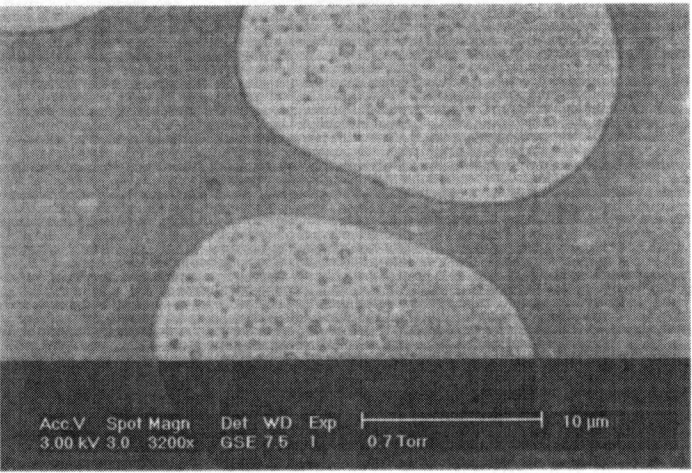

Figure 3a ESEM image showing phase separation in a PFB/F8BT blend; image taken with an accelerating voltage of 3 keV.

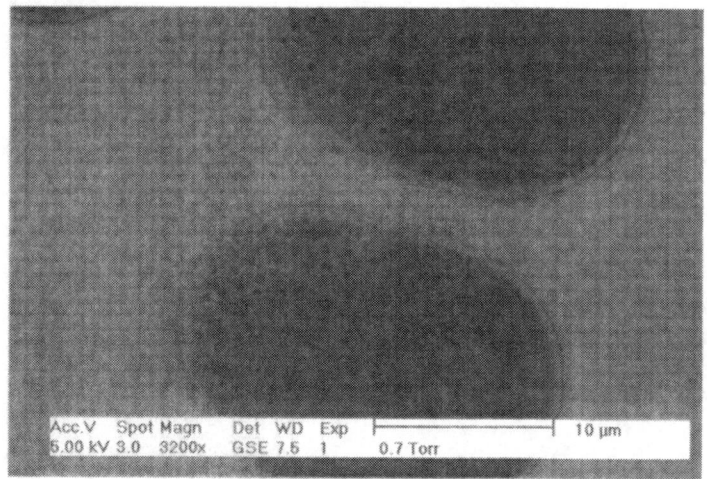

Figure 3b. As above, but the accelerating voltage now increased to 5 keV.

In order to characterise the structure of the film in the third dimension, samples were also fractured at low temperature. This enabled cross-sectional imaging, and hence the morphology through the thickness of the film could be explored. Figure 4 shows an example of such a cross-section. To achieve this cross-section the sample was cooled in liquid nitrogen after spin-casting onto silicon, and fractured immediately prior to imaging. Since silicon can readily be cleaved, this fracture can readily be carried out with minimal damage to the polymer film itself.

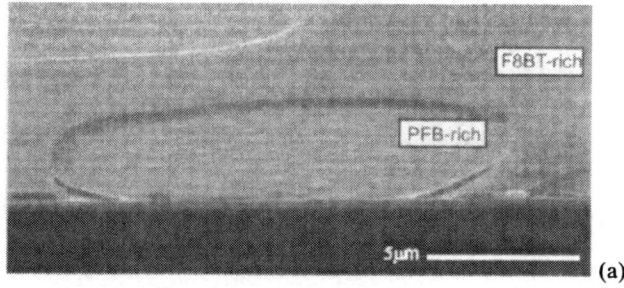 (a)

Figure 4. ESEM image of a fractured cross-section of a blend at two magnifications. (a) Shows the general morphology.

45

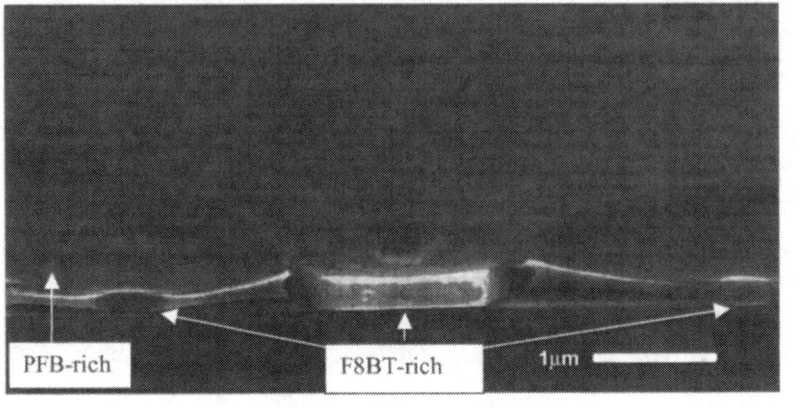

(b)

Figure 4. (b) Shows a close-up, indicating that F8BT domains exist within.

This cross-section demonstrates that both the continuous F8BT-rich phase, and the included PFB-rich phase span the film thickness. In addition the droplets of the PFB-rich phase are thinner than the matrix, with a clear step at the rim. However, within the F8BT-rich phase, the finer dispersed phase (just visible in Figure 4b at the edge) does not appear to extend through the thickness of the film. Further work is required to establish how any surface modification of the silicon (with its native oxide layer) affects the morphology.

From these images it therefore appears that there is effectively a columnar structure within these films, providing a path for efficient charge transport. The morphology both provides fairly large interfacial areas for exciton diffusion and also a path for both positive and negative charges to move to the relevant electrode. This structural information is consistent with the relatively high efficiencies which have been found for photovoltaic devices using these films (2% peak external quantum efficiency) [2].

CONCLUSIONS

ESEM clearly has significant advantages for studying polymer systems relevant to devices. The cross section analysis is particularly important for the complete understanding of conjugated polymer-based optoelectronic devices, such as photodiodes and LED's. Strong contrast can be seen between components in blends, without any need for staining, and a conductive coating is not required to prevent charging. Further work is required to establish quantitatively how contrast can be interpreted to provide information on phase composition.

ACKNOWLEDGEMENTS

The authors would like to thank the EPSRC for funding. ACA thanks the Brazilian Government for funding under a CNPq scholarship.

REFERENCES

1. A C Arias, J MacKenzie, R Stevenson et al., "Photovoltaic Performance and Morphology of Polyfluorene Blends: A Combined Microscopic and Photovoltaic Investigation," *Macromolecules* **34**, 6005-13 (2001).
2. A-C Arias, "Conjugated polymer phase separation and three dimensional thin film structure for photovoltaics," PhD, Cambridge University, 2001.
3. G D Danilatos, "Review and outline of environmental SEM at present," *J. Microscopy* **162**, 391-402 (1990).
4. D J Stokes, B L Thiel, and A M Donald, "Direct observation of water-oil emulsion systems in the liquid stated by environmental SEM," *Langmuir* **14**, 4402-8 (1998).

Mat. Res. Soc. Symp. Proc. Vol. 707 © 2002 Materials Research Society   AA10.11/DD11.11

Structural Characterization of Segmented Polyurethanes by Small Angle Neutron Scattering

Loren I. Espada[‡], Joseph T. Mang[+], E. Bruce Orler[†], Debra A. Wrobleski[‡], David A. Langlois[†] and Rex P. Hjelm[‡]
[‡]Los Alamos Neutron Science Center (LANSCE-12)
[+]Dynamic Experimental Division (DX-2)
[†]Materials Science and Technology Division (MST-7)
Los Alamos National Laboratory
Los Alamos, NM 87545

ABSTRACT

The beneficial mechanical properties of segmented polyurethanes derive from microphase separation of immiscible hard and soft segment-rich domains at room temperature. We are interested in the structure of the domains, how these are affected by hydrolytic aging, and how the structure is modified by low molecular weight plasticizers. To assessed the distribution of the plasticizer in polyurethane, we did small-angle neutron scattering measurements on mixtures of 23% hard segment poly(esterurethane) with different amounts of either non-deuterated or deuterated plasticizer. We analyzed the results using a simple model in which the contrast, $\Delta\rho = \rho_H - \rho$, between the hard and soft segment-rich domains is varied by the amount of deuterated or hydrogenated plasticizer, using the fact that $I(Q) \sim \Delta\rho^2$. The result demonstrated that the plasticizer is largely associated with the soft segment rich domains. The structure of PESU with the chain extender of the hard segment was assessed after aging under hydrolytic conditions. The results show that the microphase structure coarsens and segregates and that the hard and soft segments segregated as a result of the loss of constraints from hydrolytic soft segment chain scission. The results on plasticizer distribution and the effects of hydrolytic aging give insight on the loss of mechanical properties that occur in each case.

INTRODUCTION

It is well known that the beneficial properties of segmented polyurethanes at room temperature are a result of microphase separation of the hard (crystalline) and soft (rubbery) segments into hard and soft segment-rich domains[1-4]. We are interested in understanding what structural characteristics change during the aging processes that lead to the degradation of these mechanical properties. Plasticizers, used as processing aids, also have profound effects on the mechanical properties of segmented polyurethanes, and we want to understand the structural basis of the properties of plasticized polyurethanes.

The effects of aging on the polymer properties are critical to predicting the service life of a polymeric material. The aging of multiphase systems such as segmented polyurethanes is complex in principle, as aging may involve more than one important interrelated chemical and/or physical processes. The chemical processes may involve chain scission or crosslinking reactions or combination of these. The physical characteristics that can change with aging are the size, distribution, interfacial bonding and mixing of the hard and soft segment-rich phases, and these can be linked to chemical processes.

Investigations of of hydrolytically-aged 23% hard segment poly(ester urethane) (PESU) in humid air showed profound changes in the mechanical and thermal properties [5]. Gel Permeation Chromatography (GPC) studies showed that the molecular weight of aged samples decreased with aging and directly correlated to the changes in the mechanical and thermal properties [6].

Plasticizers are low molecular weight compounds that modify the mechanical properties of polymeric materials. In previous work it has been observed that the addition of a plasticizer increases the strain at break and the stress at break, and decreases the Young's modulus of the polymer. Since the domain structure of polyurethane has a significant impact on the mechanical properties, we are interested in how plasticizers distribute between the two phases and how this might affect the structure of polyurethanes. Previous spectroscopic investigations have focused on the static and dynamic dichroic responses of the hard and soft segments upon orientation of a poly(ester urethane) containing a nitroplasticizer. Graff and coworkers found that the plasticizer disrupts the hydrogen bonding of both the hard and soft segments [7].

Plasticized-segmented PESU is used as a binder in high explosives to impart structural integrity to the composite and decrease sensitivity to external stimuli. Our motivation is to understand the structural effect of hydrolytic aging and plasticizers to help provide predictive models for the safety, performance, and lifetime of high explosive materials.

To investigate the effects of plasticizer and hydrolytic aging we synthesized segmented PESU with 23 % hard segment content, a model polymer for Estane® 5703, used in high explosives as a binder. Small angle neutron scattering (SANS) was utilized to determine the morphology. For investigations of hydrolytic aging, we synthesized 23% hard segment PESU with deuterated chain extender in the hard segment to increase the contrast between the hard and soft segment-rich domains. The distribution of plasticizer in plasticized polyurethane was determined using deuterated and hydrogenated plasticizer in a contrast variation experiment.

EXPERIMENTAL DETAILS

Materials
An analogue of Estane® 5703, was synthesized with approximately 23 wt% hard segments composed of 4,4'-methylenediphenyl 1,1'-diisocyanate (MDI) and a 1,4 butanediol (BDO) chain extender. The soft segments are comprised of poly(butylene adipate). For the hydrolytic aging studies, the hard segment was synthesized with deuterated BDO. The plasticizer (NP) is a 50/50 wt% eutectic mixture of bis(2,2-dinitropropyl)formal and bis(2,2 dinitropropyl)acetal. The hydrogenated-NP (h-NP) was obtained from the Dynamic Experiments Division of Los Alamos National Laboratory. The synthesis of deuterated-NP (d-NP) will be described in a subsequent publication. The calculated scattering length density values of the materials are listed on Table 1.

Film Preparation
Estane films containing between 0 and 50 wt% d-NP or h-NP were cast from solution. The solutions were prepared by dissolving the Estane in methyl ethyl ketone. After dissolution, the required amount of plasticizer was added; the solution was cast and allowed dry overnight. The films were then dried under vacuum for one week.

For studies of hydrolytic aging, pellets of PESU were compression molded into 1-2 mm thick films using a hydrolytic press (Carver) with heated platens. The pellets were preheated to 110°C for 5 min before pressure was applied. After being held at 6000 psi, 110°C ± 1°C for 5 min, the films were rapidly cooled to room temperature. To allow for equilibration of the morphology, the films were held at room temperature for no less than 2 weeks before being used for the aging studies.

Environmental Aging of Polyurethanes Samples

Compression molded films were sealed in containers over an aqueous saturated NaCl solution for the 75% relative humidity conditions in air. The containers were purged through a septum and were then placed in convection ovens at 70°C ± 5°C.

SANS Measurements

Small-angle neutron scattering measurements were performed on the time-of-flight, low-Q diffractometer, LQD, at the Manuel Lujan Jr. Scattering Center. Data were reduced by conventional methods [8] to differential scattering cross section per unit volume, $I(Q)$ (cm^{-1}), as a function of momentum transfer, $Q = 4\pi/\lambda \sin\theta$, where 2θ is the scattering angle and λ is the incident neutron wavelength.

Table 1. Scattering Length Densities of Poly(ester urethane) and Hydrogenated and Deuterated Plasticizer Components

Name	Scattering Length Density $\rho(10^{10} cm^{-2})$
Estane® 5703	Overall 1.3
	Hard Segment 2.5
	Soft Segment 0.9
Bis(2,2 dinitropropyl) formal	Hydrogenated 2.5
	Deuterated 4.3
Bis(2,2 dinitropropyl) acetal	Hydrogenated 2.6
	Deuterated 4.6

RESULTS AND DISCUSSION

Figure 1a shows the effect of d-NP content on scattering intensity from PESU. The unplasticized sample exhibits a broad scattering peak with a maxima between $Q = 0.02$ and 0.03 $Å^{-1}$. At low d-NP content a decrease in the intensity is observed. Higher d-NP content results in increased intensity with a peak at about 0.023 $Å^{-1}$. The SANS of polyurethane with h-NP (shown in Figure 1b) is significantly different from that observed with d-NP (Fig.1a): the addition of h-NP results in a monotonic decrease in the scattering intensity as the h-NP content is increased.

We have analyzed these results in terms of a simple model in which the contrast between the hard and soft segment-rich domains is varied by the presence of d-NP or h-NP. This model uses the fact that $I(Q) \sim \Delta\rho^2$, where the contrast, $\Delta\rho = \rho_H - \rho_S$, is the difference between the scattering length densities of hard and soft segment-rich domains, respectively. We calculate ρ_H and ρ_S using the known scattering length densities of polyurethane hard and soft segments, h-NP and d-NP, ρ_{np} (Table 1). In this simple model we assume that the hard and soft segments are completely segregated into their respective domains, and that the NP is partitioned into one or the other of these domains. The contrast between the hard and soft segment domains in this model is varied by the amount of d-NP or h-NP present.

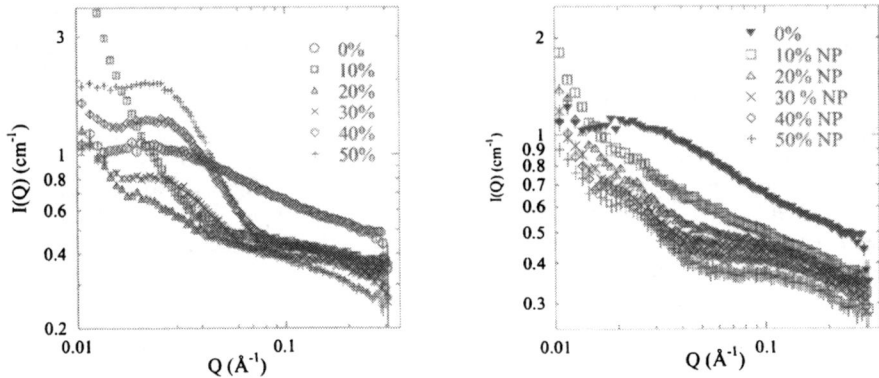

Figure 1. The effect of plasticizer content on the SANS of PESU, (a) d-NP and (b) h-NP

With these assumptions, the contrast when NP partitions into the soft segment-rich domains is, $\Delta\rho = [\rho_H - (\rho_S \phi_S + \phi_{NP} \rho_{NP})]$. Here ϕ_S is the volume fraction of soft segment and ϕ_{NP} is the volume fraction of h-NP or d-NP in the soft segment-rich domain. An analogous expression applies to the case of NP partitioning into the hard segment-rich domain.

Figure 2 shows the predicted intensity as a function of h-NP or d-NP content assuming a partition of the NP into the soft segments. The predicted intensity for h-NP decreases monotonically with increasing h-NP content. The intensity of the samples containing d-NP, on the other hand, goes through a minimum. This contrasts with the prediction assuming that the NP partitions into the hard segment-rich domains. In this case the intensity with increasing h-NP content decreases monotonically, but the intensity increases monotonically with added d-NP. Comparison of these models with the observations (Fig. 2) strongly suggests that the nitroplasticizer partitions into the soft segment-rich domains and causes minimal perturbation to the hard domains. The discrepancies between the model and the data (Fig. 2) are likely to be from incomplete segregation of the hard and soft segments, partial partitioning of the NP into the hard segment domains and heterogeneity among the domains, non of which is accounted for in the model.

Previous Differential Scanning Calorimetry data show that the nitroplasticizer displaced the glass transition temperature of the soft segments toward lower values, but had little effect on the endotherm between 30 and 80°C that is associated with the disruption of the hard domains. These results are also consistent with the model that the plasticizer preferentially interacts with the soft segment-rich domains.

SANS of BDO-deuterated PESU showed a broad scattering maximum centered about $Q = 0.04$ Å^{-1}. After 13 days of degradation under hydrolytic conditions the peak sharpens dramatically with a distinct maximum at $Q = 0.039$ Å^{-1}. With further hydrolysis the peak increases in intensity and moves to lower Q-values, the peak being at $Q = 0.036$ Å^{-1} in the sample aged for 70 days. We can qualitatively assess this result using an approximation for the scattering intensity that decouples the contribution from the average domain form facto, P(Q) and the structure factor, S(Q), namely, $I(Q) = \Delta\rho^2 P(Q)S(Q)$. Thus, the peak increase is explained as a change in contrast between the hard and soft segment-rich domains, and the shift in the maxima is due to an increase in the inter-domain spacing.

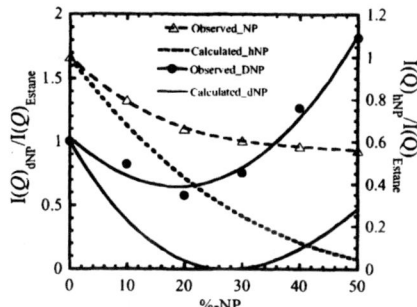

Figure 2. (a) Calculated and observed intensity of the d-Np and h-Np into the soft segment-rich domain, (b) calculated intensity for partitioning of h-Np into the hard segment-rich domain.

Degradation of the polyester soft segments decreases the molecular weight of the polymer observed using GPC [6]. The chain scission decreases the constraints imposed by the chemical linkage of the immiscible hard and soft segments. Thus, the result shown in Figure 3 can be explained by increased segregation of the hard and soft segments into the respective domains and the consequential coarsening of the domain structure of PESU.

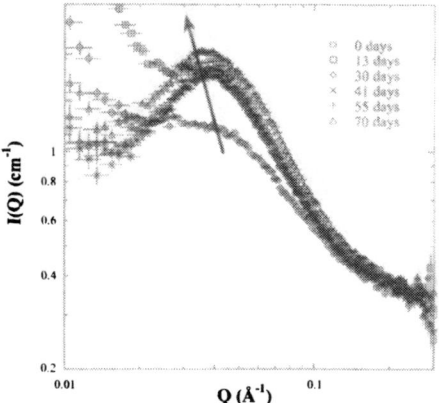

Figure 3. SANS of deuterated model Estane hydrolyzed for various times at 75% relative humidity at 70°C. The arrow indicates that aging resulted in a shift in the peak toward smaller Q with hydrolysis. Increased intensity with hydrolysis indicates stronger segregation between hard and soft domains.

CONCLUSIONS

The present work demonstrates a unique advantage in using SANS and deuteration to deconvolute complex morphologies. The SANS results show that the plasticizer partitions mostly into the soft domains. Hydrolysis causes chain scission of the soft segments. The chain scission decreases the constraints imposed by high molecular weight on the mobility of the hard and soft segments that leads to increased segregation of the hard and soft segments and to coarsening of the domain structure.

REFERENCES

1. Schneider, N.S.; Desper, C.R.; Illinger, J. L. King, A.O.; Barr, D.J. *Macromol. Sci., Phys.* **1975**, *B11*, 527.
2. Korberstein, J.T.; Stein, R.S. *J. Polym. Sci., Polym. Phys.* Ed **1983**, *23*, 1439.
3. Van Bogart, J.W.; Gibson, P.E.; Cooper, S.L. *J. Polym. Sci., Polym. Phys.* Ed **1983**, *21*, 65.
4. Miller, J.A.; Cooper, S.L.; Han, C.C.; Pruckmayr, G. *Macromolecules* **1983**, *17*, 1063.
5. Orler,. Bruce E, Wroleaski, D.A., and Smith, M. E., *Polymer Preprints,* **1998**, *39*, 763.
6. Orler,. Bruce E, Wroleaski, D.A, *manuscript in preparation.*
7. Graff, D.K.; Wang, H.; Palmer, R.A.; Schoonover, J.R. *Macromolecules* **1999**, *32*, 7147.
8. Seeger, P.; Hjelm, R P. *J. Appl. Crystallogr.* **1991**, *24*, 467.

Mat. Res. Soc. Symp. Proc. Vol. 707 © 2002 Materials Research Society   AA11.3/BB11.3

Reductive photopatterning of phenylene-vinylene-based polymers

T. Kavc[1], G. Langer[1], W. Kern[1 *)], A. Ruplitsch[1], K. Mahler[1], F. Stelzer[1], G. Hayn[1], R. Saf[1], E.J.W. List[2], E. Zojer[2], M. T. Ahmed[2 #], A. Pogantsch[2], K.F. Iskra[3], T. Neger[3], H.-H. Hörhold[4], H. Tillmann[4], G. Kranzelbinder[5], E. Toussaere[5], G. Jakopic[6]

[1] Institute for Chemistry and Technology of Organic Materials, Graz University of Technology, Graz, Austria
[2] Institute of Solid State Physics, Graz University of Technology, Graz, Austria
[3] Institute for Experimental Physics, Graz University of Technology, Graz, Austria
[4] Institute of Organic Chemistry and Macromolecular Chemistry, University of Jena, Germany
[5] Ecole Normale Superieure de Cachan LPQM, Cachan, France
[6] Joanneum Institute of Nanostructured Materials and Photonics, Weiz, Austria
[#] Affiliation: Mansoura University, Mansoura 35516, Egypt
[*)] To whom all correspondence should be adressed

ABSTRACT

Photochemical methods were developed to obtain a variation of the refractive index in aromatic polymer surfaces and a change in the photoluminescence characteristics of phenylene-vinylene-based polymers. Films of aromatic polymers, among them polystyrene (PS), poly(2-vinylnaphthalene) (PVN) and derivatives of poly(p-phenylene-vinylene) (PPV) were UV irradiated in the presence of gaseous hydrazine (N_2H_4). The photoreaction led to a strong reduction of the refractive index of the polymers due to a hydrogenation of the aromatic units. In the case of PPV, we observed reductive photobleaching. This new technique was employed to produce photogenerated patterns in PPV. The results are compared to oxidative bleaching.

INTRODUCTION

Several patterning strategies have been developed for conjugated polymers, e.g. to obtain pixelated LEDs for multicolored displays. Besides conventional photolithography, direct patterning strategies for conjugated polymers are of immediate interest. Examples of current techniques are laser ablation [1], ink-jet [2] and ion-beam deposition techniques, controlled vapour deposition [3], modified scanning force microscopy [4] and local dye diffusion [5]. UV irradiation under oxygen (i. e. photooxidative patterning) is an emerging technique to change the emissive properties of conjugated polymers [6-10]. Only recently, luminescent patterns in LED devices have been produced by oxidative bleaching [11-12]. However, detailed informations on the reaction mechanisms and photooxidation products (e.g. ketones which act as quenching centers) are seldom available. In the present work it is shown that reductive bleaching of conjugated polymers via UV illumination under gaseous hydrazine (N_2H_4) is an interesting alternative for patterning of PPV derivatives.

EXPERIMENTAL DETAILS

Films of polystyrene (PS), poly(2-vinylnaphthalene) (PVN) and MEH-PPV were cast onto Si and CaF$_2$ plates using a spincoater. The films were placed in an irradiation chamber equipped with quartz windows. Hydrazine hydrate (N$_2$H$_4$.H$_2$O) was dehydrated following a literature method [13], placed in a thermostatted vessel (50°C) and transported with a stream of nitrogen. The gas stream passed the polymer films under irradiation. UV irradiation was carried out with an unfiltered high-pressure Hg lamp (1300 W). Refractive indices were measured with spectroscopic ellipsometers from Sopra and Woolam Inc., IR spectra were taken with a Perkin Elmer Spectrum One. UV/VIS absorption spectra were recorded with a Perkin Elmer Lambda-9, emission spectra were taken with a Shimadzu RF5301 spectrofluorometer (excitation at wavelength of maximum absorption, optical spot size 3 x 10 mm, incident power in the μW range).

RESULTS AND DISCUSSION

The introduction of basic functionalities groups onto polymer surfaces such as polytetrafluoroethylene (PTFE) [14] can be achieved easily utilizing the photochemical reactions of hydrazine (N$_2$H$_4$). N$_2$H$_4$ absorbs at UV wavelengths < 300nm (n – s* transition). The photolysis of N$_2$H$_4$ has been the subject of various investigations. The photochemistry of N$_2$H$_4$, as the simplest diamine, is of fundamental interest for understandig the photodissociation mechanisms and dissociation dynamics of small nitrogen-hydrogen containing molecules. Early studies on continuous photolysis of gasous N$_2$H$_4$ have shown that the stable products are ammonia (NH$_3$), molecular hydrogen and molecular nitrogen. There are three photolysis pathways which are under discussion (see Scheme 1).

$$\text{(I)} \quad N_2H_4 \xrightarrow{\text{hv}} {}^{\bullet}N_2H_3 + {}^{\bullet}H$$

$$\text{(II)} \quad N_2H_4 \xrightarrow{\text{hv}} 2\,{}^{\bullet}NH_2$$

$$\text{(III)} \quad N_2H_4 \xrightarrow{\text{hv}} NH_3 + NH$$

Scheme 1. Photolysis reactions of N$_2$H$_4$.

Quantum yield results first led Wenner et al. [15] to suggest that the primary process might be reaction (I). The following spectral observation of NH$_2$ after the photolysis of N$_2$N$_4$ led both, Ramsey [16] as well as Hussin et al. [17] to propose reaction (II) as the primary photolysis reaction. More recent results have returned to favor reaction (I). Spectroscopic observations by Arvis et al. [18], using higher time resolution, have shown that the majority of the NH$_2$ product is rather formed by a subsequent reaction than by a primary photolysis. Vaghjiani [19] reported that the primary process in the photolysis of N$_2$H$_4$ is the formation of H atoms and not NH$_2$. Energetically, reactions (II) and (III) are possible in 248.3 nm photodissociation of N$_2$H$_4$. Therefore the formation of NH, NH$_2$ and NH$_3$ as minor products can not be excluded.

INDEX PATTERNING OF AROMATIC POLYMERS

In a previous investigation we aimed at the surface functionalization of aromatic polymers. UV irradiation of polystyrene (PS) in the presence of gaseous hydrazine were carried out and it was evidenced by XPS that amino and hydrazinyl groups were attached onto the surface. Surprisingly, we observed the disappearance of aromatic C-H signals in the FTIR spectrum of PS during irradiation in hydrazine atmosphere. The assumed reduction of phenyl units to cyclohexyl groups was backed by experiments with cumene. A stepwise photoreduction of this molecule to isopropylcyclohexane was proved by gas chromatography coupled with mass spectrometry (GC/MS).This novel photoreaction, whose mechanism is not clear in detail as yet, was utilized to change the refractive index of films of polystyrene (PS) and poly(2-vinylnaphthalene) (PVN) simply by UV illumination in an atmosphere containing N_2H_4. Ellipsometric data showed that the refractive index n of PS is lowered by 0.04 after prolonged UV irradiation in the presence of hydrazine (error limit for n: ±0.0025). In the case of PVN, the refractive index n drops by 0.11 (Figure 1). Such large changes of n are promising for the setup of waveguides and for the tuning of optical properties (patterned modification). The hydrazine photoreaction also gave significant changes of the film thickness for PS and PVN. The data presented in Figure 1 refer to a film thickness of ~ 100 nm.

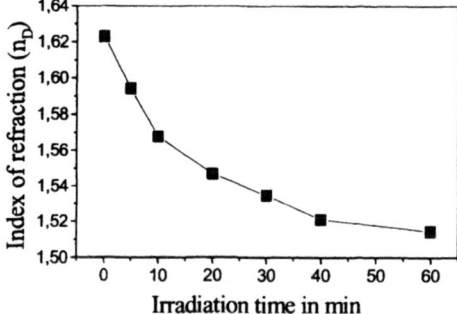

Figure 1. Refractive index of poly(2-vinyl-naphthalene) as a function of the UV irradiation time in N_2H_4 atmosphere (data from ellipsometry).

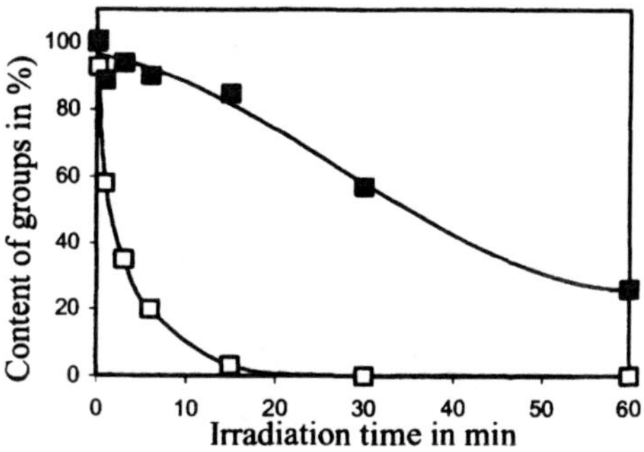

Figure 2. Bleaching of a PPV derivative: consumption of C=C double bonds (□) and aromatic units (■) during UV irradiation in N_2H_4 atmosphere (FTIR).

The hydrazine assisted UV reaction was also applied to conjugated polymers such as derivatives of PPV. Figure 2 displays FTIR data which show that in films of PPV (200 nm thickness), first, the double bonds are saturated. At a later stage of the UV process, the aromatic units are transformed.

PHOTOBLEACHING OF DERIVATIVES OF PPV

We found that UV irradiation in the presence of hydrazine is suitable to change the emissive properties of PPV derivatives. UV-VIS absorption and photoluminescence emission spectra of films of poly(2-methoxy,5-(2'-ethyl-hexoxy)-p-phenylenevinylene) (MEH-PPV) were taken prior to and after reductive photobleaching with N_2H_4. For comparison, we performed the same studies also for oxidative photo-bleaching processes. The results are depicted in Figure 3. It can be seen that - for equal exposure times - the reductive bleaching by hydrazine is much more effective in reducing the conjugation in the system (as is evidenced by the strong blue shift in the absorption spectra).

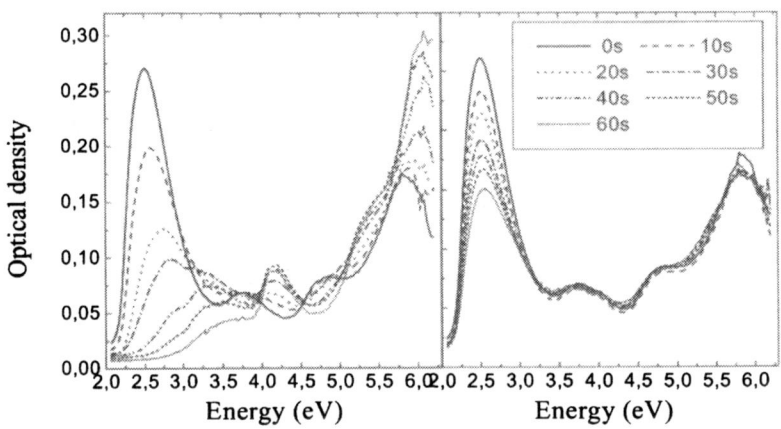

Figure 3. Optical densities of films of MEH-PPV after UV induced bleaching. Left: reductive bleaching under N_2H_4, right: oxidative bleaching under air (as a function of the exposure time).

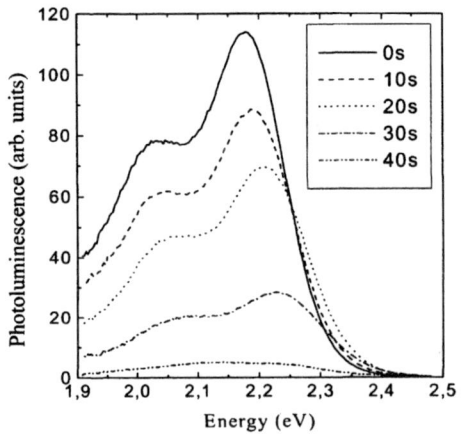

Figure 4. Patterning of the luminance of a PPV derivative by mask projection under N_2H_4 atmosphere.

Figure 5. Photoluminescence spectra of MEH-PPV after UV irradiation under N_2H_4 atmosphere.

Figure 4 shows the photoluminescence of a PPV film which was patterned by projection of a mask under N_2H_4 atmosphere. Figure 5 displays data on photoluminescence of MEH-PPV after UV irradiation under N_2H_4. The emission spectrum remains fairly constant during the initial stages of the photoreaction. This is explained by an energy transfer to the longest conjugated segments prior to emission.

The relative quantum yield (PLQY) of photoluminescence for MEH-PPV was recorded after different UV irradiation times. The experiment was carried out both under hydrazine atmosphere and under air (see Figure 6). For the data presented, an error limit of ± 5% (relative error) is estimated due to slight variations in the optical setup. The increase in relative quantum yield at low exposure times can be attributed to the shortening and increased spatial separation of conjugated segments within the material, hindering excited state migration processes to non-radiative quenching centers. After longer exposure times sites created by the photobleaching process can obviously also act as quenching centers and reduce the quantum efficiency of the material. In this context it has to be pointed out that the maximum quantum yield in the reductively bleached samples is reached after 20 s, when the optical density of the film has already dropped by 50%, whereas for the oxidative bleaching a strongly *reduced* PLQY is obtained already when the optical density has dropped by only about 20%. At this point it is not clear, whether the drop in PLQY for the reductive bleaching may not be due to the transfer of the irradiated samples from the "bleaching" set-up to the spectrometer (temporary exposure to air).

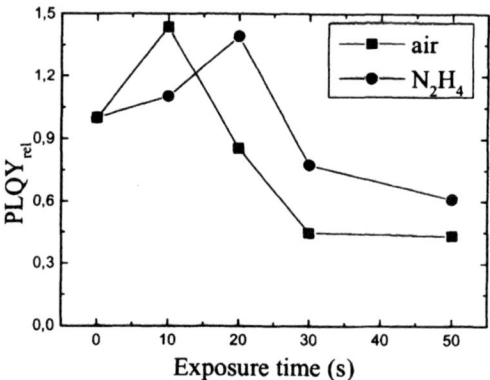

Figure 6. Relative photoluminescence quantum yields (PLQY) for films of MEH-PPV bleached reductively under N_2H_4 (•) and oxidatively under air (■) as a function of the exposure time. PLQY was measured for excitation at the maximum of the absorption spectra (cf. Figure 3).

CONCLUSIONS

The UV irradiation of unsaturated / aromatic polymers under gaseous hydrazine leads to the hydrogenation of olefinic / aromatic structural units. The process is useful for index patterning of polymers. With conjugated polymers of the PPV type, reductive photobleaching was observed. This process is a promising alternative to common oxidative photobleaching and provides a novel patterning strategy.

ACKNOWLEDGMENTS

The authors would like to thank "SFB Electroactive Materials" (TU Graz, projects F917 and F921) and FWF Vienna (project 13962-CHE) for financial support.

REFERENCES

1. S. Noach, E. Z. Faraggi, G. Cohen, Y. Avny, R. Neumann, D. Davidov, A. Lewis, *Appl. Phys. Lett.* **69**, 3650 (1996).
2. S.-C. Chang, J. Liu, J. Bharathan, Y. Yang, J. Onohara, J. Kido, *Adv. Mater.* **11**, 734 (1999).
3. K. M. Vaeth, K. F. Jensen, *Adv. Mater.* **11**, 814 (1999).
4. M. Granström, *Synth. Met.* **102**, 1042 (1999).
5. T. R. Hebner, J. C. Sturm, *Appl. Phys. Lett.* **73**, 1775 (1998).
6. V. Sinigersky, K. Müllen, M. Klappner, I. Schopov, *Adv. Mater.* **12**, 1058 (2000).
7. S.-H. Shin, J.-S. Park, J.-W. Park, H. K. Kim, *Synth. Met.* **102**, 1060 (1999).
8. J. A. DeAro, R. Gupta, A. J. Heeger, S. K. Buratto, *Synth. Met.* **102**, 865 (1999).
9. C. Kocher, A. Montali, P. Smith, C. Weder, *Adv. Funct. Mater.* **11**, 31 (2001).
10. O. Levi, G. Perepelitsa, D. Davidov, S. Shalom, I. Benjamin, R. Neumann, A. J. Agranat, Y. Avny, *J. Appl. Phys.* **88**, 1236 (2000).
11. K. Tada, M. Onoda, *J. Appl. Phys.* **86**, 3134 (1999).
12. K. Tada, M. Onoda, *Thin Solid Films* **363**, 195 (2000).
13. H. Bock, G. Rudolph, *Z. Anorg. Allg. Chem.* **311**, 117 (1962).
14. H. Niino, A. Yabe, *Appl. Phys. Lett.* **63**, 3527 (1993).
15. R.P. Wenner, A.O. Beckman, *J. Am. Chem. Soc.* **54**, 2787 (1932).
16. D.A. Ramsay, *J. Phys. Chem.* **57**, 415 (1953).
17. D. Hussin, R.G.W. Norrish, *Proc. R. Soc. London, Ser. A* **273**, 145 (1963).
18. M. Arvis, C. Devillers, M. Gillois, M. Curat, *J. Phys. Chem.* **78**, 1356 (1974).
19. G.L. Vaghijani, *J. Chem. Phys.* **98** (3), 2123 (1993).

Mat. Res. Soc. Symp. Proc. Vol. 707 © 2002 Materials Research Society

Formation of Ordered Silica–Organic Hybrids by Self-Assembly of Hydrolyzed Organoalkoxysilanes with Long Organic Chains

Kazuyuki Kuroda[1, 2] and Atsushi Shimojima[1]

[1]Department of Applied Chemistry, Waseda University,

Ohkubo-3, Shinjuku-ku, Tokyo 169-8555, Japan

[2]Kagami Memorial Laboratory for Materials Science and Technology, Waseda University,

Nishiwaseda-2, Shinjuku-ku, Tokyo 169-0051, Japan

ABSTRACT

Various layered hybrid films prepared from organoalkoxysilanes with long organic chains, based on the self-assembly of the hydrolyzed species, are reviewed. Morphological control of transparent and oriented films was achieved by cohydrolysis and polycondensation with tetraalkoxysilanes, followed by dip- or spin-coating. In addition to alkyltrialkoxysilanes, alkyldimethylmonoalkoxy- and alkylmethyldialkoxy-silanes were also used as the structural units, implying that the inorganic–organic interface can be designed at a molecular level. In these cases, co-condensation in the precursor solution plays an essential role in the formation of homogeneous and ordered films. Alkenyltriethoxysilanes with terminal C=C bonds were also employed to prepare layered hybrid films. Interlayer chains were polymerized upon UV irradiation, and the resulting films exhibited a significant increase in the hardness if compared with the films before polymerization. Hybrid films thus obtained are a new class of materials and of great interest for a wide range of materials chemistry.

INTRODUCTION

Organoalkoxysilanes are widely used as structural units to construct a variety of silica-based hybrid materials [1–3]. Because simple sol–gel reactions of alkoxysilanes usually result in the formation of amorphous materials [4], much efforts have been made in the structural control on a nanometer-length scale [5–8]. Such an attempt is of great interest from the possibility to produce novel hybrid materials with unique structures and properties.

Self-organization of organosilane molecules is a promising technique for the construction of ordered hybrid materials. The process relies on the amphiphilic nature of hydrolyzed organosilane molecules containing both hydrophilic silanol groups and hydrophobic alkyl chains. While many studies have focused on the interfacial deposition of alkylsiloxane monolayers [9], researchers have recently shown that the process can be extended to the formation of multilayered hybrids by the reaction in solution states [10–12]. We reported the formation of multi-bilayer aggregates by hydrolysis and polycondensation of alkyltrialkoxysilanes in homogeneous solutions [12]. These are a new class of layered materials consisting of organic two-dimensional arrays and siloxane networks linked by covalent Si–C bonds.

Although the layered hybrids derived from alkyltrialkoxysilanes are obtained as powders, transparent and oriented thin films can be obtained by cohydrolysis and polycondensation with tetraalkoxysilane under well-controlled conditions [13]. This synthetic approach can be extended to the development of a variety of hybrid materials by the molecular design of starting organoalkoxysilane with varied alkyl chain lengths, controlled numbers of the organic groups, and various functionalities in the organic groups.

In this paper, we present a review of the recent developments conducted in our laboratory on the formation of ordered hybrid films derived from the organoalkoxysilane–tetraalkoxysilane systems [14–16]. In addition to the alkyltrialkoxysilane with various chain lengths, alkyldimethylalkoxysilane and alkylmethyldialkoxysilanes as well as alkenyltrialkoxysilane were used as the starting molecules (figure 1). This approach will lead to the formation of highly organized architectures by a simple sol–gel route using various organoalkoxysilanes.

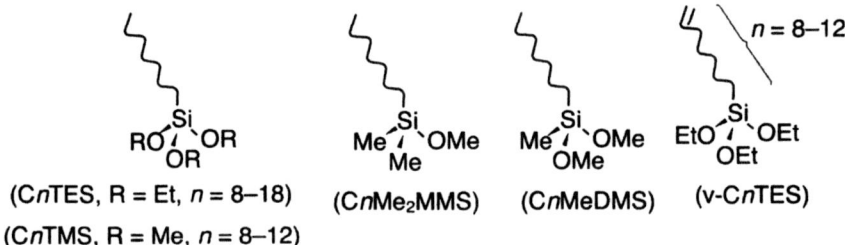

(CnTES, R = Et, n = 8–18) (CnMe$_2$MMS) (CnMeDMS) (v-CnTES)
(CnTMS, R = Me, n = 8–12)

Figure 1. Organoalkoxysilanes used in this study.

LAYERED HYBRID FILMS DERIVED FROM C*n*TES–TEOS SYSTEMS

Precursor solutions were prepared by cohydrolysis and polycondensation in the C*n*TES–TEOS–EtOH–H_2O–HCl system [14]. The ^1H NMR spectra of the reaction mixtures revealed that the signals due to ethoxy groups (SiOC$\underline{H}_2$CH$_3$, at around 3.8 ppm) almost disappeared within the first period of 15 min. The ^{29}Si NMR spectra of the solution in the C18TES–TEOS system revealed that the signals assigned to the monomeric species of both alkoxysilanes disappeared after the reaction at 40 °C for 90 min. Although the signals due to oligomeric species were not clearly resolved, the co-condensation between C*n*TES and TEOS should occur because sufficient evidences have already been reported in similar systems [17,18].

Thin films were deposited on glass substrates by dip-coating. However, the structural ordering and the macroscopic homogeneity of the films depended largely on the alkyl chain length as well as the solution temperature during the deposition. In the C10TES–TEOS system, transparent films exhibiting a sharp diffraction peak ($d = 3.48$ nm) were formed at around room temperature (20–25 °C). The layered structure of the film was confirmed by TEM (not shown). Although the films prepared at lower temperatures (10 and 15 °C) also exhibited sharp diffraction peaks, they were not homogeneous and the d values were rather variable. In contrast, the deposition at higher temperatures up to 30 °C resulted in the formation of homogeneous films with a disordered nanostructure. Similar trends were also observed for the systems with $n = 8$ and 12 .

In the case of $n \geq 14$, neither transparent nor well-ordered films were deposited from the solution at around room temperature. Figure 2 shows the variation in the XRD pattern of the hybrid films in the C14TES–TEOS system. The films obtained at around room temperature exhibited very broad peaks. However, the intensity of the peaks progressively increased as the solution temperature increases, and transparent and highly ordered films were obtained at 30–35 °C. The effect of the solution temperature was more clearly observed for the systems with $n = 16$ and 18. In these cases, well-ordered films with sharp and intense peaks were formed at 40–45 °C and 55–60 °C, respectively.

The incorporation of the Q units in the C*n*TES-derived hybrids contributed to the thermal stability. The layered hybrid derived from C18TES alone melted into an amorphous state upon heating above ~110 °C [12]. In contrast, the layered product

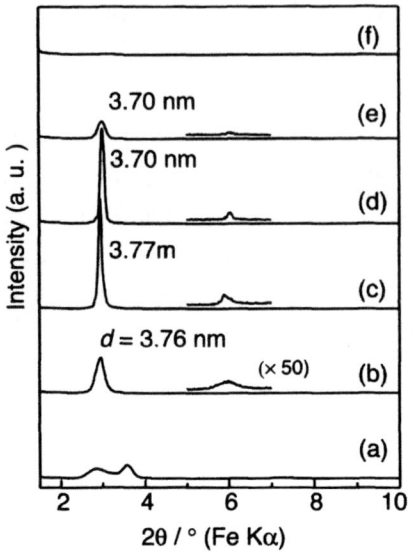

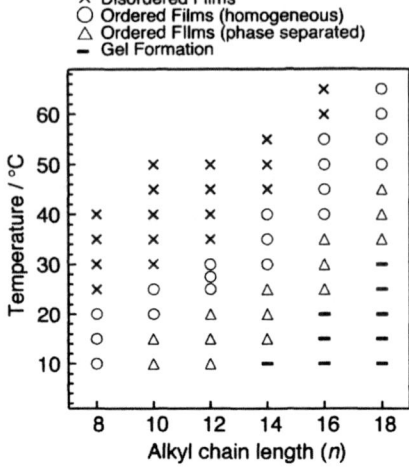

Figure 2. XRD patters of the hybrid films in the C14TES–TEOS system prepared at various temperatures: (a) 20 °C, (b) 25 °C, (c) 30 °C, (d) 35 °C, (e) 40 °C, and (f) 45 °C.

Figure 3. Variation of the products derived from the CnTES–TEOS systems (n = 8–18) depending on the solution temperatures.

derived from the C18TES–TEOS system retained the structure even at 170 °C, suggesting that the product was comprised of more stable siloxane networks.

Figure 3 summarizes the variation of the nanostructure and the macroscopic morphology of the resulting hybrids depending on the solution temperature during the film formation. Transparent and ordered films are formed at higher temperatures with increasing chain length, suggesting that the self-assembly of alkylsiloxane oligomers depend essentially on the solution temperature. When a substrate is withdrawn from the reaction mixtures, silicate species in the solutions are concentrated by the evaporation of the solvent, and finally polycondensed to form siloxane networks. The formation of oriented multilayered films requires uniform and continuous organization, presumably from solid–liquid and liquid–vapor interfaces, during this coating process. In the cases of CnTES with longer chains ($n \geq 14$), the films deposited at lower temperatures exhibited a less-ordered structure. This behavior is probably attributed to the random nucleation of the layered aggregates caused by a slight decrease in the temperature as well as the evaporation of the solvent during the film formation. The suppression of the segregation of the layered aggregate appears to play an important role in the formation of oriented films. In all the systems, further increase in the deposition temperature caused the structural disordering and eventually led to amorphous films. This is explained by the general behavior of amphiphilic assemblies that become isotropic states at higher temperature. The films may be solidified by the siloxane formation prior to the self-organization.

Figure 4 represents the basal spacings of the well-ordered films as a function of the alkyl chain length. For comparison, the d values for the layered hybrids derived from CnTES (n = 12–18) alone are also shown. As we reported previously, the products derived from CnTES alone are comprised of a bilayer structure with *all-trans* chains almost perpendicular to the siloxane layers [12]. In contrast to the linear relationship in these products, the d spacing of the films increases continuously in the range of n = 8 to 14, but decreases between n = 14 to 16. It is reasonable to assume that the thickness of the siloxane layer is almost constant independent of the chain lengths, because the TEOS/CnTES ratio is identical in all the systems. Therefore, the above behavior should be attributed to the variation in the conformation and/or the packing arrangement of the interlayer alkyl chains depending on the alkyl chain length.

The difference in the conformations of the alkyl chains in the films was confirmed by ^{13}C CP/MAS NMR. The ^{13}C signals ascribed to the interior methylene carbons appear at 30 ppm for n = 8–14, suggesting the presence of *gauche* conformers. However, the

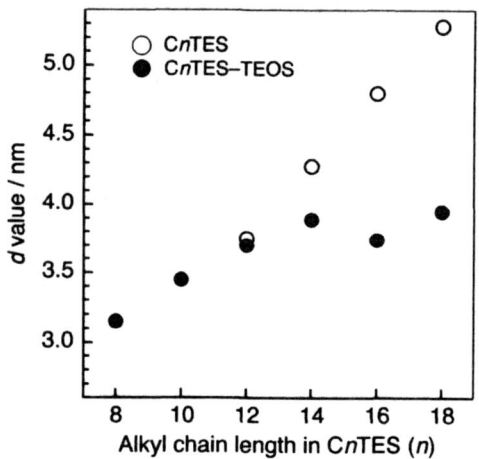

Figure 4. Variation in the d values as a function of the number of carbon atoms in the alkyl chain: The layered hybrid films derived from the CnTES–TEOS systems (filled circles) and the layered hybrids derived from CnTES alone (open circles).

relative intensity of the signal at 33 ppm, indicative of the chains in *all-trans* conformations, increases as the chain length increases to $n = 16$ and 18.

The packing arrangements of the interlayer chains can be divided into two types, i.e. interdigitated monolayer and bilayer structures, according to the representative models for amphiphilic lamellar phases [19]. In the cases of $n = 8$–12, the alkyl chains are supposed to adopt a bilayer structure with disordered state, based on our study on the layered hybrid films derived from CnTMS ($n = 8$–12)–TMOS systems [13]. However, in the cases of $n = 16$ and 18, the observed d spacings appear to be too small for the similar bilayer structures, considering that the chains are longer and take more extended states. Indeed, the d value of the film in the C18TES–TEOS system is considerably smaller than that of a bilayer structure derived from C18TES alone (figure 4), although there is no significant conformational difference in both of the products. It is considered that the alkyl chains are arranged in an interdigitated monolayer with a rather ordered state. Such interlayer structures are also reported for the alkylsilylated derivatives of crystalline layered polysilicates [20,21], and may arise from the increases in the distance between alkylsilyl groups due to the presence of co-condensed Q units.

LAYERED HYBRID FILMS DERIVED FROM CnTMS–, CnMeDMS–, AND CnMe$_2$MMS–TMOS SYSTEMS

In these systems, hybrid films were prepared by cohydrolysis and polycondensation of CnTMS ($n = 8$, 10, 12)–TMOS–THF–H$_2$O–HCl mixtures, followed by spin-coating on glass substrates [15]. As shown in figure 5, the XRD patterns of the hybrid films derived from the C10TMS–, C10MeDMS–, and C10Me$_2$MMS–TMOS systems showed sharp diffraction peaks with the d spacings of 3.56, 3.67, and 4.06 nm, respectively. The structural ordering and the macroscopic morphology of the films are strongly affected by the reaction time in the precursor solutions. The solutions at the earlier stages of the reactions afford inhomogeneous films with phase separated morphologies. On the other hand, further reactions in the precursor solutions caused the disordering of the nanostructures, and finally led to the formation of amorphous films. Controlled reaction in the precursor solutions was needed for the formation of well-ordered and transparent hybrid films.

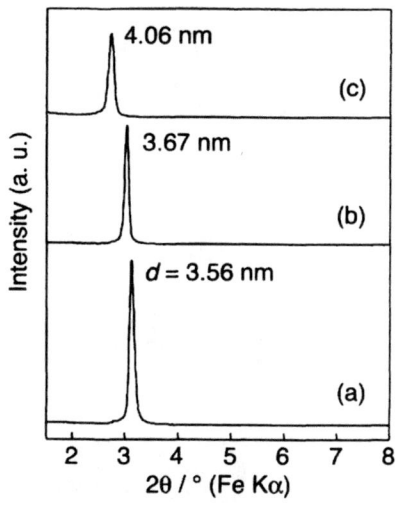

4.06 nm

(c)

3.67 nm

(b)

$d = 3.56$ nm

(a)

Figure 5. XRD patterns of the hybrid films derived from (a) C10TMS–, (b) C10MeDMS–, and (c) C10Me$_2$MMS–TMOS systems.

The TEM images of the ordered films showed well-defined stripes due to the lamellar structures, and the periodicities agree closely with the d spacings measured by XRD. The layered structures of these films were also confirmed by the structural collapse upon calcination at 450 °C for 6 h to remove organic constituents.

The formation of siloxane networks in the films was confirmed by ^{29}Si MAS NMR. The spectra for all the systems showed the signals assigned to the Q^2 (–90 ppm), Q^3 (–100 ppm), and Q^4 (–110 ppm) units derived from TMOS. In addition, the signals due to the organosiloxane units derived from C10TMS, C10MeDMS, and C10Me$_2$MMS are observed at the T^2 (–55 ppm) and T^3 (–65 ppm), D^1 (shouldered peak at –10 ppm) and D^2 (–16 ppm), and M^1 (13, 7 ppm) regions, respectively.

Similar hybrid films were prepared when the alkyl chain lengths in alkylmethoxysilanes were 8 and 12. The formation of ordered films required longer reaction time in the precursor solution with increasing chain length. This is in part ascribed to the difference in the condensation rate depending on the chain length. In this manner, the reaction time in the precursor solution was optimized depending on the number of methoxy groups and alkyl chain lengths. Figure 6 shows the relationship between the d values of the resulting hybrid films and the alkyl chain lengths of the alkylmethoxysilanes used. The d value increases continuously with increasing chain lengths and exhibit larger value as the number of methoxy groups decreases.

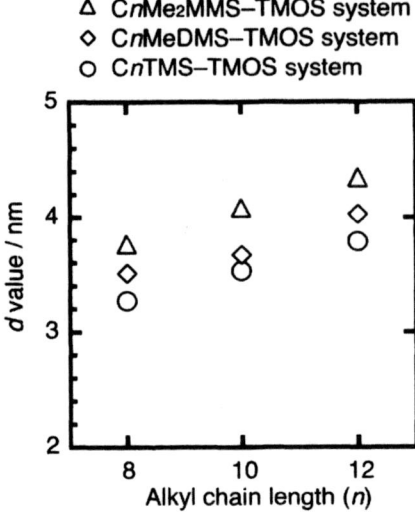

△ C*n*Me₂MMS–TMOS system
◇ C*n*MeDMS–TMOS system
○ C*n*TMS–TMOS system

Figure 6. Variation in the *d* values of the hybrid films as a function of the alkyl chain length in the alkylmethoxysilanes.

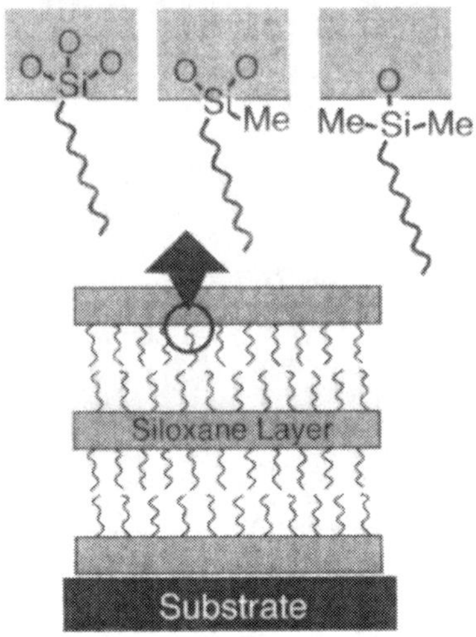

Figure 7. Proposed structural models for the hybrid films derived from C*n*TMS–, C*n*MeDMS–, and C*n*Me₂MMS–TMOS systems.

The structural models for the hybrid films are schematically illustrated in figure 7. The films have multilayered structures consisting of a bilayer arrangement of alkyl chains that are covalently attached to siloxane layers. The observed difference in the *d* value with varying the number of methoxy groups in the starting alkylmethoxysilanes (figure 6) is probably attributed to the variation in the interfacial structures. It is supposed that the trialkoxysilyl groups are in part integrated into the siloxane networks due to their cross-linking abilities, while the substitution of methyl groups for methoxy groups results in the grafting on the external surface of silica layers.

In the present system, co-condensation between alkylmethoxysilanes and TMOS in the precursor solution plays a crucial role in the formation of ordered hybrid films. This is supported by the fact that neither transparent nor ordered films were obtained by the hydrolysis and polycondensation of alkylmethoxysilane alone. In the ^{29}Si NMR spectra of the precursor solutions, the signals due to monomeric species almost disappeared and those of the oligomeric species including substantial amounts of co-condensed units were observed. These results provide a strong evidence that the origin of the self-assembly in this system is quite different from that of the conventional single-component systems using alkyltrialkoxysilanes alone [12]. The alkylsiloxane oligomers formed by the co-condensation can be regarded as amphiphilic molecules containing both hydrophobic alkyl chains and hydrophilic silanol groups. We suppose that the construction of ordered films relies on the self-assembly of these oligomeric species by the rapid evaporation of the miscible solvent during the spin-coating procedure.

LAYERED SILICA/ORGANIC POLYMER HYBRID FILMS DERIVED FROM v-CnTES–TEOS SYSTEMS

Multilayered hybrid films were also prepared from v-CnTES–TEOS–EtOH–H$_2$O–HCl mixtures [16]. The films before UV irradiation showed sharp diffraction peaks with the d spacing of ca. 3.0–3.6 nm, depending on the chain lengths. The films are considered to have a bilayer arrangement of alkenyl chains, because the d spacings of the films are similar to those of the films derived from CnTES–TEOS systems (see figure 4). The IR spectrum of the v-C8TES-derived film before the UV irradiation (figure 8a) shows the absorption bands due to the siloxane framework (1000–1200 cm^{-1}, 450 cm^{-1}) and the terminal double bonds in the organic groups at 3079 cm^{-1} (=CH stretching), 1642 cm^{-1} (C=C stretching), and 910 cm^{-1} (=CH$_2$ out-of-plane deformation). These results suggest that the films are composed of lamellar structures containing vinyl functionalities within the interlayers.

The film exposed to the UV light still retained the structural order with a slight decrease in the d value in the XRD pattern. As shown in figure 9, the TEM image of the film derived from the v-C8TES–TEOS system shows well-defined stripe patterns whose periodicity corresponds to the d spacing determined by XRD . The polymerization of organic chains was evidenced by the substantial decrease in the IR absorption bands ascribed to the terminal double bonds (figure 8b). Because the interlayer organic chains in this system appear to adopt a bilayer structure, polymerization should have proceeded either within layers or between adjacent layers. The resulting hybrid film exhibited a remarkable increase in the hardness and the scratch resistance if compared with the film before the organic polymerization.

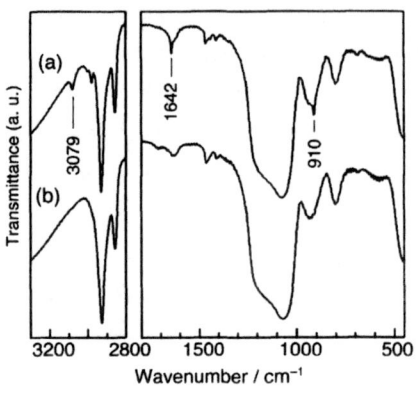

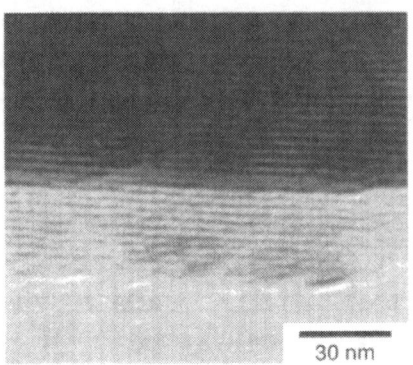

Figure 9. TEM image of the hybrid film derived from the v-C8TES–TEOS system after the UV irradiation. (The film was powdered for the observation.)

Figure 8. FT-IR spectra of the films derived from the v-C8TES–TEOS system (a) before and (b) after the UV irradiation.

All these results indicate the formation of inorganic–organic hybrid films with a lamellar structure in which the siloxane layers and the organic polymer are linked by Si–C bonds. With regard to the silica-organic polymer hybrids, intercalation of organic polymers into the interlayer spaces of crystalline layered silicates has attracted much attention for the possibility to provide new properties [22]. Covalent linking between organic moieties and silica interlayers can also be attained by using a silane coupling agent such as γ-methacryloxypropyltrimethoxysilane [23]. However, the present approach is quite different in utilizing the molecular self-assembly of organoalkoxysilane–tetraalkoxysilane system, which enabled the morphological control into the transparent and oriented thin films.

CONCLUSIONS

We have presented the formation of multilayered hybrid films consisting of alternating organic and siloxane layers by the sol–gel reaction of a series of organoalkoxysilanes and tetraalkoxysilane. This approach is quite simple and effective for the construction of ordered hybrids with precisely controlled nanostructures, macroscopic homogeneities, and compositions. The introduction of tetraalkoxysilane contributes to the morphological variation and increases the thermal stability that could not be attained by hydrolysis and condensation of alkyltrialkoxysilane alone. It is of great interest that the ordered hybrid films were also obtained from alkyldimethylmethoxysilane and alkylmethyldimethoxysilane in the presence of TMOS. In these systems, co-condensation between alkylmethoxysilane and TMOS in the precursor solutions is essential for the self-organization. The overall results provide an access to the generalized synthesis of ordered hybrid films utilizing various organoalkoxysilanes with hydrophobic organic groups.

ACKNOWLEDGMENTS

This work was financially supported by the Grant-in-Aid for COE research and for JSPS Fellows from the Ministry of Education, Culture, Sports, Science and Technology.

REFERENCES

1. U. Schubert, N. Hüsing and A. Lorenz, *Chem. Mater.* **7**, 2010 (1995).
2. C. Sanchez, F. Ribot and B. Lebeau, *J. Mater. Chem.* **9**, 35 (1999).
3. D. A. Loy, B. M. Baugher, C. R. Baugher, D. A. Schneider and K. Rahimian, *Chem. Mater.* **12**, 3624 (2000).
4. C. J. Brinker and G. W. Scherer, "Sol–Gel Science", Academic Press, San Diego (1990).
5. C. J. Brinker, Y. Lu, A. Sellinger and H. Fan, *Adv. Mater.* **11**, 579 (1999).
6. A. Stein, B. J. Melde and R. C. Schroden, *Adv. Mater.* **12**, 1403 (2000).
7. B. Boury, R. J. P. Corriu, V. L. Strat, P. Delord and M. Nobili, *Angew. Chem. Int. Ed.* **38**, 3172 (1999).
8. Y. Lu, H. Fan, N. Doke, D. A. Loy, R. A. Assink, D. A. LaVan and C. J. Brinker, *J. Am. Chem. Soc.* **122**, 5258 (2000).
9. A. Ulman, *Chem. Rev.* **96**, 1533 (1996).
10. Y. Fukushima and M. Tani, *J. Chem. Soc., Chem. Commun.* 241 (1995).
11. Q. Huo, D. I. Margolese and G. D. Stucky, *Chem. Mater.* **8**, 1147 (1996).
12. A. Shimojima, Y. Sugahara and K. Kuroda, *Bull. Chem. Soc. Jpn.* **70**, 2847 (1997).
13. A. Shimojima, Y. Sugahara and K. Kuroda, *J. Am. Chem. Soc.* **120**, 4528 (1998).

14. A. Shimojima and K. Kuroda, *Langmuir,* in press.

15. A. Shimojima, N. Umeda and K. Kuroda, *Chem. Mater.* **13**, 3610 (2001).

16. A. Shimojima and K. Kuroda, *Chem. Lett.* 1310 (2000).

17. L. Delattre and F. Babonneau, *Mater. Res. Soc. Symp. Proc.* **346**, 365 (1994).

18. S. A. Rodríguez and L. A. Colón, *Chem. Mater.* **11**, 754 (1999).

19. G. J. T. Tiddy, *Phys. Rep.* **57**, 1 (1980).

20. M. Ogawa, S. Okutomo and K. Kuroda, *J. Am. Chem. Soc.* **120**, 7361 (1998).

21. A. Shimojima, D. Mochizuki and K. Kuroda, *Chem. Mater.* **13**, 3603 (2001).

22 E. P. Giannelis, *Adv. Mater.* **8**, 29 (1996).

23. K. Isoda, K. Kuroda and M. Ogawa, *Chem. Mater.* **12**, 1702 (2000).

Mat. Res. Soc. Symp. Proc. Vol. 707 © 2002 Materials Research Society   BB10.46

Hole Transport In Self-Organized Oligosilane Thin Films With Highly Ordered Hopping Sites

H. Okumoto, T. Yatabe, A. Richter, M. Shimomura, A. Kaito, N. Minami
Nanotechnology Research Institute, National Institute of Advanced Industrial Science and
Technology (AIST), Tsukuba, Ibaraki 305-8565, Japan

ABSTRACT

Self-organized oligosilane thin films possess molecular orientation normal to substrates with multilayered structure. This unique order of σ-conjugated molecules results in good hole transport properties. In the present work, carrier transport properties at low temperature are studied for 1,10-diethyldecamethylsilane polycrystalline films. Even at a temperature as low as 173 K, a time-of-flight transient photocurrent waveform showed a clear plateau and a sharp decay, whose shape is similar to that at room temperature. Their hole mobility followed Arrhenius type temperature dependence with a small activation energy of 0.09 eV. The hole mobility of 6.3×10^{-5} cm^2/Vs at 193 K was more than 2 orders of magnitude higher than that of typical polysilanes, which inevitably contain disordered structures hindering smooth carrier transport.

INTRODUCTION

The self-organization of molecules is one of key factors in developing functional molecular materials for device fabrications. For example, π-conjugated discotic [1] and smectic [2] liquid crystalline materials attract attention as carrier transport materials with mobilities exceeding 10^{-3} cm^2/Vs. On the other hand, σ-conjugation [3] is another key factor in improving carrier transport materials. Polysilanes, possessing delocalized electrons along silicon main chains, have been extensively studied because they show high hole mobilities among conducting polymers on the order of 10^{-4} cm^2/Vs at room temperature [4, 5].

Oligosilanes can combine the above two key factors. In our previous papers, we proposed self-organized oligosilanes as a new class of hole transport materials [6, 7]. Self-organization of oligosilanes occurs as a smectic B (S_B) phase [8, 9, 10], but its high molecular order is well- preserved even in a polycrystalline phase. In these ordered states, molecules are oriented almost normal to the substrate surface, forming multi-layer structure with exceptionally large domain sizes ($> 20\ \mu$m). Such a structural peculiarity results in good carrier transport properties in crystalline phase as well as in S_B phase. In permethyldecasilane ($CH_3[Si(CH_3)_2]_{10}CH_3$; MS10) polycrystalline films, hole mobility exceeded 10^{-3} cm^2/Vs at room temperature. Moreover, the time-of-flight photocurrent showed nondispersive waveform having a clear plateau and a sharp decay, reflecting very low density of carrier traps that would hinder smooth carrier transport. This is an advantage over the π-conjugated liquid crystalline materials. While these materials show high carrier mobilities in the liquid crystalline phase, their carrier transport properties deteriorate accompanying crystallization due to many carrier traps created at grain boundaries.

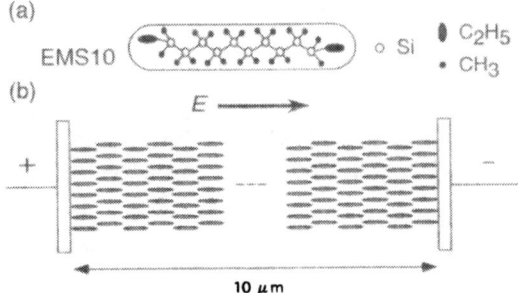

(a)

EMS10   ○ Si   ❙ C₂H₅ ● CH₃

(b)

$E \longrightarrow$

+   —

10 μm

Figure 1. (a) Molecular structure of 1,10-diethyldecamethylsilane (EMS10). (b) A schematic of molecular orientation of capillary-filled EMS10 in a sample cell for the time-of-flight transient photocurrent measurement.

In the present report, we show well-defined high molecular order in polycrystalline films of self-organized 1,10- diethyldecamethylsilane ($C_2H_5[Si(CH_3)_2]_{10}$ C_2H_5, EMS10), whose molecular structure is depicted in Figure 1(a). Like MS10, self-organized EMS10 is found to possess good hole transport properties, namely, high mobility and nondispersive TOF waveform. Different from MS10, EMS10 has a more well-defined molecular order. The effects of the high order are demonstrated by exploring the properties especially in low temperature range.

EXPERIMENTAL

EMS10 was synthesized and purified following Ref. [9]. To fabricate sample cells for photocurrent measurements, EMS10 powder was melted at 293 K and capillary-filled into a gap (∼ 10 μm) between two quartz substrates coated with indium–tin oxide (ITO) electrodes. The samples were slowly cooled from 293 K to room temperature at a rate of 2 K/min. The exact gap width of each sample was determined from the pattern of interference fringes due to multiple reflections between the two substrates. The expansion and shrinkage of the gap width caused by temperature change over a wide range was monitored *in situ* by capacitance change as measured with a precision LCR meter. The lowest temperature accessible in the present work was limited by the occurrence of poor electrical contact between the sample and ITO, which could be monitored as an abrupt drop of the capacitance. We presume that this is due to thermal shrinkage of EMS10 films. The molecular order of capillary-filled EMS10 films was evaluated by wide angle X-ray diffractometry at room temperature.

Photocurrent transients were measured by the time-of-flight (TOF) technique. The light source was an optical parametric oscillator pumped by a Nd:YAG laser (pulse width ∼ 10 ns,

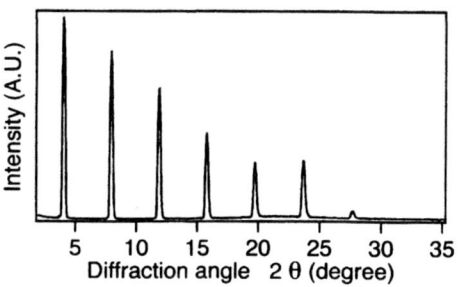

Figure 2. Wide angle X-ray diffraction pattern of an EMS10 polycrystalline film.

energy density ~ 0.3 $\mu J/mm^2$), which directly excited oligosilane molecules having strong absorption at a wavelength of 230 nm. The light pulses impinged on sample cells with an incident angle of about 45° to the substrates. This optical alignment was chosen to ensure that molecules absorbed excitation light even when their transition dipoles were aligned perpendicularly to the substrates. The TOF signals thus generated were detected using a digitizing oscilloscope. Sample cells were kept in a helium closed-cycle cryostat and the temperature was controlled within a stability of 0.3 K.

RESULTS AND DISCUSSION

An X-ray diffraction pattern of an EMS10 polycrystalline film is shown in Figure 2. A series of sharp peaks were observed, yielding an interlayer distance of 22.7 Å close to the value of 23.2 Å for the S_B phase [9]. The close proximity of these interlayer distances indicates that they possess very similar layered structures. Since EMS10 molecules in the S_B phase are known to align normal to the layer plane, those in the capillary filled films should have an ordered structure as shown in Figure 1(b). It is interesting to note that EMS10 has a well-defined single interlayer distance while MS10 showed two in their capillary filled films [6]. This is the indication of the effect of terminal groups on the self-organization behavior of oligosilanes, which should be the subject of further studies.

A series of TOF wave forms measured at different temperatures under a bias electric field of 1×10^5 V/cm are displayed in Figure 3. At 293 K, a clear plateau and a sharp decay were observed, whose inflection point (transit time) yielded a hole mobility of 4.1×10^{-4} cm^2/Vs. This excellent wave form results from the highly-ordered layer structure of the EMS10 film, where holes can move fast along σ-conjugated molecules and hop at the end of molecules. Presumably, this intermolecular hopping rate-limits the whole carrier transport process. Since the molecules are highly ordered in EMS10 films, so are the hopping sites located at the end of

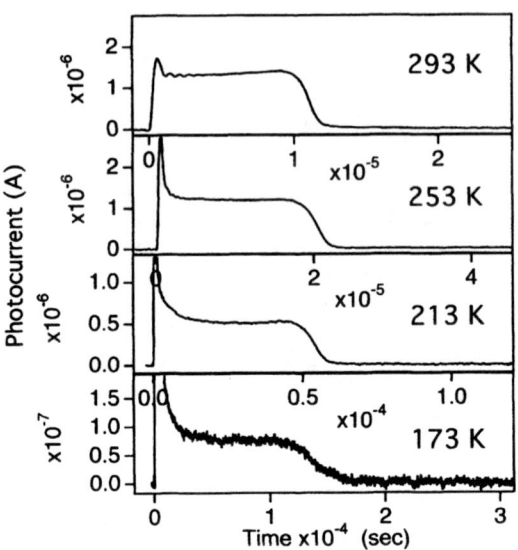

Figure 3. Time-of-flight photocurrent transients for an EMS10 polycrystalline film at different temperatures (293, 253, 213, and 173 K) under an electric field of 1×10^5 V/cm. Both current and time scales of graphs are adjusted to compare plateaus and decays of the transients.

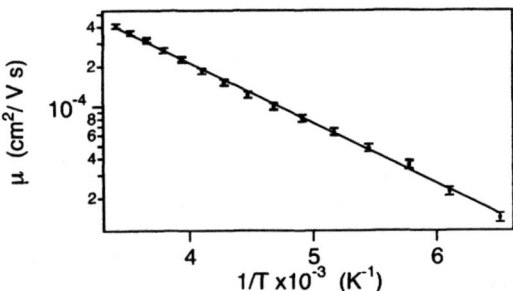

Figure 4. Temperature dependent hole mobility of an EMS10 polycrystalline film under an electric field of 1×10^5 V/cm.

each molecule. This should result in narrow distribution of both hopping distance and barrier height, contributing to the reduction in the dispersion of carrier mobility. The lower mobility of EMS10 as compared with MS10 (vs.   /Vs) can be explained by the increase in the hopping distance (ethyl vs. methyl). This is again a reflection of the highly ordered structures of the self-organized oligosilane films.

The transit time becomes longer for lower temperatures (Figure 3). However, all the photocurrent transients are nondispersive and the wave form is retained even at 173 K. In contrast, for disordered materials such as polysilanes [11] and moleculary doped polymers [12, 13], nondispersive to dispersive transition of wave forms and/or considerable lengthening of the tailing is observed as the temperature is lowered. To estimate the tailing, the parameter   is often used [14], where   is the transit time and   is the time for photocurrent to drop to the half of that at   . For EMS10,   is almost constant at ca. 0.07 throughout the entire temperature range. This fact indicates quite a small change of thickness of the carrier sheet, which is generated by a light pulse. In addition, one may suppose that only shallow traps can exist of energy level lower than 15 meV corresponding to 173 K. This temperature independence as well as the very short tailing is a distinguished feature of EMS10.

Temperature dependence of the hole mobility   is plotted in Figure 4. An activation energy of 90 meV was obtained by the Arrhenius (vs.   ) plot. This small activation energy prevents the deterioration of the mobility upon cooling. Even at temperature as low as 153 K, a mobility of   /Vs was achieved. In case of the typical polysilane (poly(methylphenylsilane)), the reported hole mobility was   /Vs at 193 K under an electric field of   V/cm [11]. In the same condition, the hole mobility of EMS10 reached   /Vs, more than 2 orders of magnitude higher.

For the disordered materials,   dependence is commonly observed. Although the distinction between   and   dependences is difficult [13],   dependence is well-fitted to the data for the EMS10 in Figure 4. Moreover, the high-temperature limit of the mobility derived by extrapolation is a reasonable value of   /Vs. Thus the highly ordered hopping sites in EMS10 enables us to apply the simple model for hopping transport as   , where   denotes the electron charge,   the hopping distance, the field and   the activation energy [6]. The details of electric field dependence of   will be discussed elsewhere.

CONCLUSION

The well-defined molecular order in capillary-filled polycrystalline films of self- organized EMS10 was established. The high molecular order generated the highly ordered hopping sites located at the end of each molecule, which resulted in good hole transport properties even at low temperature; relatively high hole mobility, narrow distribution of the mobility and small effects of traps. The self-organized oligosilanes proved to be clear carrier transport systems and showed promise for controlling hopping transport.

REFERENCES

1. D. Adam, P. Schuhmacher, J. Simmerer, L. Haussling, K. Siemensmeyer, K. H. Etzbach, H. Ringsdorf, and D. Haarer, Nature **371**, 141 (1994).

2. M. Funahashi and J. Hanna, Phys. Rev. Lett. **78**, 2184 (1997).

3. M. Pope and C. Swenberg, in *Electronic Processes in Organic Crystals and Polymers (2nd Ed.)* (Oxford University Press, New York, 1999), Chap. XI.

4. R. G. Kepler, J. M. Zeigler, L. A. Harrah, and S. R. Kurtz, Phys. Rev. B **35**, 2818 (1987).

5. M. Abkowitz, H. Bässler, and M. Stolka, Philos. Mag. B **63**, 201 (1991).

6. H. Okumoto, T. Yatabe, M. Shimomura, A. Kaito, N. Minami, and Y. Tanabe, Adv. Materials **13**, 72 (2001).

7. H. Okumoto, T. Yatabe, J. Peng, A. Kaito, and N. Minami, Synth. Met. **121**, 1507 (2001).

8. T. Yatabe, A. Kaito, and Y. Tanabe, Chem. Lett. 799 (1997).

9. T. Yatabe, T. Kanaiwa, H. Sakurai, H. Okumoto, A. Kaito, and Y. Tanabe, Chem. Lett. 345 (1998).

10. T. Yatabe, N. Minami, H. Okumoto, and K. Ueno, Chem. Lett. 742 (2000).

11. H. Bässler, P. M. Borsenberger, and R. J. Perry, J. Polym. Sci.: Part B: Polym. Phys. **32**, 1677 (1994).

12. H. Bässler, Phys. Stat. Sol. (b) **175**, 15 (1993).

13. P. M. Borsenberger, E. H. Magin, M. Auweraer, and F. C. Schryver, Phys. Stat. Sol. (a) **140**, 9 (1993).

14. P. M. Borsenberger, L. T. Pautmeier, and H. Bässler, Phys. Rev. B **48**, 3066 (1993).

Mat. Res. Soc. Symp. Proc. Vol. 707 © 2002 Materials Research Society   HH3.36

Formation of Supramolecular Assemblies by Modulating Self-Assembling Properties of Diacetylenic Phosphocholines

Alok Singh, Eva M. Wong, Mark S. Spector and Joel M. Schnur
Center for Bio/Molecular Science & Engineering, Naval Research Laboratory,
Washington, DC 20375 USA

ABSTRACT

Diacetylenic phospholipid dispersions in water produce tubules (500 nm diameter) and helices from their initial vesicular morphology as a function of temperature and concentration. A binary mixture consisting of diacetylenic phospholipid, 1,2 bis (tricosa-10, 12-diynoyl)-sn-glycero-3-phosphocholine and a short chain phospholipid, 1,2-dinonanoyl –sn-glycero-3-phosphocholine was studied to explore the morphological transformation of lipids into tubules to develop an approach to control and produce tubules of different diameters. Circular dichroic spectra not only indicated the chiral nature of these tubules, but also provided distinct spectral signatures differentiating micro- and nanotubules. The effects of temperature and lipid concentration on the formation and stability of tubules were also explored. An equimolar lipid mixture provided structures with uniform morphology, which were stable for several hours up to 36 °C. The thermal stability of nanotubules makes them an attractive candidate for many practical applications including controlled release technology.

INTRODUCTION

Phospholipids by virtue of their unique self-assembling properties belong to an important class of biomaterials. Their pivotal role in biological membranes to facilitate multiple, complex functions inside, within and on the membranes makes them technologically attractive. In non-biological environments, phospholipids typically form multi-lamellar vesicles upon dispersion in water. The formation of tubular structures from diacetylenic lipids is an important example of synthetic material derived from natural precursor, which permits extension of self-assembly for making technologically relevant novel structures [1,2]. Tubules, helices and ribbons have been the subject of numerous scientific studies due to their potential applications in the areas of chemistry, biology and material science [3-5]. Initial reports on diacetylenic phospholipids described the formation of microtubules with a fixed internal diameter of 500 nm [6]. Subsequent studies reported the formation of other forms of self-assembling structures, such as helices, ribbons and tubules [7] and the means to produce tubules with variable diameters [8,9].

The overall goal of the current research is to construct tubules with diameters ranging from nano to micrometers. The system must be reproducible, and tolerant to minor fluctuations in temperature and pH of the dispersion medium. We have focused on inserting a short chain lipid whose chain length was equal to the number of methylene groups present between diacetylene and ester group on glycerol molecule, particularly 1,2 tricosadiyn-10,12-oyl)-sn-glycer-3-phosphocholine ($DC_{8,9}PC$) and 1,2 dinonanoyl-sn-glycero-3-phosphocholine (DNPC) [10-11]. The mixed lipid system produced 28 nm wide ribbons in water and salt solution depending on the lipid concentration. These results were followed by a report on the formation of nanotubule by careful thermal manipulation of aqueous dispersions of equimolar mixture of lipids [12]. Recently, we reported on the formation of nanotubules using the binary lipid system and observed chiral-optical signature of nanotubule, twisted ribbon, and microtubule morphologies that can be used to monitor their temporal and thermal stability [13]. This report focuses on the formation of nanostructures by varying the mole fraction of DNPC in binary lipid mixture, modifying diacetylenic lipid with additional methylene groups in the distal segment of diacetylene chain, and examining the thermal stability of structures made from equimolar mixtures of lipids.

EXPERIMENTAL DETAILS

Lipid dispersions of $DC_{8,9}PC$ and 1,2-Bis (pentacosadi-10,12-ynoyl)-sn-glycero-3-phosphocholine ($DC_{8,11}PC$) with DNPC (Avanti Polar Lipids) were prepared by hydrating vacuum dried lipid films from stock chloroform solutions of each lipid in ultra pure water (18.2 MΩ.cm) at 60 °C for 3 hours. Lipids were dispersed by intermittent vortex mixing while maintaining a temperature at 60 ºC until a homogeneous dispersion was formed. The samples were slowly cooled to room temperature over a period of approximately 2 hours and placed in cold chamber maintained at 4 °C for least 10 hours before characterization. For exploring the influence of lipid ratio 7 samples were prepared keeping the $DC_{8,9}PC$ concentration fixed at 1 mM. The following molar ratios of $DC_{8,9}PC$: DNPC lipids were used: 4:6, 4.5:5.5, 1:1, 5.5:4.5, 6:4 and 1:1 (2 mM) $DC_{8,11}PC$: DNPC mixture.

A 400 MHz Bruker DRX-400 spectrometer was used for recording the nuclear magnetic resonance (NMR) spectra of mixed lipids present in the solution and in the nanostructures. The ratio of lipids was determined by comparing the number of terminal methyl protons (triplet at 0.88 ppm,) from both lipids to the total number of methylene protons in the α-position to the diacetylenic group (δ 2.27 ppm, multiplet). In an equimolar mixture, this ratio should be 12:16. Transmission electron microscopic (TEM) images were taken on a Zeiss EM-10C electron microscope operating at 60 KV. Stained samples were prepared by placing few microliters of aqueous 1% (w/v) uranyl acetate onto a drop of tubule dispersion placed on carbon-coated grid, removing the excess solvent by wicking. Circular dichroism studies were carried out on a Jasco J-720 spectropolarimeter. Samples were removed from the refrigerator and immediately placed in a water-jacketed quartz cell (0.5 mm path length) that had been precooled to 4 °C by a water circulator (Neslab).

RESULTS AND DISCUSSION

The technological relevance of hollow cylindrical structures with high aspect ratios has been well documented [1,3,14]. Tubules with smaller diameters at the same length provide higher aspect ratio structures. Numerous published reports are emerging that focus on the formation of nanotubules from a variety of materials. We have previously reported that the bilayer assemblies prepared by the addition of a short chain phosphocholine (DNPC) to diacetylenic phosphocholine, $DC_{8,9}PC$, showed enhanced polymerization of diacetylenic functionality [15] without interrupting the formation of regular bilayer assemblies [16]. However, the formation of gels consisting of 28 nm wide ribbons at room temperature and the formation of nanoribbons and nanotubules in water at 4 °C after careful cooling of the mixed lipid dispersion [11,12] led us to examine the mixed lipid system using primarily Circular Dichroic (CD) spectrometry [13]. Results revealed that the CD technique is a sensitive tool for screening the lipid dispersions for the presence of nanotubules by monitoring characteristic signals at 195 and 202 nm. Unlike microtubules (500 nm diameter) where both of these peaks were positive for naturally occurring enantiomers [17], the 202 nm peak in nanotubules was negative. Comparison between the nanostructure morphologies revealed that intensity of negative CD signal decreases with decreasing nanotubules concentration and/or increasing nanotubules diameter. A shift of the negative peak from 204 to 202 nm was also observed as nanotubules transformed into nanohelices a transformation influenced by temperature or due to the aging of the dispersion.

Effect of Lipid Ratio on Nanotubule formation

All samples studied were screened by monitoring CD spectra to confirm the presence of nanotubules before proceeding with electron microscopic observations. Table 1 summarizes the results from this study. Structures of lipids used in the study and a TEM image of nanotubules are illustrated in Figure 1. Samples with 10 and 90 mole fractions $DC_{8,9}PC$ showed only the presence of vesicles and microtubules. The rest of the samples showed the presence of nanotubules at 4 °C. However, the mixed-lipid samples stored at room temperature for more than six hours produced differing morphologies as shown in Figure 2. Mixtures containing 40 mole fraction of $DC_{8,9}PC$ (figure 2A) revealed the presence of nanoribbons (125 nm dia.) while samples containing 60 mole fraction of diacetylenic lipid (figure 2B) produced nanoribbons of 640 and 80 nm diameters. Circular dichroic spectra of both of these samples displayed only a negative peak between 196-198 nm. Mixtures with 45 (data not shown) and 55 mole fractions of $DC_{8,9}PC$ (figure 2C) showed the presence of helices (145 nm diameter) in TEM and the CD spectrum showed only a negative peak at 196 nm. Equimolar mixtures (figure 2D) produced 56 nm diameter ribbons stacked in parallel fashion. These samples showed a negative CD signal at 198 nm. A typical CD spectrum of nanotubule is shown in figure 3.

Table 1. Results observed for DC$_{8,9}$PC:DNPC mixtures held at room temperature.

Lipid Ratio	Diameter (nm)	Morphology	CD Observations
40:60	125	ribbons	negative peak at 196-198 nm
45:55	145	helices	negative peak at 196 nm
10:10	56	ribbons	negative peak at 198 nm
55:45	145	helices	negative peak at 196 nm
60:40	640/80	ribbons	negative peak at 196-198 nm

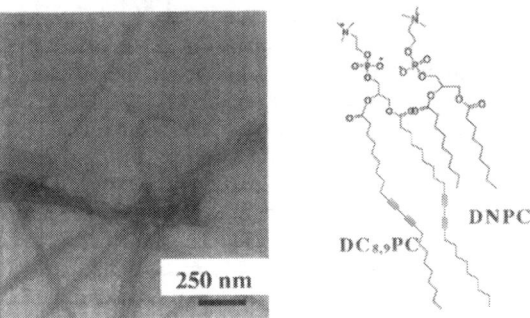

Figure 1. TEM of Nanotubule at 4 °C and Structure formula of diacetylenic phospholipid and DNPC.

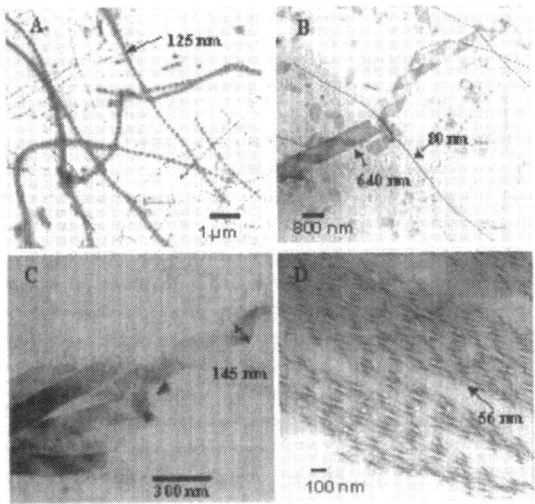

Figure 2A-D. TEM of Mixed lipid dispersions at room temperature; DC$_{8,9}$PC : DNPC molar ratio A. 40:60; B. 60:40; C. 55:45; and D. 1:1.

Effect of Temperature on Nanotubules

Equimolar mixtures of phospholipids with 2 mM total lipid concentration consistently produced nanotubules in abundance. Storage at 4 °C resulted in the retention of nanotubule morphology and storage at room temperature resulted in the formation of ribbons and helices. Nanotubules prepared from equimolar mixture of lipids upon slow heating transform into helical ribbons, which upon storing at room temperature produce microtubules of 500 nm diameter. We immersed a solution of nanotubules stored at 4 °C in hot water baths maintained at 15, 25, and 34°C. Examination of the tubules by TEM, and CD techniques after 2 hours of incubation period showed only the presence of nanotubules and vesicles and CD spectra showed a decrease in the peak intensity, but the peak position remained unchanged (see Figure 3). NMR analysis showed the presence of both lipids in equimolar ratio in the tubules. These results indicated that these nanotubules are sufficiently stable to withstand the processing conditions required for applications development.

Effect of Acyl Chain Length on Nanotubules

In order to examine the influence of acyl chain length on nanostructured properties, we examined a 2 mM, equimolar dispersion of and DNPC mixture in water. This lipid contains 2 additional methylene groups in its hydrocarbon chain. This mixed lipid dispersion again showed two peaks, a positive peak at 200 nm and a negative peak at 207 nm with molar ellipticities of 1.2×10^5 and 6.9×10^4 deg.cm^2/dmol respectively. TEM image showed the average diameter of tubules to be 63 nm (Figure 4). This diameter is slightly larger than those observed for the DC$_{8,9}$PC:DNPC mixture.

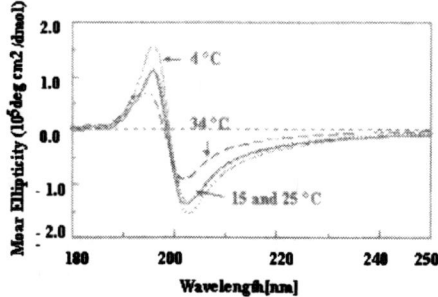

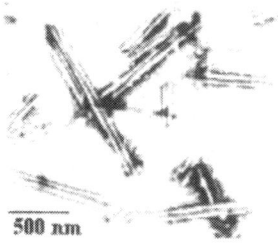

Figure 3. CD Spectra of nanotubules from 2mM, equimolar mixture of DC8,9PC and DNPC held at 4,15, 25 and 34 °C for two hours.

Figure 4. Nanotubules from equimolar mixture of DC$_{10,11}$PC and DNPC.

CONCLUSIONS

Mixed lipid systems consisting of diacetylenic phospholipid and short chain saturated phospholipids may provide a key for the development of novel, simple and efficient strategies for making nanotubules and controlling their diameters. Stability of nanotubules over a wider range of temperatures makes them suitable for developing practical applications.

ACKNOWLEDGEMENTS

This work was supported by the Office of Naval Research. We thank Robert Hendel for TEM assistance. Eva Wong is a NRC Research Associate.

REFERENCES

1. J. M. Schnur, *Science* **262**, 1669 (1993).
2. A. Singh and J. M. Schnur, "Polymerized Phospholipids" in *Handbook of Phospholipids*, G. Cevc, ed. (Marcel Dekker, 1995), pp 233-291.
3. D.T. Bong, T. D. Clark, J. R. Granja, and M. R. Ghediri, *Angew. Chem. Int. Ed.* **40**, 988 (2001).
4. K. Ariga, J. Kikuchi, M. Naito, E. Koyama, and N. Yamada, *Langmuir* **16**, 4929 (2000).
5. J.H. Fuhrhop and W. Helfrich, *Chem. Rev* **93**, 1565 (1993).
6. P. Yager, P. E. Schoen, Mol. Cryst. Liq. Cryst. **106**, 371 (1984).
7. J. H Georger, A. Singh, R. R. Price, J. M. Schnur, P.Yager, and P. E Schoen, *J. Am. Chem. Soc.* **109**, 6169-6175 (1987).
8. M. A. Markowitz, A. Singh, and J. M. Schnur, *Chem. Phys. Lipids* **62**, 193 (1992).
9. A. Singh, M. A. Markowitz, and L. Tsao, *Chem. Phys. Lipids* **63**, 191 (1992).
10. D. G. Rhodes and A. Singh, *Chem. Phys. Lipids* **59**, 215 (1991).
11. M. A. Markowitz, E. L. Chang, and A. Singh, *Bichem. Biophys. Res. Commun.* **203**, 296 (1994).
12. S. Svenson and P. B. Messersmith, *Langmuir* **15**, 4464 (1999).
13. M. S. Spector, A. Singh, P. B. Messersmith, and J. M. Schnur, *Nano Letters* **7**, 375 (2001).
14. N. Nakashima, R. Ando, T. Muramatsu, and T. Kunitake, *Langmuir* **10**, 232 (1994).
15. A. Singh and B. P. Gaber, *Applied Bioactive Polymeric Materials*, C. G. Gebelein, C. E. Carraher, and V. R. Forster eds. (Plenum Press, 1988) p.239.
16. D. G. Rhodes and A. Singh, *Chem. Phys. Lipids* **59**, 215 (1991).
17. J. M. Schnur, B. R. Ratna, J. V. Selinger, A. Singh, G. Jyothi, and K. R. K. Easwaran, *Science* **264**, 945 (1994).

Bio-Inspired Nanomaterials

Mat. Res. Soc. Symp. Proc. Vol. 707 © 2002 Materials Research Society   AA4.5/W4.5

FABRICATION OF TWO- AND THREE-DIMENSIONAL STRUCTURES OF NANOPARTICLES USING LB METHOD AND DNA HYBRIDIZATION

Takayuki Takahagi, Shujuan Huang, Gen Tsutsui, Hiroyuki Sakaue, and Shoso Shingubara
Graduate School of Advanced Sciences of Matter, Hiroshima University,
1-3-1 Kagamiyama, Higashi-Hiroshima 739-8526, Japan

ABSTRACT

In this paper we describe fabrication methods for two types of nanostructures, two- and three-dimensional arrays of gold nanoparticles. Large-scale and high-ordered monolayers of alkanethiol-encapsulated gold particles were fabricated by using Langmuir-Blodgett (LB) method. Three-dimensional nanoparticle arrays composed of gold nanoparticles of two different sizes, which were encapsulated by complementary thiol-capped DNA oligonucleotides, were fabricated by using DNA hybridization. DNA hybridization occurred upon mixing these particles, which resulted in the assembly of three-dimensional nanostructure of gold particles. Scanning electron microscopy observations and UV spectroscopy measurement were performed to confirm the construction of the nanostructures.

INTRODUCTION

Recently, nanostructured materials made of nanoparticles have attracted much research attention for the novel electrical, optical, magnetic, and chemical properties [1-4]. The construction of nanoparticles in an ordered structure is the key issue for their potential application in nanoelectronics devices and magnetic storages. In our previous work [5], we developed a self-organization method for fabricating a planar ordered array of gold particles encapsulated by alkanethiol. We have observed the Coulomb blockade phenomena in the nanostructures consisting of the encapsulated gold particles of 9 nm in diameter [6], which suggests that alkanethiol monolayer surrounding the gold particles can serve as a stable tunnel barrier and the proposed nanostructure is very useful for developing nanoelectronics devices. In this paper, we report the fabrication of a large-scale highly ordered monolayer of gold particles by using LB method. In addition, in order to fabricate more complex nanostructures, one feasible way is DNA-assisted fabrication. DNA oligonucleotides have hybridization information encoded in their sequence. In this study, by utilizing this property of DNA, we fabricated a three-dimensional structure of gold nanoparticles with two particle sizes, which were encapsulated by complementary DNA oligonucleotides. Scanning electron microscopy (SEM) observations and UV spectroscopy measurement were carried out to confirm the construction of the nanostructures.

EXPERIMENT

LB preparation for gold nanoparticle monolayers

The gold particles used in this work were synthesized by using a mixture of trisodium citrate and tannic acid for the reduction of chloroauric acid ($HAuCl_4$) [7]. It is reported that gold particles synthesized by this method is homodispersed. The size of the particles is determined by the quantity of the tannic acid in the reducing solution. Because alkanethiols are water insoluble, the aqueous gold colloidal solution synthesized by this method is not suitable for preparing alkanethiol-encapsulated gold particle solution in high concentrations [5]. In addition, the preparation of the LB monolayer of encapsulated gold particles requires the suspension of encapsulated gold particles in a water-insoluble volatile solvent such as chloroform. For these reasons, we made use of a mediating solvent of ethanol, which is soluble in both of water and chloroform, to prepare the LB spreading solution [8]. First, gold colloidal particles were dissociated by centrifugation and were then dispersed in fresh ethanol using an ultrasonic processor. By mixing the above-prepared gold particle suspension in ethanol with dodecanethiol solution, the suspension of alkanethiol-encapsulated gold particles was prepared [9]. The mixed solution was kept at room temperature overnight, and was then centrifugalized to remove the unreacted dodecanethiol molecules. This centrifugation caused the dodecanethiol-encapsulated gold particles to precipitate at the bottom of the tube, which were dispersed in fresh chloroform by an ultrasonic processor. The concentration of gold particles was determined by the amount of chloroform infused.

The LB trough we used was a KSV mini-trough manufactured by KSV Instruments. To prepare LB monolayer, minute droplets of the encapsulated gold particle suspension prepared above were very carefully cast on the surface of the pure water in the LB trough. After the solvent evaporated, the hydrophobic dodecanethiol-encapsulated gold particles remained on the water's surface. These particles were then compressed by moving the barriers at a speed of 5~10 mm/min. The surface pressure isotherm of water was recorded throughout the compression. The gold particles were transferred by retracting a hydrogen-terminated silicon substrate at a speed of 0.5 mm/min, which was immersed vertically in the trough before the particle suspension was spread.

DNA-assisted fabrication of three-dimensional array of gold nanoparticles

Figure 1 shows the fabrication strategy for three-dimensional structures of gold nanoparticles using DNA hybridization. Two colloidal gold solutions containing particles of 9 nm and 20 nm in diameter and oligonucleotides that are functionalized by heptanethiol at their 5' termini were used in this study. The thiol group at the end of the oligonucleotides was adsorbed to the particle

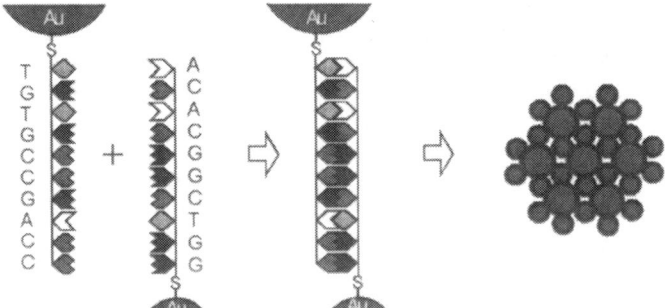

Figure 1. Fabrication strategy for a three-dimensional nanostructure based on complementary property of DNA.

surfaces by a chemical bond upon mixing with the colloidal gold solution, resulting in oligonucleotide monolayers on the particles. More specifically, 0.2 mL of 9 nm particle solution was mixed with 5 μL of 100 μM thiol-5'-GGTCGGCACA-3', and 0.2 mL of 20 nm particle solution was mixed with 5 μL of 100 μM thiol-5'-TGTGCCGACC-3'. After keeping for 48 h at room temperature, the two mixture solutions were blended and diluted with 0.5 mL of 1 M NaCl with pH 7.

UV spectroscopy measurement of above-prepared solution was carried out at 80 °C and 30 °C to confirm the hybridization of DNA oligonucleotides. Also, for SEM observation, a droplet of above-prepared solution was then cast on a silicon substrate and dried at room temperature in air. SEM observations were performed on a Hitachi S-5000.

RESULTS AND DISCUSSION

LB monolayer of alkanethiol-encapsulated gold particles

The monolayer of the gold particles was deposited onto a hydrogen-terminated silicon substrate over an area of 1 cm^2. SEM observations demonstrated that it was composed of two-dimensional domains of ordered, close-packed gold particles. Figure 2 shows the SEM micrograph of part of a domain in LB monolayer prepared from 8.3 nm gold particle solution of 0.06 mg/mL and transferred at water's surface pressure of 10 mN/m. The inset shows a high-magnification SEM image, which demonstrates that a very high level of orderliness of the particle arrangement had formed locally. However, at longer ranges, it can be seen that the gold particles have not formed a perfect order. The particle arrangement looks like a polycrystal, which has many "crystal grains". Some particle vacancies and dislocations were observed in

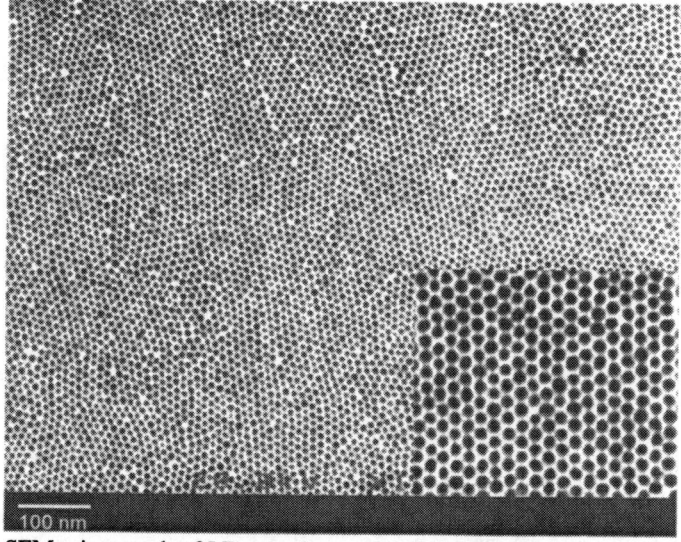

Figure 2. SEM micrograph of LB monolayer of dodecanethiol-encapsulated gold particles of 8.3 nm in diameter. The inset shows a high-magnification SEM image.

SEM observations. This is due to arrangement defects that were mainly caused by the variation of the particle sizes.

Forming mechanism of the domain structure

It is known that a homogenous ordered monolayer of amphiphilic molecules can be fabricated using the LB method. However, in the case of gold particles, the monolayer that was formed was not a homogenous monolayer, but rather contained many domains made up of locally ordered particles. In order to elucidate the mechanism of the formation of domain structures, LB monolayers of encapsulated gold particles were prepared using a variety of gold particle densities on the surface of the water without moving of the barriers. Particle density d on the surface is defined as $d = N/A$, where N is the amount of encapsulated gold particles spread over the surface of the water, and A is the effective area of water's surface in the trough. Figure 3 shows SEM images of encapsulated gold particles transferred under different particle density conditions. It should be noted that although gold particles were not compressed by the barriers, the domain structures of the ordered close-packed particles had formed under each of the density conditions. In the case of the low particle density, crystal nucleus-like structures were formed as shown in Figure 3(a). When the particle densities on the water's surface increased, larger and larger domains were observed as shown in Figure 3(b) and 3(c). These results suggest that the ordered domain structures of gold particles were not a result of barrier compression, but rather a result a self-organization process that occurred on the water's surface.

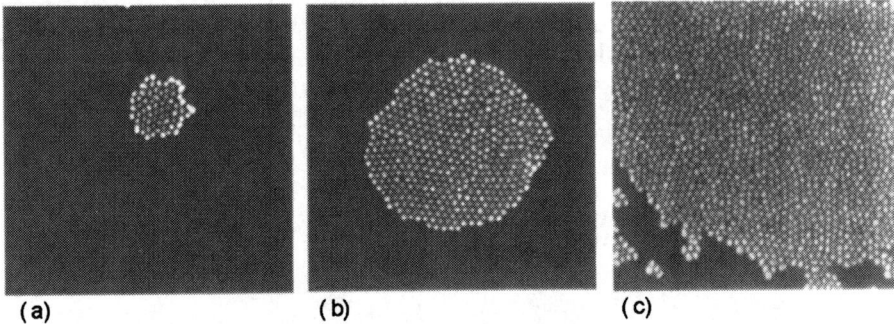

(a)   (b)   (c)

Figure 3. SEM images of domain structures formed under different particle densities on the water's surface: (a) a low density of 2.74×10^{13} /m^2, (b) a medium density of 1.37×10^{14} /m^2, and (c) a high density of 2.74×10^{14} /m^2.

In the LB preparation, the encapsulated gold particle solution was spread on the surface of the water in the trough. Because the solvent used is water-insoluble, and because alkanethiol-encapsulated gold particles are hydrophobic, the sample evaporation process on the water's surface is similar to that on a solid substrate [5]. In the concrete, when a droplet of encapsulated gold suspension was cast onto the surface of the water in a trough, its water-insolubility caused it to spread into a thin layer. As the solvent evaporated, convective flows of the solvent appeared. These flows carried the particles toward the solvent surface, whereby a particle assembly arose. After the solvent became a thin layer with a thickness equal to the particle size, the attractive force between particles induced by the surface tension of the solvent increased significantly and caused the particles to pack together into ordered arrays. When the sample first began to be spread, the particle density on the water's surface was quite low, so that only small domains were formed. When the particle suspension was spread and evaporated in succession, more and more particles were transported to these nucleus-like structures driven by the same attraction, so that the nuclei grew larger and larger. Finally, these self-organized domains were compressed to a densely packed monolayer on the water's surface by moving the barriers.

In this study, we discovered that the degree of long-range order of the monolayer made from a low particle concentration of 0.06~0.3 mg/mL is better than that made from a high particle concentration of 0.6 mg/mL. In other words, a low particle concentration of 0.06~0.3 mg/mL is more suitable for the LB preparation of gold particle monolayers. Also, We found that a surface pressure of π = 10 mN/m is the optimal condition for a good quality monolayer of 8.3-nm-diameter gold particles. The particle coverage at this pressure is about 85%. When the gold particles were compressed beyond 10 mN/m, some patches of bilayer had formed and the monolayer collapsed

DNA Hybridization

The results of UV spectroscopy measurement on the blended solution of DNA-encapsulated gold particles are shown in Figure 4. The absorbance peak at 260 nm at 80 °C, 24 °C above the dissociation temperature (T_m = 56 °C), is obvious higher than that at 30 °C. As known, the intensity of optical absorbance of DNA is dependent on the state of hybridization. Unhybridized DNA shows higher absorbance than hybridized one. Therefore, the different intensity of absorbance in Figure 4 gives an evidence for the DNA hybridization of the encapsulated particles at 30 °C.

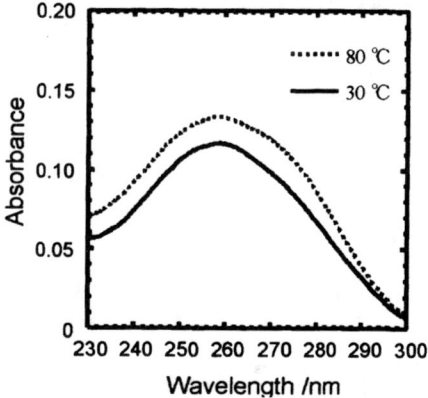

Figure 4. UV spectrum for blended solution of thiol-5'-GGTCGGCACA-3'-capped and thiol-5'-TGTGCCGACC-3'-capped gold particles.

SEM observation of a three-dimensional array

Figure 5 shows the SEM micrograph of DNA-encapsulated gold nanoparticles of 9 and 20 nm in diameter. Obviously, a three-dimensional structure has formed with alternating superposition of the big and small particles. Though the internal structure cannot be observed with SEM, the 20 nm gold particles can be seen to exclusively connect with the 9 nm particles on their surfaces. Hundreds of similar particle aggregations were observed. Comparatively, using uncomplimentary oligonucleotides, the encapsulated gold particles never aggregated in this way, but rather in a way of the aggregation of the same size. These results suggest that the internal structure of DNA-linked particles might consist of alternating small and large particles, and three-dimensional particle networks had formed.

CONCLUSION

We reported the fabrication of a highly ordered monolayer of alkanethiol-encapsulated gold particles using LB method. The LB monolayers were composed of domains of ordered close-packed particles, some of which extended to about 100 μm^2 in size. Our study of the mechanism by which such an LB monolayer was formed demonstrates that the domain structure was a result of a self-organization process induced by evaporation of the solvent before LB compression began. The LB compression aggregated these domains to form a densely packed monolayer on the water's surface. A low particle concentration of 0.06~0.3 mg/mL resulted in a highly ordered monolayer of gold particles.

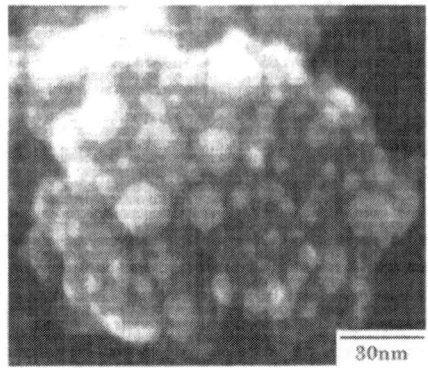

Figure 5. SEM image of three-dimensional nanostructure composed of 9 nm and 20 nm gold particles linked by DNA hybridization.

Also, we described the fabrication of three-dimensional particle structures based on DNA hybridization. Our results indicate that DNA hybridization is a very useful method for assembling nanoparticles into a three-dimensional nanostructure. By using both methods proposed, it is possible to fabricated more complex nanostructures of various particle materials and particle sizes.

REFERENCE

1. B. Korgel, S. Fullam, S. Connolly and D. Fitzmaurice, *J. Phys. Chem.* **B102**, 8379 (1998).
2. G. Markovich, D. V. Leff, S.-W. Chung, H. M. Soyez, B. Dunn and J. R. Heath, *Appl. Phys. Lett.* **70**, 3107 (1997).
3. S. Sun, C. B. Murray, D. Weller, L. Folks and A. Moser, *Science* **287**, 1989 (2000).
4. V. P. Roychowdhury, D. B. Janes, S. Bandyopadhyay and X. Wang, *IEEE Trans. Electron Devices* **43** 1688 (1996).
5. S. Huang, H. Sakaue, S. Shingubara and T. Takahagi, *Jpn. J. Appl. Phys.*, Part 1 **37**, 7198 (1998).
6. S. Huang, G. Tsutsui, H. Sakaue, S. Shingubara and T. Takahagi, *J. Vac. Sci. Technol.* **B18**, 2653 (2000).
7. J. W. Slot and H. J. Geuze, *Eur. J. Cell Biol.* **38**, 87 (1985).
8. S. Huang, G. Tsutsui, H. Sakaue, S. Shingubara and T. Takahagi, *J. Vac. Sci. Technol.* **B19**, 115 (2001).
9. C. D. Bain, E. B. Troughton, Y. T. Tao, J. Evall, G. M. Whitesides and R. G. Nazzo, *J. Am. Chem. Soc.* **111** 321 (1989).

Mat. Res. Soc. Symp. Proc. Vol. 707 © 2002 Materials Research Society   AA5.5/HH5.5

Bio-Inspired Nanocomposites: From Synthesis Toward Potential Applications

Tewodros Asefa, Neil Coombs, Hiltrud Grondey, Mietek Jaroniec[1], Michal Kruk[1], Mark J. MacLachlan, and Geoffrey A. Ozin*
Materials Chemistry Research Group, Department of Chemistry, University of Toronto, Toronto, Ontario M5S 3H6, Canada
[1]Department of Chemistry, Kent State University, Kent, Ohio, 44242, USA

ABSTRACT

In recent years, the extraordinary properties of bio-inspired nanocomposites have stimulated great interest in the development of bottom-up synthetic approaches to organic-inorganic hybrid materials in which molecular scale control is exerted over the interface between the organic and inorganic moieties. These developments have led to advanced materials with novel properties and potential use in catalysis, sensing, separations and environmental remediation. Periodic mesoporous organosilica (PMO) materials are an entirely new class of nanocomposites with molecularly integrated organic/inorganic networks, high surface areas and pore volumes, and well ordered and uniform size pores and channels. We recently have extended the approach to include novel PMO materials incorporating chiral and heteroatom-containing organic functional groups inside the inorganic framework that may be useful in asymmetric catalysis, enantiomeric separations and heavy metal remediation.

INTRODUCTION

Over evolutionary time scales, Nature has mastered the process of making several complex and hierarchal structures that can perform various tasks in living organisms [1]. Most of these form through biomineralization of inorganic compounds within membrane-delineated space and/or organic matrix and scaffolding composed of specialized macromolecules [2]. Examples of such structures include calcium phosphates in bones and teeth of vertebrates, calcium carbonates of mollusk shells and statoliths of squids, biogeomagnetic compass of magnetotactic bacteria, and exoskeleton of coccolith [2,3]. Control over the crystal nucleation, polymorph type, crystal texture and morphology in all these materials are the result of the build-up of the hierarchal structures starting at the molecular level to form composite organic-inorganic biomaterials [2]. Understanding of the mechanism of biomineralization and mimicking this process from the natural world has been an enjoyable and rewarding area of research in materials science, biomimetics, bioinorganic chemistry, biomedical science, bioelectronics and bio-inspired self-assembly in the last few decades [3-7]. This inspiration, for instance, has led self-assembly material chemists to use surfactant-templating self-assembly approaches to discover ordered mesostructured silicas (M41S materials) in 1992 [8]. These materials have periodic organic-inorganic hierarchical structures and their synthesis mimics nature's extraordinary ability in a number of ways. In 1999, similar approaches led us and two other groups to discover new types of surfactant-templated periodic mesoporous organosilica (PMO) materials that have self-assembled structures with organic-inorganic hybrid frameworks [9-12]. Since this time, a number of papers have been published on the synthesis of PMO materials and on their chemical, surface and mechanical properties [9-30]. The discovery of ordered mesoporous materials with organosilica frameworks is believed to offer many advantageous features compared to the earlier silica-based and terminal organic group functionalized mesoporous materials. The presence of the organic moiety in the framework could result in unusual physical, electrical, optical, mechanical, surface and chemical properties in the materials. The organic functional groups could also help immobilize biological molecules that can display similar properties as they do in their natural environment. A good article in this area has recently been published by Tertykh and Yanishpolskii, which details the adsorption and chemisorption of enzymes and natural macromolecules immobilized on organosilica surfaces, their physicochemical properties and many of their analytical and biocatalytic applications [31]. Furthermore, heteroatom functionalized organosilica materials have been reported to have important non-covalent interactions that help selective intake and release of proteins, immobilize enzymes and proteins, [32,33] and consequently play selectivity roles for mixtures of compounds for separation and sensing applications. For instance, amorphous organosilica functionalized materials can be used to separate Heme from cytochrome c [32].

In this report, we highlight some of the contributions that we made in the last three years in the area of surfactant-templated periodic mesoporous organosilica (PMO) materials. The vast choice of functional groups that can be anchored onto the bridging framework organic group of these materials potentially includes biologically active, electroactive, enzymes, proteins, aromatic and organometallic, catalytic, proteins, heteroatoms and coordination complexes.

EXPERIMENTAL DETAILS

For the synthesis of PMO materials, surfactant-mediated sol-gel hydrolysis and condensation of bis(trialkoxysilyl)organic $((R'O)_3Si)_nR)$ compounds were used [9-30]. Co-condensation with tetraethoxysilane (TEOS) was employed when control over the relative amount of organosiloxane groups in the materials was required [9,13]. The PMO precursors were either commercially obtained or synthesized by polysilylation of various organic molecules through Grignard, hydrosilylation, lithiation and silylation-alcoholysis reactions (Scheme 1). Most of the synthetic reactions were carried out under inert atmosphere in pre-purified solvents [10,19,29].

The PMO materials were synthesized from the corresponding precursors under acid, base, or fluoride catalyzed [9-12,29,34] aqueous surfactant solutions. For a typical base-catalyzed synthesis, a solution of cetyltrimethylammonium bromide (CTABr) (0.67g, 1.84 mmol), NH_4OH (14.18 g, 35 wt%, 0.14 mol) and H_2O (26.73 g, 1.48 mol) was prepared at room temperature, and 2.94 mmol of bis- or tris(trialkoxysilylated)organic PMO precursor (or a mixture of PMO precursor and TEOS with a total of 5.88 mmol Si) was added. The mixture was stirred for 30 mins during which time a precipitate was formed. After aging at 80 °C for 4 d to increase the condensation process and subsequently obtain structurally integrated material, the product was filtered, washed with water, and dried under ambient conditions to give a fine powder.

For a typical acid-catalyzed synthesis, a solution of CTABr (0.34 g, 0.93 mmol), HCl (7.18 g, 36 wt%, 70.8 mmol) and H_2O (13.4 g, 0.74 mol) at room temperature was prepared, and 3.85 mmol (or 7.70 mmol Si) of PMO precursor was added. The mixture was stirred for 30 mins and then aged at 80 °C for 4 d. The product was isolated by filtration, washed with water, and dried under ambient conditions.

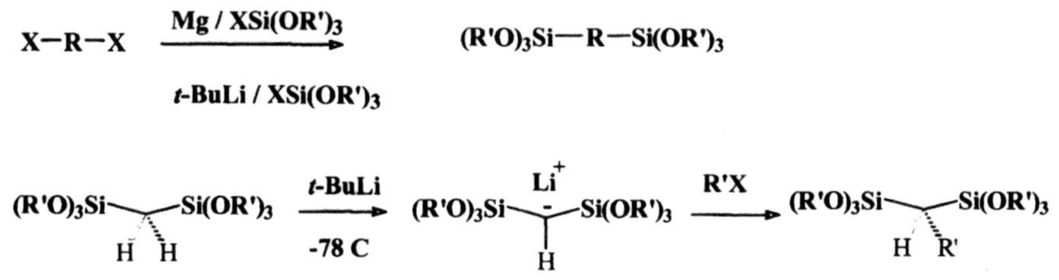

$$X-R-X \xrightarrow[t\text{-BuLi / }XSi(OR')_3]{Mg\,/\,XSi(OR')_3} (R'O)_3Si-R-Si(OR')_3$$

Scheme 1. Synthesis of PMO precursors through Grignard and lithiation reactions.

For fluoride mediated synthesis, a similar composition as the base-catalyzed procedure was used, but with *ca.* 3.86 mmol NH_4F (per 7.72 mmol precursor or 15.44 mmol Si) instead of NH_4OH. The samples were aged for 3-24 h at 80 °C [34].

The surfactant from the PMO materials was removed either by solvent-extraction in an HCl/MeOH solution or calcination in air at 300-400 °C for 3 h. Generally, the solvent-extraction procedure was effective for most PMOs when about 0.3 g of as-synthesized powder was stirred for 6 h at 55 °C in a solution of 4 g of conc. (36 wt %) HCl and 170 g MeOH. Typical weight losses ranged from *ca.* 20-30 % after a single solvent extraction. Calcination was effective for PMOs containing $Si-C(sp^3)$ bonds and bridging aromatic groups.

For preparation of oriented PMO thin films, a hydrolyzed PMO precursor in CTABr/EtOH/H_2O solution [22] or a non-ionic liquid crystalline solution [17] was prepared and the film was made either by spin or dip coating or drop casting on a glass or silicon substrate. The film in CTABr templated materials dried at ambient conditions or aged between 40-60 °C while for those containing non-ionic templates, the hydrolysis product (ethanol) was pumped off at a moderate rate right after the film was made.

The PMO materials were characterized with powder X-ray diffraction (PXRD), transmission electron microscopy (TEM), ^{13}C, ^{29}Si CP MAS, ^{29}Si MAS, ^{11}B MAS and ^{79}Br MAS NMR spectroscopy, N_2 adsorption, thermogravimetric analysis (TGA), FT-Raman spectroscopy and elemental analysis (EA). The experimental details for each measurement were described in references [13,17-20].

RESULTS AND DISCUSSION

Before the discovery of PMO materials in 1999, various inorganic and terminal organic-functionalized mesoporous materials were reported. The synthesis of many of these inorganic materials were performed from the hydrolysis of metal alkoxides ($M(OR)_4$) in an appropriate surfactant solution. However, the synthesis of organic containing mesoporous materials were carried out either through co-condensation of organotrialkoxysilane ($RSi(OR)_3$) compounds with tetraalkoxysilanes (the former contributing *ca.* 43 mol % or less silicon atoms in the overall material) or through grafting the organotrialkoxysilane compounds onto silanol (Si-OH) groups of a preformed material [35-37]. Unfortunately, both of these methods usually give materials with a lower fraction of organics than those achievable in PMOs and grafting often results in a less uniform distribution of organics [35-37].

The recent discovery of PMO materials from the self-assembly of bridged silesesquioxane ($\{(RO)_3Si\}_nR'$, n=2,3,etc.) precursors in a surfactant solution has allowed for the first time the incorporation of organic groups in the framework of ordered mesoporous materials with uniform distribution and an ordered materials to be obtained from 100% bridged organosilanes without co-condenastion. The synthesis occurs from the hydrolysis and condensation of the polyalkoxysilylated PMO precursors around the head groups of surfactant-templates mediated by acid, base, or fluoride catalysts. Upon completion of hydrolysis and condensation, the surfactant templates were removed under solvent-extraction or calcination, leaving ordered porous PMO materials with uniformly distributed organic groups inside the framework. In the past three years, we and many other groups have reported a number of 2-D and 3-D hexagonal and cubic ordered PMO materials that have various kinds of bridging organic groups [9-30]. This review will only briefly highlight and discuss the contributions from our group in this area in the last three years.

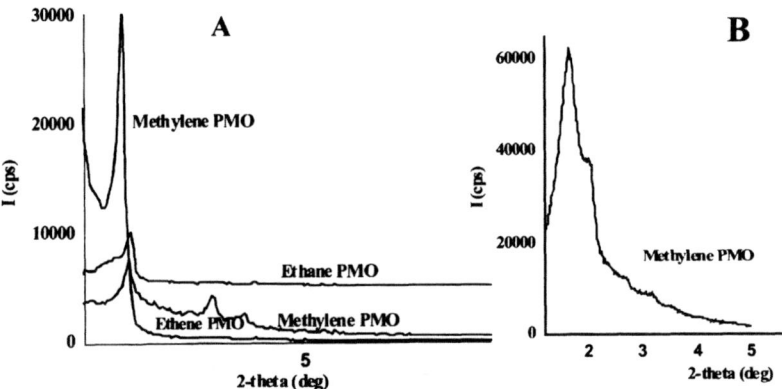

Figure 1. PXRD patterns of surfactant-extracted hexagonal ordered methylene, ethylene and ethane PMOs (A) and surfactant-extracted cubic ordered methylene PMO (B).

Among the first PMO materials we reported, most contained short and rigid organic groups such as methylene, ethylene and ethane (Scheme 2) [9,13,34]. All these three materials showed ordered 2-D hexagonal structures with unit cell dimensions a_o = 45 - 47 Å (d_{100} at $ca.$ 39 - 41 Å). The PXRD patterns of 2-D hexagonal-ordered ethane, ethylene and methylene PMOs are shown in Figure 1A. The d_{100} values of the materials are similar when the same surfactant (CTABr) was used. This suggests that the structures of such short and rigid organic group containing PMO materials seem to depend heavily on the nature and size of the templating surfactant micellar species rather than the organic groups. Similarly, the synthesis of ethane PMOs catalyzed with NH_4F gave well-ordered mesostructured materials [34]. By changing the surfactant concentration in the preparation, we recently succeeded in making cubic ordered methylene PMO materials that were indexed in the space group Pm3m or Pm3n (Figure 1B). [20] Typical transmission electron microscopy (TEM) images for hexagonal and cubic ordered

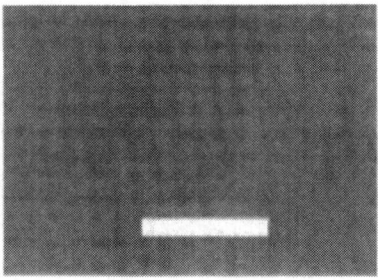

Figure 2. Transmission electron micrographs of 2-D hexagonal ordered (left) and cubic ordered (right) methylene PMO materials (scale bar = 25 nm).

methylene PMO materials are shown in Figure 2. The solid state NMR spectra of these materials confirmed that the organic groups remained intact with no Si-C bond cleavage after synthesis and surfactant-extraction (Figure 3).

As few PMO precursors are commercially available, we have devoted a great effort to synthesizing known and new polyalkoxysilylated PMO precursors and PMOs that have many potential applications. Scheme 1 shows some synthetic routes that we have been following to make various PMO precursors. Among the methods, lithium or magnesium based metallation have been the most routinely used in which commercially available polyhalogenated reagents were converted to the corresponding organomagnesium or organolithium compounds, which were subsequently coupled with chloro-triethoxysilane [10,19,29,30]. Among these precursors, many of them contained aromatic rings and were found to give ordered mesoporous materials. These include 1,4-benzene, 1,3,5-benzene, 1,4-tolyl, 1,4-xylyl, 1,4-phenylmethylene, and 1,4-dimethoxybenzene and even 2,5-thiophene, 2,2'-bithiophene and 1,1'-ferrocene containing PMOs (Scheme 2) [10,17,19,29,30]. However, in some of these materials a relatively low degree of order and cleavage of Si-C bonds were observed and much work was required to optimize the syntheses [10]. In some cases, the difficulties were minimized or overcome by allowing the reaction to take place under basic conditions followed by immediate neutralization of the base catalyst or by using non-aqueous solvents [10,19]. Overall, most of the aromatic PMOs have a significant fraction of well-ordered material. The aromatic groups also survive the hydrolysis, condensation, and solvent-extraction processes as shown from aromatic peaks at ca. 129-140 ppm in the ^{13}C CP and NQS CP MAS NMR spectra [13] of 1,3,5-benzene PMO (Figure 4A). Most of these peaks also remained in the material after calcination at 350 °C for 6 hrs. The ^{29}Si CP MAS NMR spectra (Figure 4B) of the material showed only slight cleavage of Si-C bonds. In general, the aromatic PMOs, like methylene PMOs, were found to be more thermally stable than ethylene PMOs [9,13,18].

Significant loss of uniform mesoporosity or formation of amorphous materials were observed when we attempted the synthesis of PMOs with precursors containing longer and more flexible organic groups such as bis(triethoxysilyl)hexane bis(triethoxysilylethyl)benzene and bis(triethoxysilyl)propylethylenediamine [29]. Nonetheless, solid state NMR spectroscopy of the resulting materials indicated that the organic groups remained intact. The Si-C bonds in these materials are hydrolytically stable since the silicon atoms are bound to sp^3 carbon atoms [10]. Our experiments have shown that the Si-C(sp^3) bonds are also more stable than Si-C(sp or sp^2

carbons) during thermolysis [13,18,20,30]. For instance, the methylene groups in both hexagonal and cubic methylenesilica PMOs are stable up to *ca.* 600 °C during calcination in air and N_2 for a few hours. However, they undergo an unprecedented bridging-to-terminal organic transformation during thermolysis before being expelled, the ordered mesoporous structure remaining intact at all stages. This transformation has been monitored using ^{13}C non-quaternary suppression (NQS) CP MAS and ^{29}Si NMR experiments [13,20]. This thermal process allowed the formation of terminal organic group functionalized mesoporous materials to be synthesized from PMOs by a novel route.

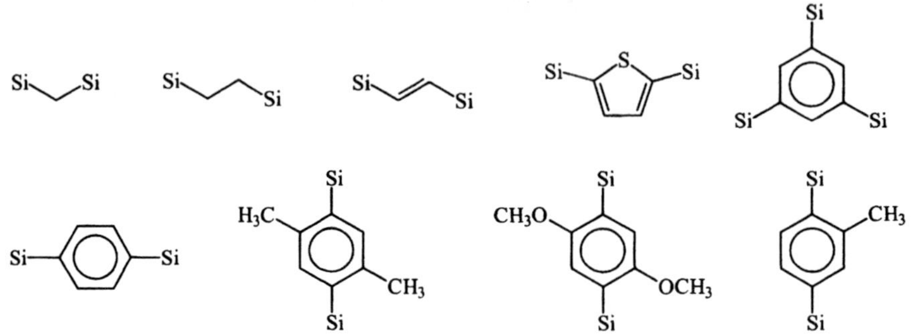

Scheme 2. Typical alkyl and aromatic PMOs.

The co-condensation of two or more bridging and/or terminal alkoxysilane precursors results in bi- or multi-functionalized PMO materials where the electroactive, surface, mechanical and catalytic properties of the materials can be further tuned [18]. For instance, we synthesized bridging and terminal organic group containing bi-functional PMO material (BPMO) that has its surface, mechanical and chemical properties modified due to the presence of two different organic groups in the material. These include vinyl-ethylene (Figure 3), methylene-ethylene-methyl bi- and trifunctional PMO materials [18]. Nitrogen adsorption isotherm for vinyl-ethylene BPMO synthesized using 25% of the vinyl-functionalized precursor (evaluated as a percentage of silicon atoms from a given precursor in the synthesis mixture) is shown in Figure 5A. This BPMO exhibited the BET specific surface area of 690 m^2 g^{-1}, pore diameter of about 2.8 nm (pore size distribution shown in Figure 5B) and primary mesopore volume of about 0.48 cm^3 g^{-1}. It exhibited one strong peak on its XRD pattern (Figure 6A). The surfactant was successfully removed from the as-synthesized material under solvent extraction, as seen from relatively small weight loss corresponding to surfactant at about 200°C under air for the solvent-extracted BPMO material (Figure 6B). A small weight gain following this weight loss can be attributed to a reaction of vinyl groups with oxygen. The subsequent significant weight loss is related to removal of organics from the material, and its magnitude is correlated well with the formula of the BPMO product predicted on the basis of the composition of the synthesis mixture, indicating a stoichiometric incorporation of both of the organosilanes used in the synthesis [18].

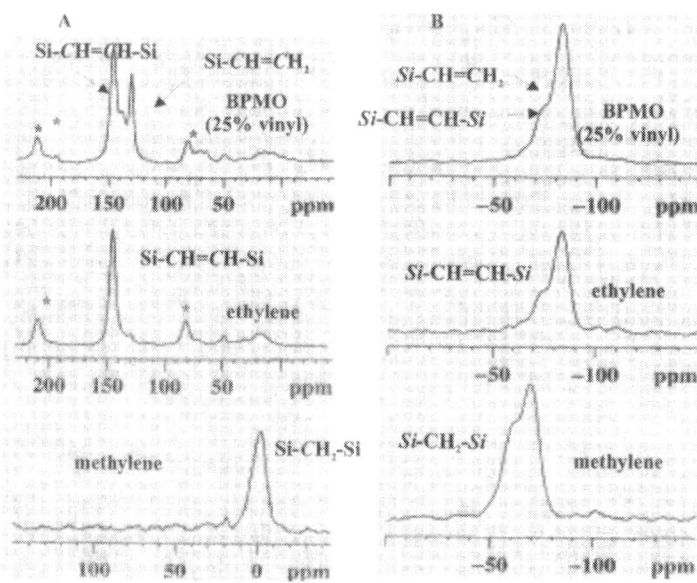

Figure 3. (A) ^{13}C CP MAS NMR spectra of solvent-extracted methylene, ethylene and bifunctional (25% vinyl/75% ethylene) PMO materials. (B) ^{29}Si CP MAS NMR spectra of solvent-extracted methylene, ethylene and bifunctional (25% vinyl/75% ethylene BPMO) PMO materials ($^{\bullet}$ denotes spinning side bands).

In-situ reactivity of organic groups of PMO materials or what we call "*chemistry of the channels*" has been an area of interest in which we investigate the accessibility, reactivity and kinetics of the various organic groups in PMOs [9,12,18,29]. The fundamental studies in this area lead to further functionalization of the materials. This way, various catalytic reagents and functional groups can be introduced into PMOs either through coordination or covalent bonds onto the functional sites on the walls of the PMOs. It was reported that the functional groups on the frameworks are accessible and reactive enough to allow further modification but showed some differences compared to similar functional groups within the pores or in solution phase [9,12,18,29]. These differences could be advantageous for carrying out selective reactions important for separating molecules. For instance, ethylene groups in the framework are less reactive than pendent vinyl groups in an ethylene-vinyl bifunctional PMO (BPMO). Furthermore, the hydrophobicity caused by the bridging organic groups could help less polar reactant and product species to access the channels and reactive sites. We recently demonstrated this through hydroboration-alcoholysis and epoxidation-alcoholysis reactions in BPMO materials [18]. Both of these reactions are very practical in organic chemistry as intermediate steps for making diverse functional and chiral groups. Chiral group functionalized PMO and BPMOs could in turn be useful for heterogeneous asymmetric catalysis and chiral separation applications. Figure 7 shows typical ^{13}C CP MAS NMR spectra of BPMO and ethylene PMO after hydroboration. Peaks corresponding to sp^2 organic groups (vinyl : 130 and 137 ppm and ethylene : 146 ppm) undergo hydroboration forming alkyl sp^3 carbons (7 ppm). However, these reactions depend on the structure and pore size of the materials and the solvent used. In some cases reaction conditions have to be modified to leave the organic groups intact and avoid Si-C bond cleavage. Moreover, the hydroboration of ethylene carbons required much longer reaction time than vinyl groups [29]. Subsequent reactions of hydroborated and epoxide functionalized materials resulted in alcohol and chiral diol functionalized BPMOs, respectively. We also recently reported similar sequential transformation of vinyl groups in vinyl-functionalized mesoporous materials that were prepared in a one-pot synthesis [38].

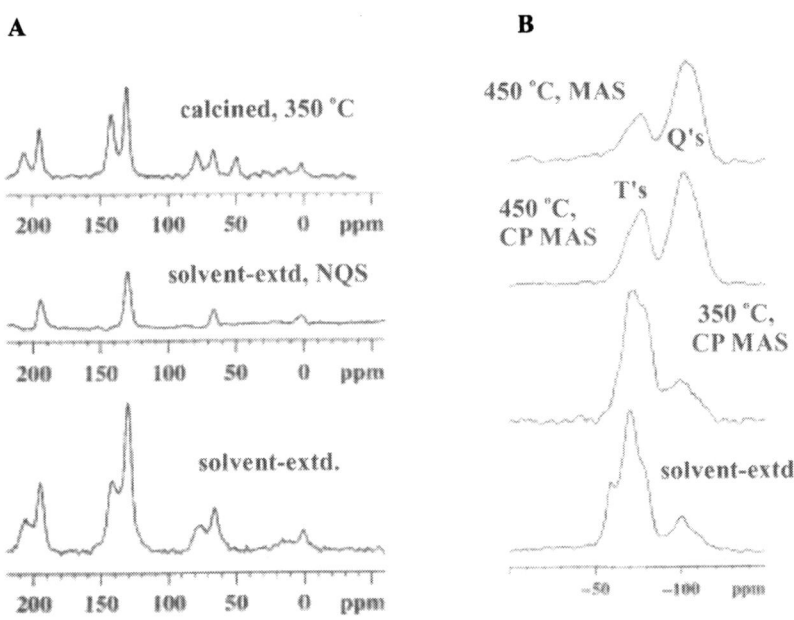

Figure 4. (A) ^{13}C CP and NQS CP MAS NMR spectra of solvent-extracted and calcined 1,3,5-benzene PMO. (B) ^{29}Si CP MAS spectra of solvent-extracted and calcined 1,3,5-benzene PMO.

Preliminary results indicated that the presence of bridging organic groups in the framework seems to improve the hydrothermal stability of the mesoporous structures. For instance, the PXRD pattern and adsorption isotherm of a BPMO material seems to have changed not very significantly after soaking in boiling water at 95 °C for 1d, conditions deleterious to many siliceous MCM-41 materials. This enhanced structural stability was also evident in BPMOs vs. VIN-MCM-41 materials after *in-situ* hydroboration-alcoholysis reactions. The former containing bridging organic groups was observed to remain structurally more stable after hydroboration than the latter that contains Si-O-Si linkages in its backbone [18,37].

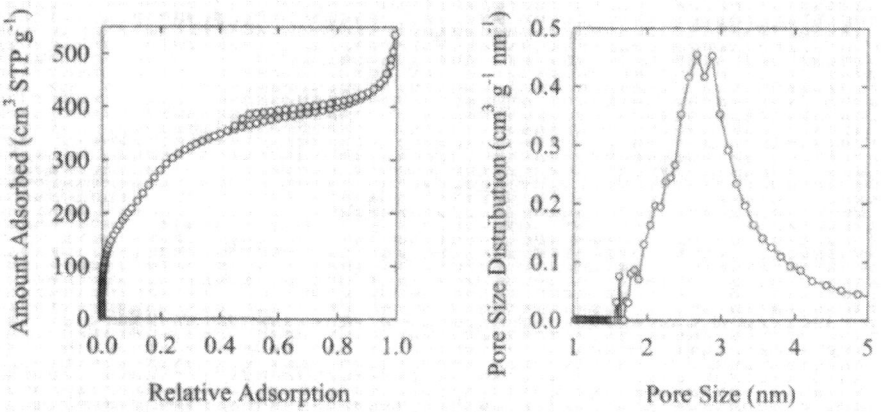

Figure 5. Nitrogen adsorption isotherm and pore size distribution for surfactant-extracted vinyl-ethylene BPMO synthesized using 25% of the vinyl-functionalized precursor.

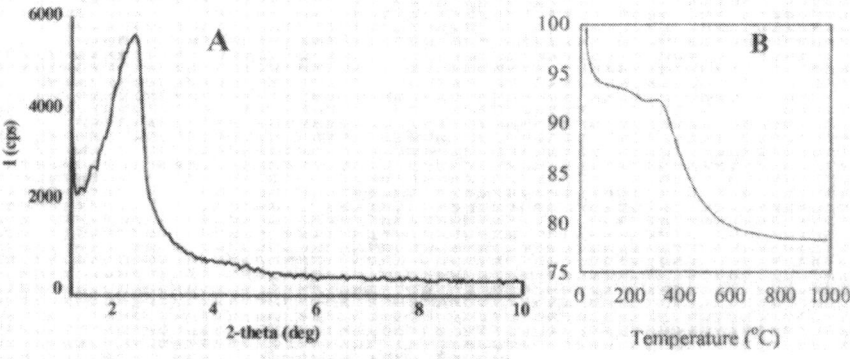

Figure 6. (A) PXRD pattern and (B) TGA plot of surfactant-extracted vinyl-ethylene BPMO synthesized using 25% of the vinyl-functionalized precursor.

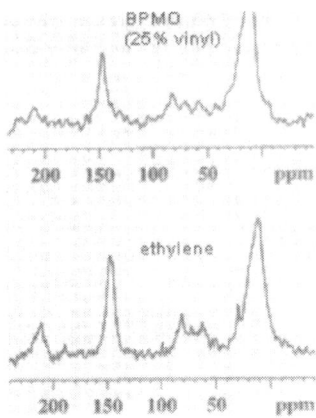

Figure 7. ^{13}C CP MAS NMR spectra of hydroborated ethylene and bifunctional (25% vinyl/75% ethylene) PMO materials.

We recently also synthesized various bromo-functionalized PMO materials which are anticipated to have a wide range of potential applications [29]. These organobromine functionalized mesoporous materials can be used as important starting materials for making other functionalized materials, as organobromine groups are known to undergo nucleophilic, electrophilic and radical reactions easily. Moreover, the bulky bromine atoms could change the surface and adsorption properties of the materials.

Preparation of thin films is important for many applications and devices. Lyotropic liquid crystalline templating approaches under optimized conditions yielded the first oriented PMO films [17]. Figure 8 shows an example of a highly oriented mesostructured thiophene PMO film made by dip coating from a solution containing hydrolyzed benzene precursor in oligoethyleneoxide $(C_{12}H_{25}(EO)_{10}H)$ non-ionic surfactant template.

Figure 8. Polarized optical microscope image of a PMO film showing fan-like optical birefringence.

CONCLUSIONS

Biology has inspired the generation of new classes of materials. In this brief review, we have highlighted some of the contributions that our group has made to the discovery and development of a new class of bio-inspired nanocomposite materials called periodic mesoporous organosilicas (PMOs). In particular, this new class of materials offers exciting opportunities for fundamental materials investigations (precursor synthesis, assembly mechanism, organic transformation inside PMOs, instrument development) and hold promise for applications as advanced materials.

ACKNOWLEDGEMENTS

The financial support of NSERC (Canada) and the NSF (Grant CHE-0093707) are acknowledged. GAO is a Canada Research Chair in Materials Chemistry. The authors also thank Dr. Masakatsu Kuroki, Prof. Ömer Dag, Stephen Knauer, Chiaki Yoshina-Ishii, Galina Temtsin, Prof. Shmuel Bittner, Terry Fedorkiw and Dr. Srebri Petrov for their contributions to the development of PMO materials.

REFERENCES

1. S. Weiner, L. Addadi, and H. D. Wagner, *Mater. Sci. Eng. C* **11**, 1 (2000).
2. Y. Levi-Kalisman, G. Falini, L. Addadi, and S. Weiner, *J. Struct. Biol.* **135**, 8 (2001).
3. W. Becker, J. Marxen, M. Epple, and O. Reelsen, *J. Appl. Physiology* **89**, 1601 (2000).
4. M. Louloudi, Y. Deligiannakis, and N. Hadjiliadis, *J. Inorg. Biochem.* **79**, 93 (2000).
5. M. Louloudi, Y. Deligiannakis, and N. Hadjiliadis, *Inorg. Chem.* **37**, 6847 (1998).
6. C. Nicolini, *Biosensors & Bioelctronics* **10**, 105 (1995).
7. B. L. Zhou, *Mater. Sci. Eng. C* **11**, 13 (2000).
8. C. T. Kresge, M. E. Leonowicz, W. J. Roth, J. C. Vartuli, and J. S. Beck, *Nature* **359**, 710 (1992).
9. T. Asefa, M. J. MacLachlan, N. Coombs, and G. A. Ozin, *Nature* **402**, 867 (1999).
10. C. Yoshina-Ishii, T. Asefa, N. Coombs, M. J. MacLachlan, and G. A. Ozin, *Chem. Commun.* 2539 (1999).
11. S. Inagaki, S. Guan, Y. Fukushima, T. Ohsuna, and O. Terasaki, *J. Am. Chem. Soc.* **121**, 9611 (1999).
12. B. J. Melde, B. T. Holland, C. F. Blanford, and A. Stein, *Chem. Mater.* **11**, 3302 (1999).
13. T. Asefa, M. J. MacLachlan, H. Grondey, N. Coombs, and G. A. Ozin, *Angew. Chem. Int. Ed. Engl.* **39**, 1808 (2000).
14. M. Kruk, M. Jaroniec, S. Guan, and S. Inagaki, *J. Phys. Chem. B* **105**, 681 (2001).
15. M. J. MacLachlan, T. Asefa, and G. A. Ozin, G. A. *Chem. Eur. J.* **6**, 2507 (2000).
16. T. Asefa, C. Yoshina-Ishii, M. J. MacLachlan and G. A. Ozin, *J. Mater. Chem.* **10**, 1751 (2000).
17. Ö. Dag, C. Yoshina-Ishii, T. Asefa, M. J. MacLachlan, H. Grondey, and G. A. Ozin, *Adv. Funct. Mater.* **3**, 213 (2001).

18. T. Asefa, M. Kruk, M. J. MacLachlan, N. Coombs, H. Grondey, M. Jaroniec, and G. A. Ozin, *J. Am. Chem. Soc.* **123**, 8520 (2001).
19. G. Temtsin, T. Asefa, S. Bittner, and G. A. Ozin, *J. Mater. Chem.* **11**, 3202 (2001).
20. T. Asefa, M. Kruk, N. Coombs, S. Petrov, M. Jaroniec, and G. A. Ozin, *J. Mater. Chem.*, submitted.
21. S. Guan, S. Inagaki, T. Ohsuna, and O. Terasaki, *J. Am. Chem. Soc.* **122**, 5660 (2000).
22. Y. Lu, H. Fan, N. Doke, D. A. Loy, R. A. Assink, D. A. LaVan, and C. J. Brinker, *J. Am. Chem. Soc.* **122**, 5258 (2000).
23. A. Sayari, S. Hamoudi, Y. Yang, I. L. Moudrakovski, and J. R. Ripmeester, *Chem. Mater.* **12**, 3857 (2000).
24. S. Hamoudi, Y. Yang, I. L. Moudrakovski, S. Lang, and A. Sayari, *J. Phys. Chem. B* **105**, 9118 (2001).
25. S. S. Park , C. H. Lee , J. H. Cheon, and D. H. Park, *J. Mater. Chem.* **11**, 3397 (2001).
26. K. Yamamoto, Y. Nohara, and T. Tatsumi, *Chem. Lett.* 648 (2001).
27. M. C. Burleigh, S. Dai, E. W. Hagaman, and J. S. Lin, *Chem. Mater.* **13**, 2537 (2001).
28. O. Muth, C. Schellbach, and M. Fröba, *Chem. Commun.* 2032 (2001).
29. T. Asefa, G. A. Ozin, H. Grondey, M. Jaroniec, and M. Kruk, *Stud. Surf. Sci. Catal.* **141**, 1 (2002).
30. M. Kuroki, and G. A. Ozin, Manuscript in preparation.
31. V. A. Tertykh, and V. V. Yanishpolskii, *Surfactant Sci. Ser.* **90**, 523 (2000).
32. M. S. Rao, and B. C. Dave, *J. Am. Chem. Soc.* **120**, 13270 (1998).
33. T. I. Denisova, G. F. Karpenko, T. A. Khalyavka, D. I. Shvetz, *Adsorpt. Sci. Technol.* **17**, 139 (1999).
34. T. Fedorkiw, T. Asefa, and G. A. Ozin, Unpublished results.
35. M. H. Lim, C. F. Blanford and A. Stein, *Chem. Mater.* **10**, 467 (1998).
36. C. E. Fowler, S. L. Burkett and S. Mann, *Chem. Commun.* 1769 (1997).
37. T. Asefa, M. Kruk, M. J. MacLachlan, N. Coombs, H. Grondey, M. Jaroniec and G. A. Ozin, *Adv. Funct. Mater.* **11** (2001) in press.

Semiconductor Quantum Dots

Mat. Res. Soc. Symp. Proc. Vol. 707 © 2002 Materials Research Society   AA3.2/N2.2

Self-assembling and Ordering of Ge/Si Quantum Dots on Flat and Nanostructured Surfaces

N. Motta[1,2*], A. Sgarlata[1], A.Balzarotti[1], F. Rosei[1,3]
[1]INFM Dipartimento di Fisica, Università di Roma Tor Vergata,
Via della Ricerca Scientifica 1, 00133 Roma, Italy
[2]INFM Dipartimento di Fisica, Università di RomaTre,
via Vasca Navale 84, I-00146 Roma, Italy
[3]Institute of Physics and Astronomy and CAMP, University of Århus, 8000 C Århus, Denmark

ABSTRACT

We have studied by Scanning Tunneling Microscopy (STM) the effect of step bunching on Ge/Si(111) epitaxy. We have verified that self-organization of Ge islands is greatly influenced by "step bunching" which arises from the flash-annealing procedure used to reconstruct the Si surface. Two different growth regimes arise: initially islands nucleate and evolve only at steps, up to complete ripening; subsequently the same evolution is observed on flat areas of the sample. The average distance between islands and steps is nearly constant, originating a single row of equally spaced islands, followed by other rows of islands in between. The exploitation of this phenomenon, which is governed by the surface diffusion length of Ge on Si (estimated from our data) and by the terrace width, constitutes one possible path to achieve self-organization of quantum dots.

INTRODUCTION

It has been shown that during resistive heating of Si(111) a bunching of natural surface steps can form, yielding a simple way to obtain nanostructured Si surfaces. Several authors have studied this phenomenon [1,2], demonstrating that the step configuration at a vicinal surface (with a small misorientation angle to the (111) plane) during sublimation of a Si crystal depends on the direction of the heating current flowing through the crystal as well as on temperature. The temperature dependence of this effect is not simple, but general agreement exists on the fact that, for Si sublimation at T > 1220° C, step bunching occurs in step-down direction, while a regular step distribution occurs in step-up direction. In this way we have obtained both regular (i.e. with steps naturally distributed) and step-bunched Si(111) surfaces on which we have grown epitaxially Ge at T = 450° C and 530° C. We have analyzed by STM and Atomic Force Microscopy (AFM) the evolution and distribution of the islands on different surfaces. An evident self-ordering on step-bunched surfaces exists and the parameters of this ordering have been studied. The amount of Ge-Si intermixing has been also investigated by means of X-Ray Absorption Fine Structure (XAFS), in connection with island shape modification.

EXPERIMENTAL

Ge/Si(111) samples were grown by Physical Vapour Deposition (PVD) in a UHV chamber (base pressure 4 x 10^{-11} mbar) equipped with an e-gun evaporator and a commercial Variable Temperature STM-AFM microscope [3]. Si(111) substrates (n-type $\rho=10^{-3}$ Ω cm, miscut angle < 0.5°) were flashed at 1250 °C to form the 7x7 reconstruction. Both regular (R) and step bunched (SB) surfaces were obtained by using respectively a current flow oriented in step-up and step-down direction. Ge was evaporated at a rate of about 1 Å·min⁻¹ on these surfaces kept at a fixed temperature in the range 450 – 550 °C. A new Si sample was prepared for each Ge deposition in order to minimize the uncertainty of successive evaporations. Two groups of samples will be analyzed here, grown at 450 ± 20 °C and 530±20 °C, respectively.

STM RESULTS

The STM image of a step bunched surface obtained by flashing Si(111) in step-down direction is shown in Fig 1, along with a profile taken across the steps.

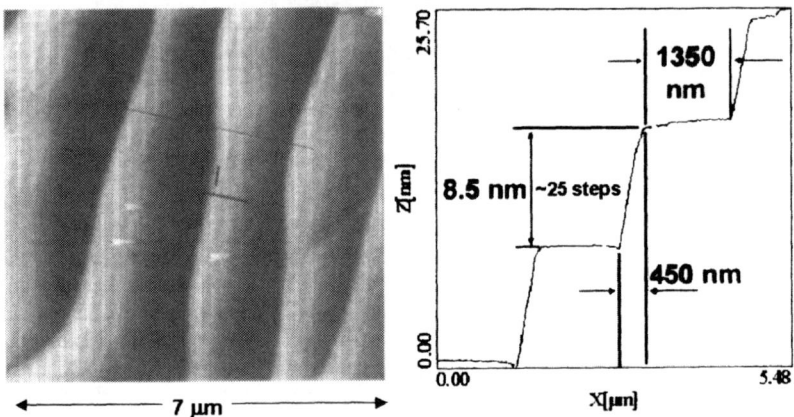

Fig.1 (a) STM topography of a Si(111) surface flashed at T=1250 °C with current flowing in step-down direction; image dimensions are (7000x7000x36) nm³. (b) Profile taken along the red line.

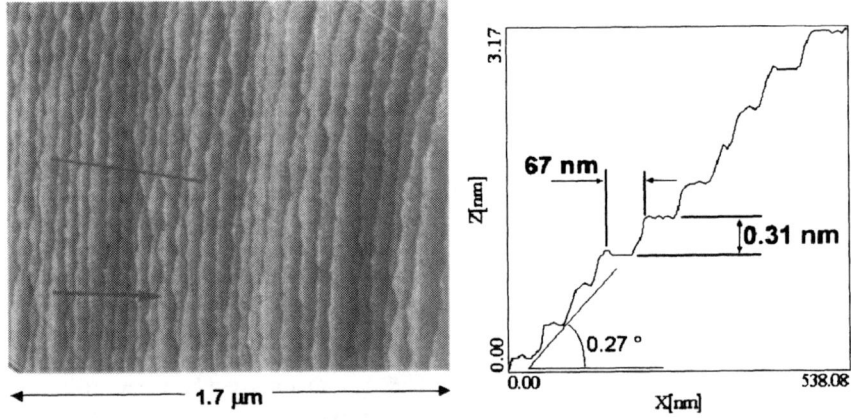

Fig 2 (a) STM topography of a Si(111) surface flashed at T=1250 °C with current flowing in step-up direction; image dimensions are (1700x1700x10) nm³. (b) Profile taken along the red line.

108

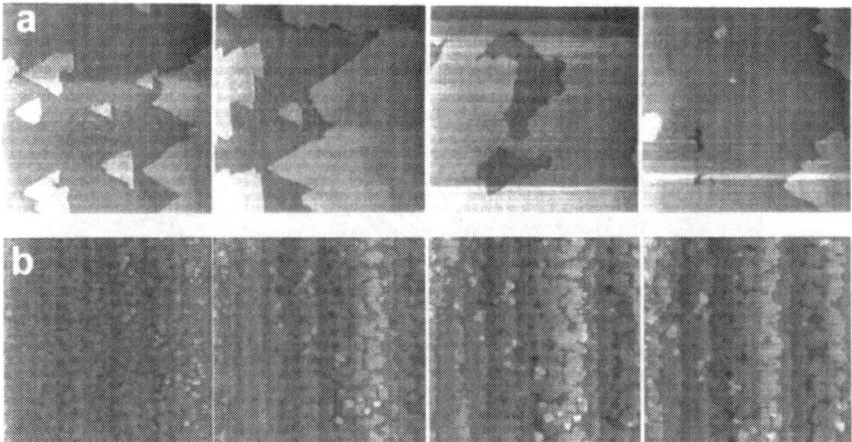

Fig.3 Ge/Si(111): growth of the wetting layer at T=400°C. a) on a step-bunched surface with large terraces, low flux (0.02 ML/min.); 0.3x0.3 μm² STM images. b) on a regular surface with small steps, higher flux (0.15 ML/min), 1x1 μm² STM images. Full movies are available at the www.roma2.infn.it/infm/nanolab web site.

Width of terraces is about 1300 nm, and height is typically 8.5 nm, corresponding to ≈25 atomic steps. The image of a regular surface obtained by flashing in step-up direction is shown in Fig. 2, where steps width is about 70 nm. From the image profile it is possible to estimate an average miscut angle <0.3°.

The growth mode of the Wetting Layer (WL) depends on both substrate arrangement and deposition flux. In Fig 3(a) we show a sequence of STM images corresponding to low flux Ge evaporation on a large terrace of a step-bunched substrate. Triangular 2D islands are observed [4], which expand and coalesce up to the full coverage of the surface. Subsequent layers (up to the third) are formed, which nucleate before the completion of the previous layer. In Fig 3(b) we show a set of images corresponding to the WL growth on a regular surface with a higher Ge flux. Here 2D islands have fractal-like borders, due to the higher growth rate, but growth occurs in the step-flow mode, due to the much shorter distance between the atomic steps. All the images of Fig.3 have been extracted from STM movies acquired during growth.

3D island nucleation starts at a 3-5 ML Ge coverage, with a clear dependence on Ge flux and substrate temperature, as reported for Ge/Si(001) by Kamins *et al.* [6]. Initially islands nucleate as truncated pyramids with one corner pointing in the <11-2> direction; the evolution proceeds by including new facets. Subsequently, by insertion of dislocations, the islands round out, often forming a hole at their center [7].

An interesting point is the islands distribution on different substrates. On the step-bunched substrate, islands first nucleate and evolve on the steps, and then on terraces (Fig 4 a-b). Islands on the steps are subjected to ripening, elongation and coalescence up to the formation of a continuous ribbon.

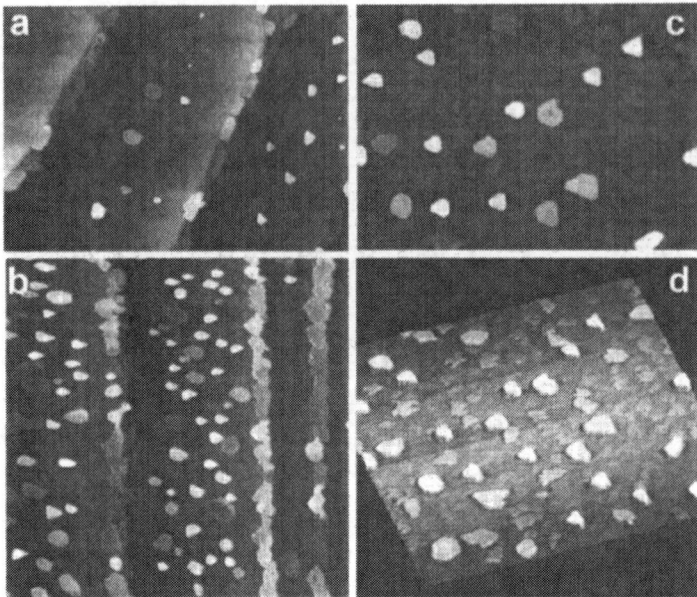

Fig 4 Island distribution on different surfaces obtained by Ge/Si(111) epitaxy at 450 °C. a) 2.5 nm Ge on a step-bunched surface: STM image (3.7x3.7) μm²; vertical scale: 62 nm. b) 6 nm Ge on a step-bunched surface: STM image (10x10) μm²; vertical scale: 75 nm. c) 5.4 nm Ge deposited at T=500 °C on a regular surface: (3x3) μm²; vertical scale: 29 nm. d) 3D representation of image (c).

When the evolution is complete, some islands at the center of the terraces begin to appear (Fig 4 a). They have a regular distribution, with a mean spacing of 0.39 μm and a mean distance from the steps of 0.62 μm. This gives a rough estimate (≈0.5 μm) of the diffusion length of Ge on these surfaces. By increasing the deposited material new islands nucleate on free areas, determining full coverage of the surface (Fig 4 b). It is very interesting to observe that a regular distribution of islands has been obtained by CVD growth on lithographically patterned substrates by Kamins et al. [8] and by Goryll et al. [9]. Our experiments show that the same is possible on Si(111) just by using the natural patterning due to step bunching.

On a regular substrate islands nucleate without any specific ordering (Fig 4 c-d). It is also notable that tetrahedral islands point in the same $<11\bar{2}>$ direction both on step-bunched and regular surfaces.

XAFS RESULTS

XAFS spectra at the Ge K-edge were recorded at the 'GILDA' beamline at ESRF, Grenoble, in order to provide a quantitative value of the intermixing level in our samples [10].

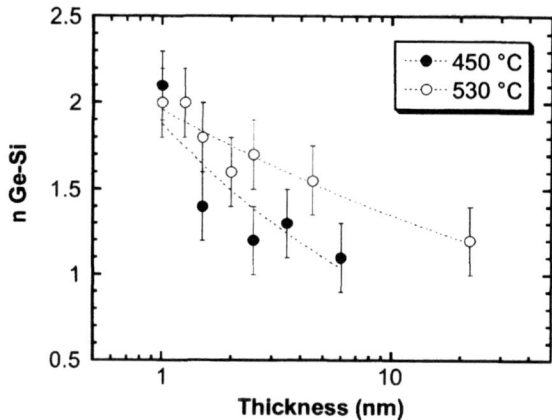

Fig 5. Average number of Si atoms around a Ge absorber obtained by fitting XAFS data taken on Ge/Si(111) samples grown at 450 and 530 °C substrate temperature.

In Fig 5 the number of Si atoms around Ge, obtained from the fit of the Ge/Si(111) XAFS spectra, is shown as a function of deposited Ge thickness. The data refers to substrate temperatures of 450±20 °C and 530±20 °C. The number of Si atoms decreases from 2 to 1 with increasing thickness. This means that intermixing decreases from 50% (as previously reported [10, 11]) to 25%. Moreover, samples grown at higher T display higher values of n, which points to a larger intermixing. Interesting experiments on this aspect have been performed by Capellini et al.[12], who have analyzed by XPS the Si content in Ge/Si(001) islands grown at different temperatures by CVD, finding an increase of intermixing in the islands with temperature. Our intermixing (ranging from about 25% for samples with 3 D islands to 50% for samples with WL only), matches well the observed lateral size of the islands (up to 200 - 300 nm). In fact this size is considerably larger than that measured on typical Ge/Si islands (40 - 80 nm) grown by fast evaporation or by CVD [13, 14]. The decrease of intermixing with thickness, shown in Fig.5, is connected to the increase of Ge in the 3D islands at the expenses of the WL. We suggest that Si content is limited both by diffusion (the height of the islands can reach 50 nm) and from the fact that in the islands the lattice is more relaxed than in the WL, so reducing the driving force for intermixing.

CONCLUSIONS

We have analyzed the effects of substrate on the formation and distribution of Ge/Si(111) islands. By using *in situ* STM we have observed the growth of the wetting layer, and the nucleation of 3D islands. The island distribution on step-bunched and regular substrates has been analyzed. A nearly ordered distribution of islands has been found along the terraces of step-bunched substrates; their spacing has been connected to the diffusion length of Ge on the surface. A random nucleation has been observed instead on regular substrates.

From XAFS analysis we measured intermixing of Si in Ge on two set of samples with substrate temperature of 450 and 530 °C and Ge thicknesses ranging from 5 to 220 Å. By increasing coverage, we found that the number of Si atoms surrounding Ge decreases from 2 to 1, which means that intermixing decreases from 50% (as reported in our previous work) to 25%. The temperature dependence indicates a larger intermixing for the samples grown at higher temperatures.

ACKNOWLEDGEMENTS.

This work has been partially supported by INFM. We are grateful to G. Capellini and F.Boscherini for their collaboration on the XAFS data acquisition and analysis.

REFERENCES

* corresponding author. e-mail: motta@roma2.infn.it

[1] K.Yagi, H.Minoda, M.Degawa *Surf. Sci. Rep.* **43** (2001) 45, and references therein
[2] J.J. Metois and S.Stoyanov, *Surf. Sci.* **440** (1999), 407
[3] N.Motta, *Proceedings of the NN2000 school, S.Margherita di Pula, (Italy)*, Ed. by S.Bellucci, M.De Crescenzi, SIF Conference proceedings series, Bologna (2001).
[4] U. Köhler, O. Jusko, G. Pietsch, B. Muller and M. Henzler, *Surf. Sci.* **248**, 321 (1991).
[5] N. Motta, A. Sgarlata, R. Calarco, Q. Nguyen, F. Patella, J. Castro-Cal, A. Balzarotti and M. De Crescenzi, *Surf. Sci.* **406**, 254 (1998). G.Capellini, N. Motta, A. Sgarlata and R. Calarco, *Solid State Comm.* **112**, 145 (1999).
[6] T.I. Kamins, E.C. Carr, R.S. Williams and S.J. Rosner, *J. Appl. Phys.* **81** 211 (1997); T.I. Kamins, G. Medeiros-Ribeiro, D.A.A. Ohlberg and R.S. Williams, *J. Appl. Phys.* **85** 1159 (1999).
[7] F.Rosei, N.Motta, A.Sgarlata and A.Balzarotti, these proceedings, pag. xxx
[8] T. Kamins and R.S. Williams *Appl. Phys. Lett.* **71**, 1201 (1997)
[9] M. Goryll, L. Vescan, K. Schmidt, S. Mesters, and H. Luth, K. Szot, *Mat. Sci. and Eng. B* 69-70 p.251 (2000).
[10] N. Motta, F. Boscherini, G. Capellini, F. Rosei, A. Sgarlata, and S. Mobilio, *Mat. Sci. and Engineering B. in press (2001)*.
[11] F. Boscherini et al., *Appl. Phys. Lett.* **76**, 682 (2000); F. Rosei et al., *Thin Solid Films* **369**, 29 (2000); F. Boscherini et al., *Thin Solid Films* **380**, 173 (2000).
[12] G.Capellini, M.De Seta, F.Evangelisti, *Appl. Phys. Lett.* **78**, (2001), 303.
[13] G. Capellini, L. Di Gaspare, F. Evangelisti and E. Palange, *Appl. Phys. Lett.* **70**, 493 (1997).
[14] H. Kajiyama *et al., Phys. Rev. B* **45**, 14005 (1992).

Mat. Res. Soc. Symp. Proc. Vol. 707 © 2002 Materials Research Society

Spontaneous Pattern Formation from Focused and Unfocused Ion Beam Irradiation

Alexandre Cuenat and Michael J. Aziz
Division of Engineering and Applied Sciences, Harvard University, Cambridge, MA 02138.

ABSTRACT

We study the formation and self-organization of "ripples" and "dots" spontaneously appearing during uniform irradiation of Si, Ge, and GaSb with energetic ion beams. Features have been produced both with sub-keV unfocused Ar^+ ions and with a 30 keV Ga^+ Focused Ion Beam. We follow the evolution of features from small amplitude to "nanospikes" with increasing ion dose. It appears that the edge of the sputtered region influences the patterns formed, an effect that may make it possible to guide the self-organization by the imposition of lateral boundary conditions on the sputter instability.

INTRODUCTION

Self-organization of surface morphological features to produce structures of potential use in nanotechnology is in its infancy. A few processes leading to self-organized patterns, particularly in heteroepitaxial growth, have been identified and studied. However, control of the self-organization process remains an issue. We have been investigating alternative processes leading to self-organized surface structures - in particular, "sputter patterning" produced by unfocused ion irradiation. Spontaneous rippling of surfaces by ion irradiation was first observed on glass [1]. Similar patterns have been obtained on a variety of materials, including semiconductors [2-5], metals [6] and insulators [7]. These self-organized patterns have been observed for a wide range of ion energy (from 200 eV to 40 keV); the orientation of the ripples depends mainly on the incidence angle of the ion beam relative to the sample normal.

The pattern formation mechanism is usually discussed within the linear stability framework of Bradley and Harper [8]. Patterning is due to the dependence of the sputter yield on the curvature of the surface. Because ion energy deposition rate increases as the ion penetrates the solid, concave regions are sputtered more than convex regions, resulting in a growing instability on the surface. This instability is opposed by a smoothening mechanism, e.g. surface diffusion or viscous flow, which has a different dependence on wavelength than the roughening effect. The interplay of these two effects selects a characteristic wavelength for the pattern. This linear theory reproduces many experimental observations [9] but not all [10] [4] [11]. At large amplitudes, nonlinear terms have been proposed [12]. The nonlinear behavior may be especially important in enhancing the uniformity of the size distribution of "quantum dots" recently produced on GaSb surfaces [4] using sub-keV irradiation at normal incidence and room temperature. In this paper, self-organized patterns on Si (001) and Ge (001) using both sub-keV unfocused Ar^+ beam and 30 keV Ga^+ Focus Ion Beam (FIB) are compared. Boundary effects are shown to have an effect on the self-organization process, opening a way to control the symmetry of the pattern.

EXPERIMENT

All unfocused beam experiments are performed in a UHV chamber working at a base pressure of $2 \cdot 10^{-10}$ Torr. The surface is bombarded with 750 eV Ar^+ from a collimated 3 cm Kaufman ion gun working at 10^{-4} Torr. The sample can be heated to 800 °C and the angle between the beam and the sample normal can vary from 0 to 70 degrees. Typical ion flux is 0.4 mA/cm^2.

Focused ion beam (FIB) experiments are done using a FEI dual beam FIB-scanning electron microscope (SEM) producing 30 keV Ga^+ and delivering it to a sample at a pressure of $2 \cdot 10^{-7}$ Torr and room temperature. The incidence angle between the beam and the sample normal can vary from 0 to 60 degrees. The ion beam current can vary from 1 pA to 3 nA and the dwell time at each point was chosen to be 1 μs. Separation between each spot hit by the FIB is set to have a 50% overlap for each adjacent point, assuming the beam profile to be gaussian.

Post-irradiated surfaces are observed in air using contact mode AFM.

RESULTS

Focused and unfocused ion beam induced self-organized patterns. Ripple patterns develop under both sub-keV unfocused Ar^+ and Ga^+ FIB irradiations. Here we compare them.

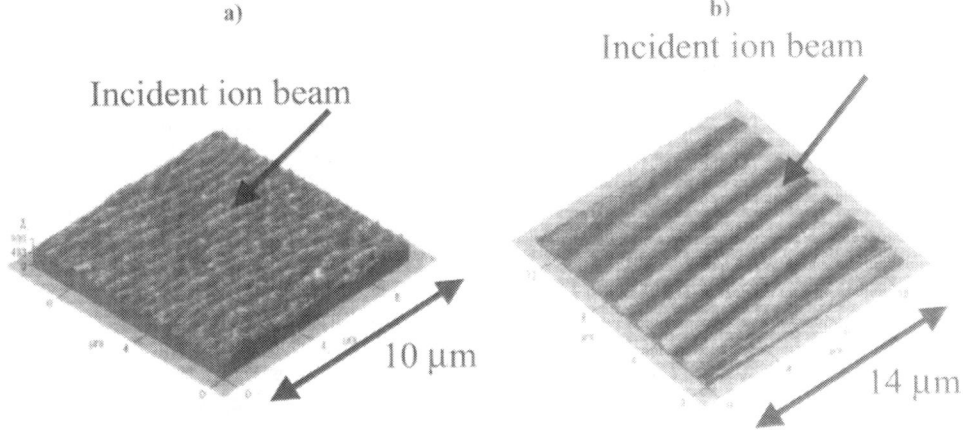

Figure 1. a) Sputter-induced ripples on Si (001) after bombardment with an unfocused 750 eV Ar^+ beam at 600 °C. Wavelength $\lambda = 520$ nm. **b)** Sputter-induced ripples on Si (001) after bombardment with a focused 30 keV Ga^+ beam at room temperature. Wavelength $\lambda = 2.15$ μm

Fig. 1a shows ripples produced using the unfocused Ar beam. The incident angle is 55° from normal and the wavevector is perpendicular to the incoming ion beam. The wavelength can be tuned by changing the temperature or the ion flux [9].

Fig. 1b shows ripples produced at room temperature using the FIB; the incident angle is 52° and the beam is rastered over a total irradiated area of 200 μm by 200 μm. The ripple wave vector is also perpendicular to the beam direction. The ion beam current is 1054 pA for a beam spot size of 40 nm and an irradiation time of 20 minutes. Changing the raster direction of the beam or the overlapping percentage between adjacent FIB points does not change the characteristics of the ripples. Figure. 2 shows the morphology resulting from normal-incidence 1 keV Ar^+ irradiation of Si(001) at room temperature for six minutes. The features are qualitatively different from those reported by Gago et al [13] under the same conditions and we cannot reproduce their results. Also, the features are not nearly as regular as the "quantum dots" produced by Facsko et al. on GaSb [4]. The underlying reasons remain a puzzle.

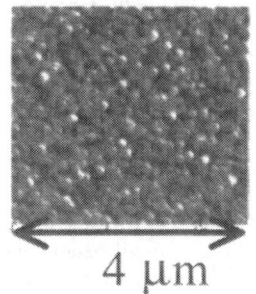

4 μm

Figure 2. Si (001) bombardment at normal incidence with unfocused 1 keV Ar^+ beam at room temperature.

Boundary effects. When the surface evolution is nonlinear, it is interesting to see whether the boundary conditions influence the spontaneously arising pattern. A square 50 μm on edge was sputtered away on a Si (001) surface using the FIB at room temperature. The ion beam current was 1082 pA for a beam spot size of 40 nm and duration of 20 minutes.

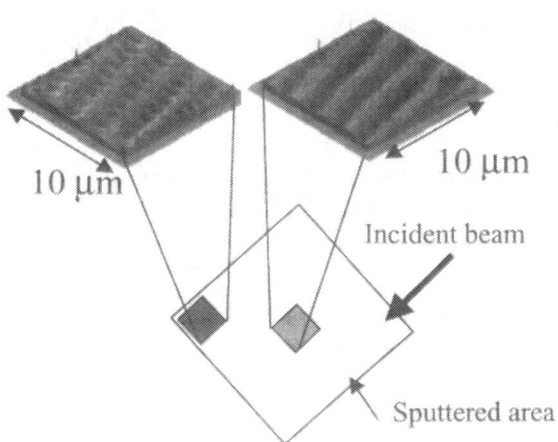

10 μm

10 μm

Incident beam

Sputtered area

Figure 3. Ion sputter-induced ripples on Si (001) after bombardment for 20 min. with a focused 30 keV Ga^+ beam at room temperature in a (50 μm)^2 box. Wavelength λ = 2.07 μm.

Fig. 3 shows AFM images of two regions of a sputtered square crater. In the center of the square, ripples in the expected orientation are produced. However, near the edge of the box a new symmetry appears. New ripples perpendicular to the previous one are produced, both wave-lengths being almost the same. Both ripple directions are parallel to the edge of the square. The amplitude of these new ripples decreases from to 4 to 0 nm, when scanning from the edge of the square to a distance of about 15 μm. Each corner of the square displays the same new symmetry.

FIB sputtering of Ge and GaSb

Using the FIB a variety of "rough" surface are observed on Ge (001). For low dose irradiation, small nanometer bumps are produced as shown on Fig. 4a and b. The morphology depends only on the total dose and is independent of the flux or the dose rate. The dots produced are constant in size, but no order is apparent on the surface. This feature can be produced in a very short time (a few seconds), for small irradiated area and large ion beam current, or on a much longer time scale. Even for irradiation as long as 10 minutes, no order is apparent in the Fourier transform of the AFM image.

Figure 4. a) SEM image of low dose ($3.75 \cdot 10^{15}$ /cm^2) focused 30 keV Ga$^+$ irradiation of Ge (001) surface at room temperature; **b)** corresponding AFM image.

The surface becomes rougher with increasing dose and, for high enough dose, a qualitatively different "nanospiked" surface is produced (Fig. 5a). The evolution is qualitatively the same for GaSb (Fig. 5 b) and the rate of morphological progression is much faster on these cases than on Si. These morphologies are reminiscent of those found in room-temperature ion implantation of Ge [14].

DISCUSSION

Focused and unfocused ion beam experiments permit the study of the self-organization process over a wide range of conditions. Comparison of the resulting morphologies should help unravel the various contributions to morphology evolution associated with this phenomenon. Room temperature irradiation of semiconductors, as is reported here for the FIB experiments, amorphizes the top ~100 nm very early in the irradiation. The higher-temperature, lower-energy experiments reported here for the Ar⁻ irradiation permitted the surface to remain crystalline throughout the experiment. One should expect different mass transport mechanisms to dominate these two regimes, perhaps leading to qualitatively different behavior once the nonlinear regime is reached. A model based on adatom surface diffusion during ion irradiation has successfully accounted for experimental observation for high-temperature sub-keV unfocused irradiation [9, 15]. For FIB conditions we are not yet able to conclude whether the smoothening mechanism is due to irradiation induced viscous relaxation of the amorphized layer, nonlinear effects in the instability induced by the ion beam, or other effects. Large variations in ion beam energy should permit us to address this question.

a)   **b)**

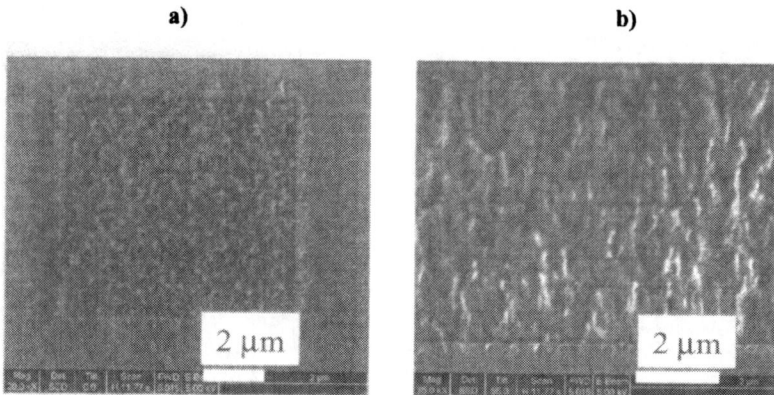

Figure 5. a) High dose ($8 \cdot 10^{17}$ /cm^2) FIB irradiation of Ge (001); plan view SEM image. **b)** Side view SEM image of high dose ($5 \cdot 10^{17}$ /cm^2) FIB irradiation of GaSb (001).

The ripples appear to be more regular when produced with the FIB than with the unfocused Ar⁺ beam. This result could be due to differences in beam characteristics or differences in ripple formation mechanisms. Our experiments indicate that it is not due to any synchronization between the pattern and the grid size used in the FIB pattern generation.

That the boundary of the irradiated region changes the pattern formed in its vicinity indicates the possibility of further morphology control in the future. This observation is a first step toward the guided self-organization of surface features using lateral templating. By combining both focused and unfocused ion beam sputtering we hope to be gain a much higher degree of control over the range of morphologies obtainable.

SUMMARY

We have produced self-organized patterns on Si (001) using both focused and unfocused ion beams. The mechanism for the formation of the ripples in the case of FIB is not completely elucidated, but raster effects can be ruled out. A boundary effect on the morphology has been observed for a small box sputtered using the FIB. Nanometer bumps have been produced using the FIB on Ge (001) surface at low dose, the induced topography depending only on the total dose, independent of dose rate. For higher dose, the surface rapidly develops a high aspect ratio "nanospike" structure. The rapid evolution of nanospikes is also observed for GaSb, but not for Si, where we observe disorganized bump.

ACKNOWLEDGMENTS

This was supported by the Harvard MRSEC, NSF-DMR-98-09363. One of us (A.C.) was supported in part by the Swiss National Science Foundation.

REFERENCES

1. Navez, M., D. Chaperot, and C. Sella. Comptes Rendus Academie des Sciences, 1962. **254**(2): p. 240.
2. Carter, G. and V. Vishnyakov. Phys. Rev. B, 1996. **54**: p. 17647.
3. Erlebacher, J.D. and M.J. Aziz. Mater. Res. Soc. Symp. Proc., 1997. **440**: p. 461.
4. Facsko, S., T. Dekorsy, C. Koerdt, et al.. Science, 1999. **285**(5433): p. 1551.
5. Frost, F., A. Schindler, and F. Bigl. Phys. Rev. Lett., 2000. **85**(19): p. 4116.
6. Rusponi, S., C. Boragno, and U. Valbusa. Phys. Rev. Lett., 1997. **78**: p. 2795.
7. Mayer, T.M., E. Chason, and A.J. Howard. J. Appl. Phys., 1994. **76**: p. 1633.
8. Bradley, R.M. and J.M. Harper. J. Vac. Sci. Technol. A, 1988. **6**: p. 2390.
9. Erlebacher, J., M. Aziz, E. Chason, et al.. Phys. Rev. Lett., 1999. **82**(11): p. 2330.
10. Erlebacher, J., M.J. Aziz, E. Chason, et al.. J. Vac. Sci. Technol. A, 2000. **18**(1): p. 115.
11. Habenicht, S., W. Bolse, K.P. Lieb, et al.. Phys. Rev. B, 1999. **60**(4): p. R2200.
12. Cuerno, R., H.A. Makse, S. Tomassone, et al.. Phys. Rev. Lett., 1995. **75**(24): p. 4464.
13. Gago, R., L. Vazquez, R. Cuerno, et al.. Appl. Phys. Lett., 2001. **78**(21): p. 3316.
14. Appleton, B.R., O.W. Holland, J. Narayan, et al.. Appl. Phys. Lett., 1982. **41**(8): p. 711.
15. Chason, E., J. Erlebacher, M.J. Aziz, et al.. Nucl. Instr. Meth. B, 2001. **178**: p. 55.

Mat. Res. Soc. Symp. Proc. Vol. 707 © 2002 Materials Research Society

3D periodic arrays of nanoparticles inside mesoporous silica films

Sophie Besson[1,2], Thierry Gacoin[1], Catherine Jacquiod[2], Christian Ricolleau[3], Jean-Pierre Boilot[1]

[1] LPMC, UMR CNRS 7643, Ecole Polytechnique, 91128 Palaiseau France
[2] LSVI CNRS/Saint-Gobain, UMR CNRS 125, 39 quai Lucien Lefranc, 93303 Aubervilliers France
[3] LMCP, UMR CNRS 7590, Universités Paris 6 et 7, 4 place Jussieu, 75252 Paris France

ABSTRACT

CdS nanoparticles were grown inside a 3D hexagonal porous silica film. The film pore size and organization allowed the perfect control of particle repartition and size (3.5 nm), leading to a 3D nanocrystal array inside the silica matrix. The method was extended to another silica porous structure with larger pores, which allowed to obtain larger particles (5.8 nm). This process was then successfully generalized to other metal sulfides.

INTRODUCTION

Since the past decade, nanocrystals below 10 nm have attracted worldwide interest as they display optical, electronic or magnetic properties which differ significantly from those of bulk materials [1]. These specific effects depend strongly on their size, shape and surface state. To elaborate nanoparticulate materials for applications such as nonlinear optics, optoelectronics, biology or catalysis, it may be important to control not only these parameters, but also the repartition of the nanocrystals which can strongly modify the properties. In this context, 3D periodic arrays of nanoparticles seem to be the optimal systems to study collective effects. Nevertheless, the production of such materials at a large scale and with good mechanical stability remains a challenge.

Periodic mesoporous materials are potentially excellent candidates to support nanocrystals. These materials, discovered by Mobil Oil Corporation researchers in 1992 [2], are synthesized via the polymerization of inorganic species, generally silica, around a surfactant micelle liquid crystal, which is then eliminated by thermal treatment. By varying the surfactant nature and the synthesis conditions, different periodic structures with pore sizes from 2 to 30 nm can be formed [3,4]. As they have pores which are perfectly controlled in size, shape and organization, these materials can act as templates for the synthesis of periodic arrays of nanoparticles.

A lot of studies have been published about nanoparticle growth inside mesoporous powders, but only few have succeeded in controlling the size, shape and organization of the nanocrystals, and usually in small domains [5,6]. Besides, for most applications, coatings filled with nanoparticles are needed instead of powders.

We report here a general method to synthesize 3D quantum dot lattices (CdS, ZnS, PbS, Ag$_2$S) in mesoporous silica films.

EXPERIMENT

Porous matrices were synthesized using a procedure previously described [7]: TEOS (Si(OC$_2$H$_5$)$_4$), ethanol and water (pH=1.25) were mixed in the molar ratio 1:3.8:5 and aged 1h at

60°C. The cationic surfactant CTAB (cetyltriethyammonium bromide) was then added to the sol with Si/surfactant molar ratio equals to 0.1. The obtained solution was diluted with ethanol (1:1) and spin-coated on glass or silica slides at 3000 rpm. The organic template was eliminated by calcination at 450°C.

The strategy to grow metal sulfide nanocrystals inside mesoporous films was first optimized for CdS quantum dots. Details about the synthesis can be seen in reference [8]. This method was then generalized to other sulfides.

Nanoparticles were grown in two steps: the adsorption of metal cations on the silica surface and the precipitation by injection of H_2S. Metal cations were chemisorbed to the silanols at the pore surface by immersing the porous film in a basic water solution of metallic salt. After 1 min the film was thoroughly washed with deionized water to eliminate cations in excess. The pH needed to avoid ion leaching during the washing depends on the metal: 9.5 for cadmium and silver ions, 7.5 for zinc ions [9]. More precisely, 0.1M solutions of metal nitrate were prepared in water and their pH was adjusted to the desired value with ammonia. To avoid precipitation of the hydroxide, sodium citrate was used as a complexing agent [8]. After the cation adsorption, films were placed under vacuum and H_2S was injected in order to homogeneously induce nanoparticle precipitation inside the pores. The two steps (adsorption and precipitation) were repeated until the saturation of the films.

Films were characterized after each step of impregnation-precipitation by UV-visible spectroscopy and X-ray diffraction in the Bragg Brentano geometry using CuKα radiation. HRTEM observations were made using a PHILIPS CM20 operating at 200kV.

RESULTS AND DISCUSSION

Porous film structure

The key points to grow nanoparticles inside a mesoporous film is the accessibility of the porous network and the quality of the periodic structure, which must be excellent on the whole film thickness. Recently, we have reported the synthesis of a new 3D organization of mesopores in CTAB templated silica films [7,10]. It is a 3D hexagonal structure consisting in an interconnected spherical pore network which is oriented with the c-axis perpendicular to the film plane. Figure 1a displays an in-plane view of a film where the 6-fold symmetry of the hexagonal structure is clearly visible, along with grain boundaries between domains, showing the high degree of organization and texture. The cross-section in figure 1b shows the excellent quality of the structure throughout the film thickness. The lattice parameters are a = 6 nm and c = 7 nm. The characterization of this structure by X-ray diffraction in the Bragg Brentano geometry gives rise to an intense and sharp peak (indexed 002) and its harmonic (no other peak can be seen because of the film texture).

3D hexagonal array of CdS nanoparticles

CdS nanocrystals were grown in the previously described 3D hexagonal porous structure. The evolution of the UV-visible spectrum during the impregnation-precipitation cycles is shown in figure 2a. The film absorbance increases until the saturation which is reached after 9 cycles. The presence of discrete transitions in the spectra due to quantum size effect attests that the particle size distribution is narrow. According to correlation curves [11], we can deduce the

 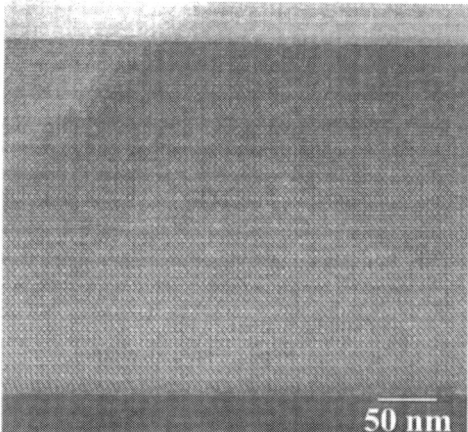

Figure 1: HRTEM images in plane view (a) and cross-section (b) of the 3D hexagonal structure.

particle size from the peak position: Figure 2b displays the evolution of the particle size during the impregnation precipitation cycles, which increases from 2.2 nm after one impregnation to 3.6 nm at the saturation. Hence, the particle size seems to be controlled by the pores, and the saturation occurs when they are totally filled. Figure 2c shows the evolution of the 002 diffraction peak during the process. The rapid extinction of the peak and its recovery (while the peak width remains constant), can be explained by an inversion of the scattering contrast between the silica and the pores during their filling by CdS. This means that the structure is preserved during the impregnations, and that the particles grow inside the pores. Besides, the figure shows a peak shift to lower 2θ values, corresponding to an increase of the lattice distance of 0.2 nm, which can be explained by the full filling of the pores which slightly deforms them. An HRTEM image of the film in cross-section is shown in figure 2d, which confirms the complete filling throughout the whole film thickness, and the periodic organization of the nanoparticles.

This growth of CdS nanocrystals inside a 3D hexagonal matrix demonstrates the feasibility of using mesoporous silica films as templates to elaborate nanoparticulate coatings with perfectly controlled nanoparticle size and organization.

Generalization to other porous host matrices and other metal sulfides

To check the ability of tuning particle size by changing the host matrix, CdS particles were grown inside a mesoporous film structured with a triblock copolymer instead of the CTAB, which allows to produce larger pores [4,12]. It is highly organized and consists in a cubic arrangement of spherical micelles. The evolution of the UV-visible spectrum of the film during the impregnation process is presented in figure 3a. Compared to the spectra obtained with the CTAB structure, we can observe first that the absorption edges are shifted toward lower energies, which corresponds to larger particles. The evolution of the particle size after each impregnation cycle deduced from correlation curves is shown in figure 3b. The particles grow from 3.3 nm to 5.8 nm at the saturation. For sizes above 5 nm the quantum size effect is weak, so that no discrete transition is observed in the spectra. The absorption edge is very close to the bulk energy gap.

This experiment shows that if we are able to tune the porosity of the matrix, we can elaborate a great variety of naoparticulate coatings.

Figure 4a displays the evolution of the UV-visible spectrum of ZnS nanoparticles grown in a 3D hexagonal mesoporous film, which was deposited on silica plates to avoid the absorption of glass substrates. This figure is very similar to the one obtained with CdS nanoparticles, with an absorption edge at higher energy, due to the fact that ZnS has a larger energy gap than CdS (respectively 3.6 and 2.5 eV for bulk materials). Here again the quantum confinement effect is clearly visible. The evolution of the 002 diffraction peak intensity is shown in figure 3b. As for CdS, there is an inversion of the scattering contrast during particle growth, which confirms that ZnS particles are periodically distributed inside the film. The phenomenon is slightly less pronounced than for CdS, as the electron density in ZnS is lower than in CdS. ZnS nanoparticles

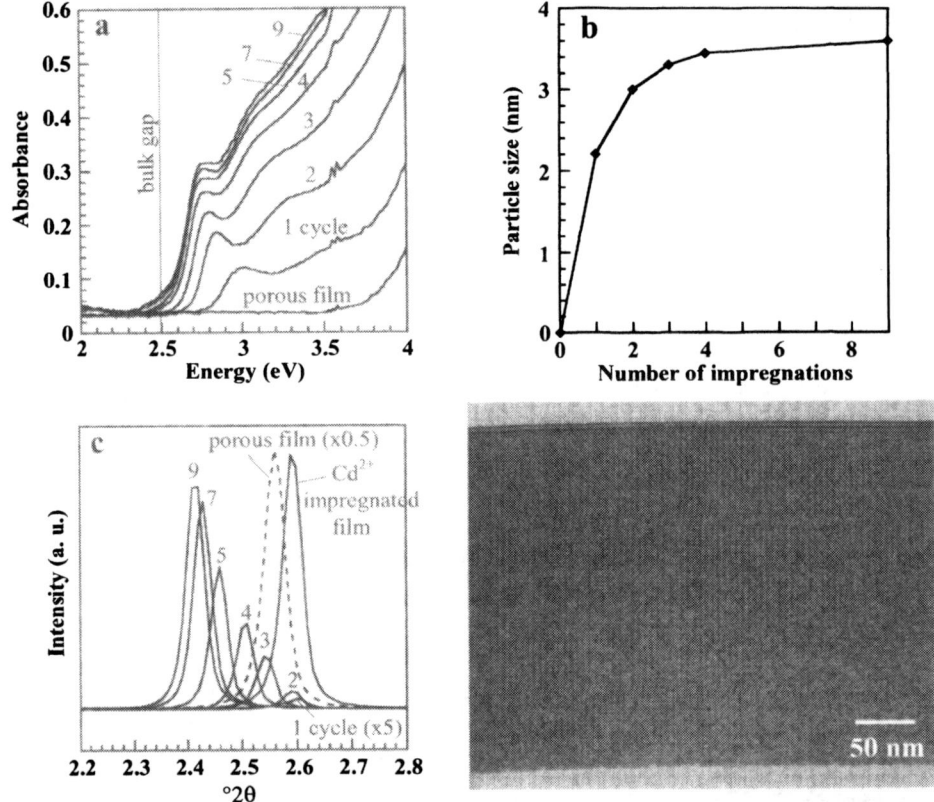

Figure 2: CdS nanoparticle growth inside a 3D hexagonal mesoporous silica film.(a) Evolution of the absorption spectrum during the impregnation cycles. (b) Nanoparticle sizes deduced from gap-size correlation curves. (c) Evolution of the diffraction peak. (d) HRTEM image in cross section of the saturated film.

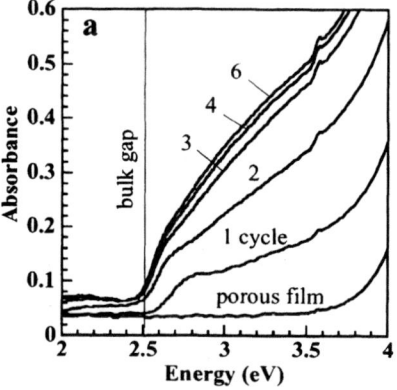 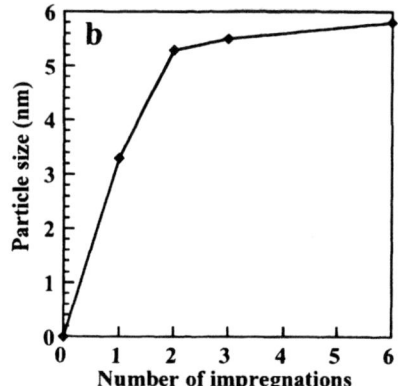

Figure 3: CdS nanoparticle growth in a mesoporous film elaborated with a triblock copolymer. (a) Evolution of the absorption spectrum. (b) nanoparticle sizes deduced from gap-size correlation curves.

are therefore periodically distributed inside the film and their size is controlled by the mesoporous structure.

These experiments demonstrate that the process optimized for CdS can be generalized to other metal sulfides. In figure 5 are presented UV-visible spectra of PbS (a) and Ag$_2$S (b) nanoparticles grown inside 3D hexagonal mesoporous films, recorded after one impregnation. As both compounds do not have direct gap at the center of their Brillouin zone, one cannot see discrete transitions in the spectra. However, the adaptability of the process is once again confirmed.

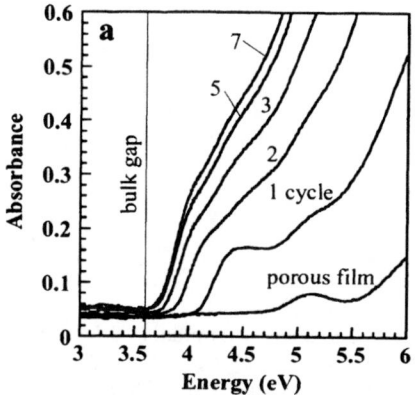 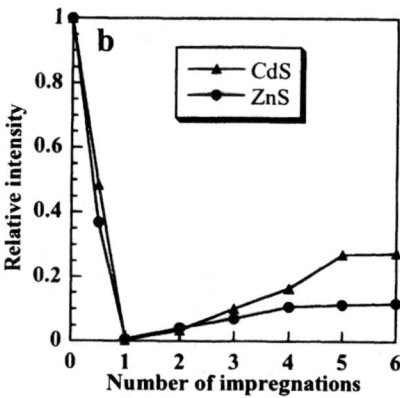

Figure 4: ZnS nanoparticle growth inside a 3D hexagonal mesoporous silica film.(a) Evolution of the absorption spectrum. (b) Evolution of the 002 diffraction peak intensity compared to the one observed for CdS.

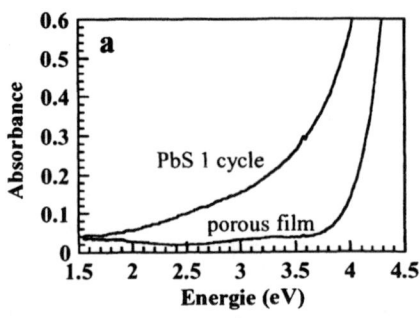
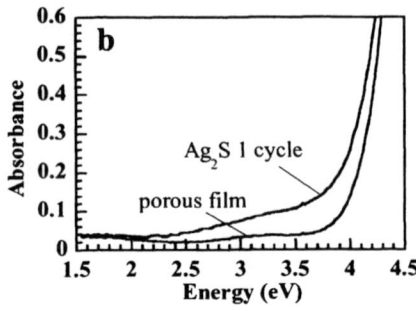

Figure 5: UV-visible spectra of PbS (a) and Ag₂S (b) grown in 3D hexagonal silica films (1 cycle of impregnation).

CONCLUSION

We have succeeded in totally filling mesoporous silica films with CdS nanoparticles, perfectly controlled in size and arrangement. The process is very simple and can be applied to various porous matrices and metal sulfides, allowing the production of a great variety of nanoparticulate coatings at a large scale. Besides, the process should be applicable to the synthesis of metal selenides and halogenides.

ACKNOWLEDGMENTS

This work was partially supported by Saint-Gobain Recherche.

REFERENCES

1. A. N. Shipway, E. Katz, I. Willner, *Chemphyschem*, **1**, 18 (2000)
2. J.S. Beck, J.C. Vartuli, W.C. Roth, M.E. Leonowiccz, C.T. Kresge, *J. Am. Chem. Soc.*, **114**, 10834 (1992)
3. Q. Huo, D. I. Margolese, G. D. Stucky, *Chem. Mater.*, **8**, 1147 (1996)
4. D. Zhao, Q. Huo, J. Feng, B. F. Chmelka, G. D. Stucky, *J. Am. Chem. Soc.*, **120**, 6024 (1998)
5. L.-Z. Wang, J.-L. Shi, W.-H. Zhang, M.-L. Ruan, J. Yu, D.-S. Yan, *Chem. Mater.*, **11**, 3015 (1999)
6. H. Kang, Y.-W. Jun, J.-I. Park, K.-B. Lee, J. Cheon . *Chem. Mater.*, **12**, 3530 (2000)
7. S. Besson, T. Gacoin, C. Jacquiod, C. Ricolleau, D. Babonneau, J.-P. Boilot, *J. Mater. Chem.*, **10**, 1331 (2000)
8. S. Besson, T. Gacoin, C. Ricolleau, C. Jacquiod, J.-P. Boilot submitted to *Adv. Mater.*
9. R. K. Iler, in *The Chemistry of Silica*, (Ed: Wiley-Interscience), 667-676 (1979)
10. S. Besson, C. Ricolleau, T. Gacoin, C. Jacquiod, J.-P.Boilot *J. Phys. Chem. B*, **104**, 51, 12095 (2000)
11. Y. Wang, N. Herron, *Phys. Rev. B*, **42**, 7253 (1990)
12. S. Besson, T. Gacoin, C. Ricolleau, C. Jacquiod, J.-P.Boilot, to be published.

Mat. Res. Soc. Symp. Proc. Vol. 707 © 2002 Materials Research Society   AA8.7/V10.7

Surfactant Templated Assembly of Hexagonal Mesostructured Semiconductors Based on [Ge₄Q₁₀]⁴⁻ (Q=S, Se) and Pd²⁺ and Pt²⁺ ions.

Pantelis N. Trikalitis, Krishnaswamy K. Rangan and Mercouri G. Kanatzidis*
Department of Chemistry, Michigan State University, East Lansing, MI 48824.

ABSTRACT

Mesostructured semiconducting non-oxidic materials were prepared by linking $[Ge_4Q_{10}]^{4-}$ (Q=S, Se) clusters with the square planar noble metal cations of Pd^{2+} and Pt^{2+} in the presence of cetylpyridinium surfactant molecules. The use of Pt^{2+} afforded materials with exceptionally high hexagonal pore order similar to those of high quality silica MCM-41. These materials are semiconductors with energy band gap in the range $1.8 < E_g < 2.5$ eV.

INTRODUCTION

The emergence of the family MCM-X of surfactant templated silica mesoporous molecular sieves with regular pore shape and adjustable pore size a decade ago sparked a flurry of activity worldwide that resulted in an abundance of oxidic mesoporous solids with promising technological properties (1). Whereas these materials will impact catalytic, separation and adsorption applications, they lack interesting electronic properties. Bulk materials that combine mesoscale features and electronic properties are envisioned for novel applications in quantum electronics (2), photonics (3) and non-linear optics (4) among others. Such characteristics may be expected in non-oxidic materials such as the chalcogenides (5). A suitable method for the construction of semiconducting mesostructured materials utilizes a self-assembly process between chalcogenido building blocks and metal cations in the presence of surfactant molecules acting as templates (6). However outstanding issues as to how these systems form remain. The mesostuctured chalcogenides reported todate mainly were synthesized by linking adamantane $[Ge_4Q_{10}]^{4-}$ (Q=S, Se) clusters (see scheme 1) with first row transition and main-group ions such as Mn, Co, Ni, Zn, Cd, In and Ga, whose coordination preference is mainly tetrahedral. In the exploration of the role of the linkage metal in the assembly process we decided to use noble metal Pd^{2+} and Pt^{2+} ions because of the strong square-planar coordination preference. Moreover these ions are less kinetically labile than the first-row transition metals and therefore may slow down the self-assembly reaction thus achieving a more ordered structure. In this work we employed the clusters $[Ge_4Q_{10}]^{4-}$ (Q=S, Se) with Pt^{2+} and Pd^{2+} in the presence of cetylpyridinium ($C_{16}PyBr$) surfactant molecules. We find that these two similar metal ions behave very differently. In the case of Pt^{2+} the materials show remarkable hexagonal pore order similar to those of high quality silica MCM-41 whereas in the case of Pd^{2+} the solids are significantly less ordered.

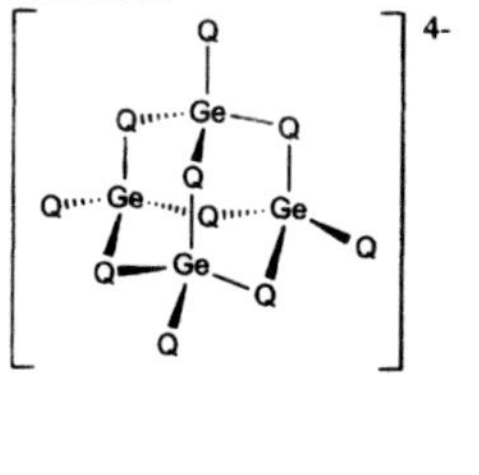

Scheme 1. Chalcogenido adamantane-type $[Ge_4Q_{10}]^{4-}$ (Q=S, Se) cluster.

EXPERIMENTAL

The syntheses of the materials were carried out as follows: 1 mmol of $TMA_4[Ge_4Q_{10}]$ (TMA=tetramethylammonium; Q=S, Se) was dissolved in 20 ml of formamide at 80 ^{0}C. To this clear solution 10 mmol of surfactant $C_{16}PyBr$ was added and the mixture stirred at 80 ^{0}C until a clear solution formed. In a flask 1 mmol K_2MCl_4 (M=Pt, Pd) was dissolved in 10 ml of formamide and added to the surfactant/$[Ge_4Q_{10}]$ solution dropwise using a pipet. The mixture was aged overnight under stirring and the product was isolated with suction filtration, washed with warm formamide and water and dried under vacuum. The yield was >80 % and the solids were in the form of light powder.

RESULTS AND DISCUSSION

The mesostructured materials are denoted as $C_{16}PyMGeQ$ were M=Pt, Pd. Unlike in previous cases (6) with other, kinetically labile linkage metals, we observe a considerable slower reaction upon addition of the Pt^{2+} metal ions. That is when the K_2PtCl_4/FM solution was added to the $C_{16}Py/[Ge_4Q_{10}]^{4-}$ solution, instantaneous precipitation did not take place. The deposition of the mesophase began 1-1.5 min after and was completed in approximately 10-15 min. As we discuss below the platinum containing mesophases exhibit remarkably good hexagonal mesoscopic order. In contrast Pd^{2+} ions, react much more rapidly giving mesophases that exhibit high degree of disorder. Figure 1 shows powder X-ray diffraction patterns of $C_{16}PyMGeQ$ materials. The platinum products show three or four well defined Bragg reflections in the $2^0<2\theta<7^0$ region, characteristic of mesostructured materials with regular hexagonal pore arrangement. Accordingly these reflections are indexed to a hexagonal $p6m$ mesophase, see Figure 1. The intense, sharp, well-defined high order reflection (110) and (200) as well as the observation of the fourth (210) reflection, betray a high degree of hexagonal order in these materials as observed directly by transmission electron microscopy (TEM) (see below). In the case of palladium however, the (100) reflection is clearly broader and the high order reflections (110) and (200) are not well resolved, indicating the formation of less ordered mesostructured phases.

Samples of the mesostructured $C_{16}PyMGeQ$ were examined by TEM. Figure 2a shows a characteristic image of $C_{16}PyPtGeSe$ looking down the pore channel axis ([100] direction) where a remarkably uniform hexagonal order is clearly visible. Figure 2b shows a view of $C_{16}PyPtGeSe$ perpendicular to the pore channel axis ([110] direction). The long, straight parallel tunnels are apparent in this image and the observed interpore distances are in good agreement with those obtained from the X-ray diffraction patterns. Similar images were observed in $C_{16}PyPtGeS$ material. Figure 2c,d shows a characteristic image of $C_{16}PyPdGeS$ were the presence of local hexagonal pore arrangement along with disordered regions is evident, as indicated by the X-ray powder patterns. The quality of $C_{16}PyPtGeQ$ solids as judged by the degree of hexagonal order, is comparable to those of high quality silica MCM-41 (7). Figure 3 shows a TEM image of a large $C_{16}PyPtGeSe$ particle where the size of coherent, hexagonally organized domain is >500 nm.

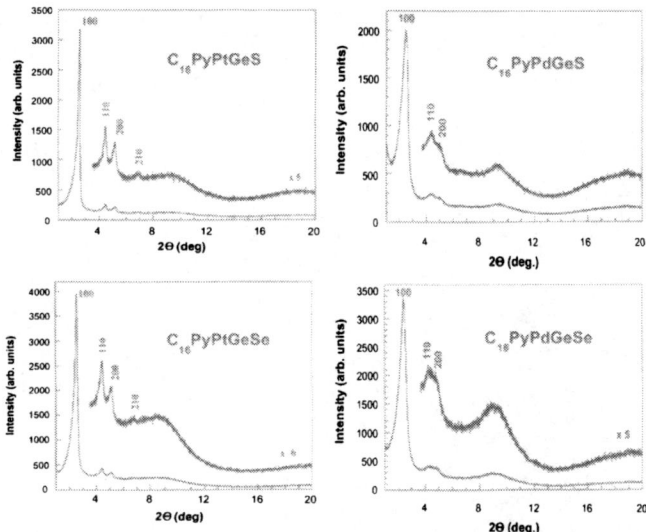

Figure 1. Powder X-ray diffraction patterns of mesostructured noble-metal chalcogenides (CuKα radiation).

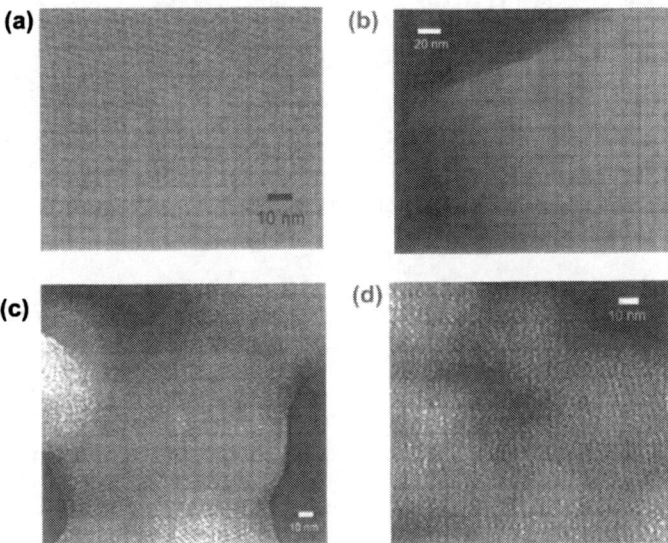

Figure 2. Representative TEM images of $C_{16}PyMGeQ$ materials. (a) $C_{16}PyPtGeSe$ down to [100] direction, (b) $C_{16}PyPtGeSe$ down to [110] direction, (c) $C_{16}PyPdGeS$ down to [100] direction and (d) highly disordered region of $C_{16}PyPdGeS$.

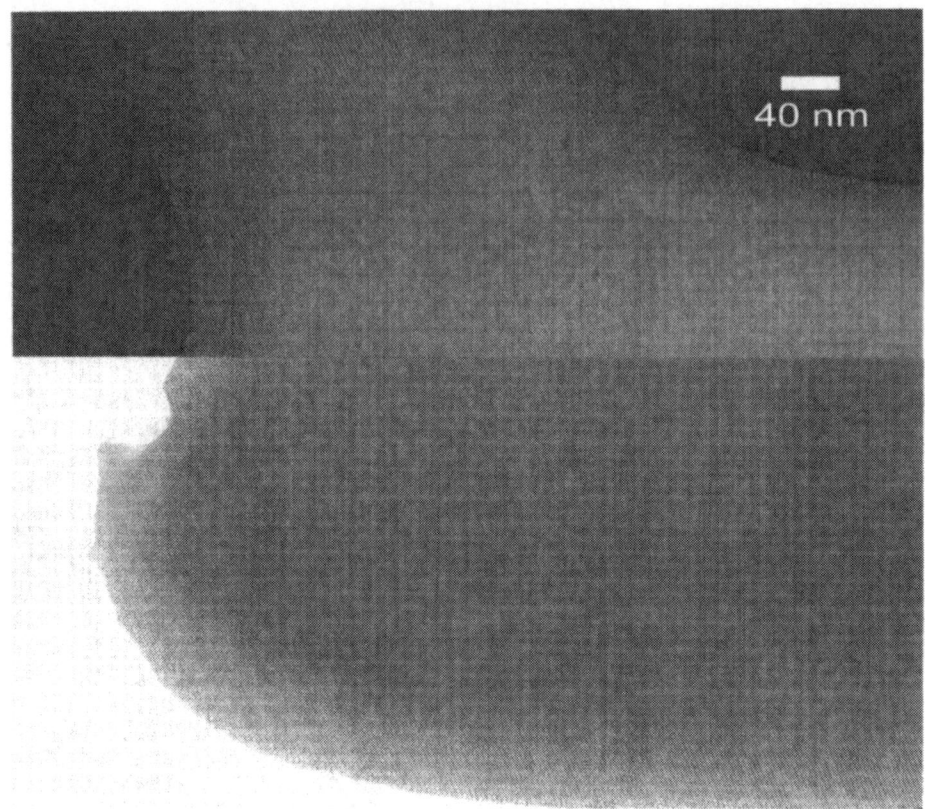

40 nm

Figure 3. Representative image of a large particle of $C_{16}PyPtGeSe$ showing the hexagonal organization extending over its full body. Particle length >500 nm.

Undoubtedly the quality of the hexagonal pore order in the platinum germanium chalcogenide materials is significantly higher than in palladium analogs. Since the synthesis of $C_{16}PyPtGeQ$ is based on simple metathesis reactions according to (1),

$$K_2MCl_4 + TMA_4[Ge_4Q_{10}] + C_{16}PyBr \longrightarrow C_{16}PyMGeQ + TMA\text{-}Cl + KBr \quad (1)$$

it is natural to expect that ligand substitution kinetics and coordination preference of the linkage metal ions will affect the quality of the final product. However noble metal ions Pd^{2+} and Pt^{2+} exhibit strong square planar coordination geometry. Therefore the striking difference found between $C_{16}PyPtGeQ$ and $C_{16}PyPdGeQ$ materials in terms of long range hexagonal pore order, is attributed to the slower reaction observed in the former. In that case, the reactants have more time to react and reach equilibrium thus leading to considerably higher quality hexagonal mesostructured phases.

Infrared (IR) and Raman spectroscopy was used to examine the inorganic framework and also confirm the presence of the surfactant molecules in these materials. In the mid-IR region we observed the characteristic absorption bands of cetylpyridinium cations. Shown in Figure 4a is typical far-IR spectrum from $C_{16}PyPtGeS$ together with the corresponding spectrum from the free adamantane cluster for comparison. The spectrum of $C_{16}PyPtGeS$ shows characteristic peaks in the same range as the free adamantane cluster, however, the peaks are much broader indicating possible different bonding environments present.

FT-Raman spectra were collected for the mesostructured $C_{16}PyPtGeS$ while for the other materials spectra could not be obtained as they decomposed under the laser beam. We observe several vibrational modes attributed to Ge-S and Pt-S stretching modes, see Figure 4b. By comparison with previous Raman studies (6), the sharp peak centered at 376 cm^{-1} can be assigned to the totally symmetric breathing mode of the "Ge_4S_6" cage, while the peak at 441 cm^{-1} to terminal Ge-S_t stretching mode. The peak at 340 cm^{-1} can be assigned to the Pt-S vibration, according to the spectroscopic analysis of Raman spectrum of PtS.

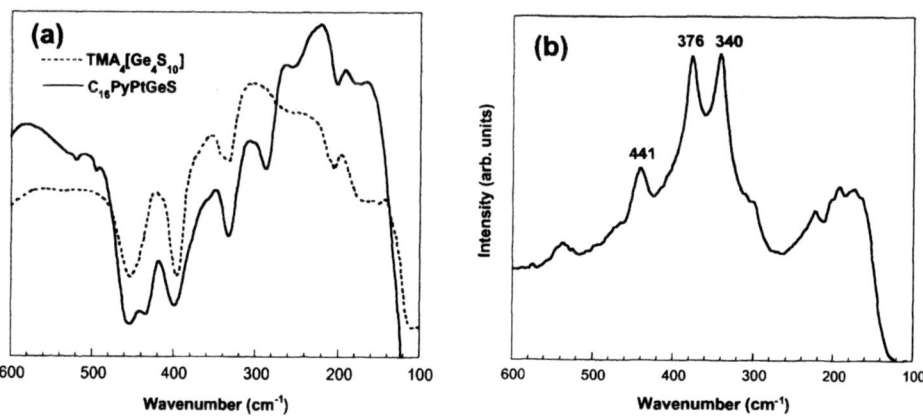

Figure 4. (a) Far-IR and (b) FT-Raman spectrum of $C_{16}PyPtGeS$.

The optical absorption properties of the $C_{16}PyMGeQ$ mesostructured materials were investigated with solid state diffuse reflectance UV-vis/Near IR spectroscopy. All solids posses well-defined, sharp optical absorptions associated with bandgap transitions in the energy range $1.8 < E_g < 2.5$ eV, see Figure 5. The bandgap narrows in going from the lighter $C_{16}PyMGeS$ (Pd:2.5 eV, Pt:2.3 eV) to the heavier $C_{16}PyMGeSe$ (Pd:2.0 eV, Pt: 1.8 eV). Moreover the effect on the bandgap in going from $C_{16}PyMGeS$ to $C_{16}PyMGeSe$ is greater (0.5 eV) rather than in going from $C_{16}PyPdGeQ$ to $C_{16}PyPtGeQ$ (0.2 eV) suggesting a more dominant role of the chalcogenide element in the valence and conduction band of the materials.

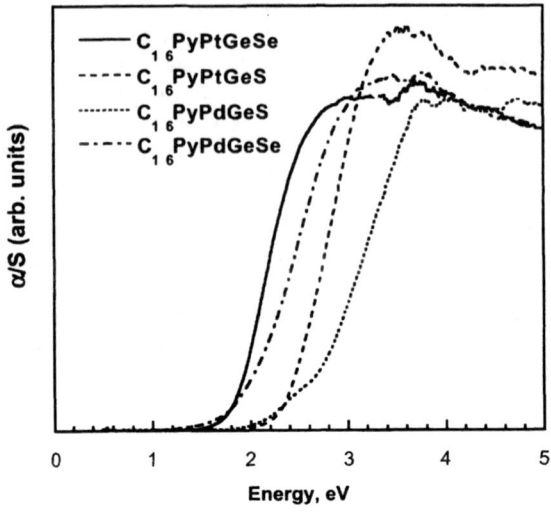

Figure 5. Solid state UV-vis absorption spectra of C_{16}PyMGeQ materials.

CONCLUSIONS

The use of Pd^{2+} and Pt^{2+} as linkage metal ions in a supramolecular assembly of adamantane $[Ge_4Q_{10}]^{4-}$ clusters leads to a new class of non-oxide mesostructure materials with noble metal ions as part of the framework. In the case of platinum the solids show exceptionally high hexagonal order similar to those in silica MCM-41, whereas in the case of Pd such high order was not observed. This suggests that slower reaction kinetics of framework assembly is an important factor. The materials posses optical bandgaps, ranging between $1.8 < E_g < 2.5$ eV. The bandgap narrows with the incorporation of heavier elements in the framework.

ACKNOWLEDGEMENTS The support of this research by NSF-CRG grant CHE 99-03706 is gratefully acknowledged. This work made use of the SEM and TEM facilities of the Center of Advanced Microscopy at MSU.

REFERENCES

1.   C. T. Kresge, M. E. Leonowicz, W. J. Roth, J. C. Vartuli, J. S. Beck, *Nature* **359**, 710 (1992); Y. Ma, W. Tong, H. Zhou, S. L. Suib, *Microporous Mesoporous Mater.* **37**, 243 (2000); A. Corma, *Chem. Rev.* **97**, 2373 (1997); A. Sayari, P. Liu, *Microporous Mater.* **12**, 149 (1997).

2.   P. V. Braun, P. Osenar, S. I. Stupp, *Nature* **380**, 325 (1996).

3.   D. J. Norris, Y. A. Vlasov, *Adv. Mater.* **13**, 371 (2001).

4.   G. A. Ozin, *Supramol. Chem.* **6**, 125 (1995).

5.   H. L. Li, A. Laine, M. O'Keeffe, O. M. Yaghi, *Science* **283**, 1145 (1999).

6.   K. K. Rangan, P. N. Trikalitis, M. G. Kanatzidis, *J. Am. Chem. Soc.* **122**, 10230 (2000); K. K. Rangan, P. N. Trikalitis, T. Bakas, M. G. Kanatzidis, *Chem. Commun.*, 809 (2001); P. N. Trikalitis, K. K. Rangan, T. Bakas, M. G. Kanatzidis, *Nature* **410**, 671 (2001); M. J. MacLachlan, N. Coombs, G. A. Ozin, *Nature* **397**, 681 (1999).

7.   Q. S. Huo, D. I. Margolese, G. D. Stucky, *Chem. Mater.* **8**, 1147 (1996).

Mat. Res. Soc. Symp. Proc. Vol. 707 © 2002 Materials Research Society

Enhanced Photoluminescence from Long Wavelength InAs Quantum Dots Embedded in a Graded (In,Ga)As Quantum Well

L. Chen, V. G. Stoleru, D. Pan, and E. Towe
Department of Electrical and Computer Engineering, University of Virginia, Charlottesville, Virginia 22904

ABSTRACT

Three sets of self-organized InAs quantum dots (QDs) embedded in an external InGaAs quantum well samples were grown by solid source molecular beam epitaxy (MBE). By modifying Indium composition profile within quantum well (QW) region, it's found the photoluminescence emission from quantum dots can be greatly enhanced when employing a graded quantum well to surround QDs. This quantum dots in a graded quantum well structure also preserves the long wavelength (1.3 µm) spectrum requirement for the future use in optoelectronics devices.

INTRODUCTION

The first GaAs-based quantum-dot laser was reported in 1994 [1]. Since then, remarkable progress has been made in the development of this device. Edge-emitting lasers operating from 1.0 to 1.3 µm with very low threshold currents have been reported [2,3,9]; in addition, vertical-cavity surface-emitting lasers (VCSELs) have also been successfully demonstrated [4].

There are currently several approaches to grow 1.3 µm (In,Ga)As quantum dots by MBE. These include (i) the alternate supply of group-III and group-V source materials to form the (In,Ga)As dots [5], (ii) very slow growth (<0.01 ML/sec) of the InAs dots at high substrate temperatures [6], (iii) burying the InAs dots with an InGaAs overlayer to reduce the strain [7], and (iv) embedding the InAs dots inside an InGaAs quantum well [8].

Most quantum-dot lasers operating at 1.3 µm appear to have two major unresolved problems: a low density of dots and a weak carrier confinement to the active region. Both of these problems imply low optical gain. In fact, the low gain is believed to prohibit the lasers from operating at the ground state without high-reflectivity coatings [3,10]. One approach to achieving tight carrier confinement is to use growth method (iv), which embeds the dots in an InGaAs quantum well. In addition to serving as a strain reducing (and hence wavelength tuning) layer, the InGaAs quantum well also serves to confine carriers within the vicinity of the quantum-dot layer, thus promoting capture within the dots [9].

In this paper, we report a modified structure with InAs quantum dots embedded in a graded InGaAs quantum well, which not only tunes the QDs emission to 1.3 µm, but also greatly enhances the photoluminescence (PL) efficiency.

EXPERIMENT

The samples in our study were grown in a Riber 32P solid source MBE system on semi-insulating (001) GaAs substrates. The basic structure for PL study consists of a 250 nm GaAs buffer grown at a substrate temperature of 580°C, a 90 nm GaAs layer with a reduced substrate temperature at 490°C, 1~2.5 ML InAs quantum dots symmetrically sandwiched in an 8 nm

In$_x$Ga$_{1-x}$As layer and a final 90 nm GaAs top layer. For Atomic-Force Microscopy (AFM), samples with similar structures, but with growth stopped after InAs deposition, were grown. The growth rate of the InAs dots was 0.05 ML/sec. The substrate holder was continuously rotated to improve the uniformity. A Coherent Argon ion laser (λ=488 nm) is used for PL and the emitted radiation was detected by a SPEX 500M spectrometer with a cooled Hamamatsu Ge detector. A Bomen MB155S FTIR is used for photocurrent (PC) measurement and a PSI AutoProbe CP is used for AFM.

RESULTS AND DISCUSSIOIN

Fig. 1 shows the PL spectra of InAs QDs embedded in an 8 nm In$_x$Ga$_{1-x}$As QW with different Indium composition x under different InAs deposition. From Fig.1, it's seen that the wavelength can be red shifted with the increase of InAs deposition. When InAs content is increased to 2.5 ML, the peak wavelength of QDs in an In$_{0.18}$Ga$_{0.82}$As QW reaches 1.3 μm but with apparent decrease of PL intensity, which indicates the possible formation of defects in this highly strained structure. Our best result is from 2.5 ML InAs QDs in an In$_{0.15}$Ga$_{0.85}$As QW.

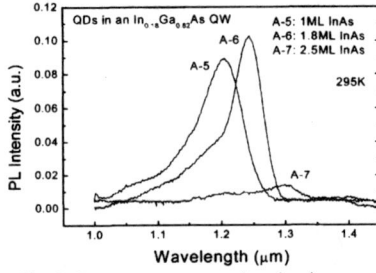 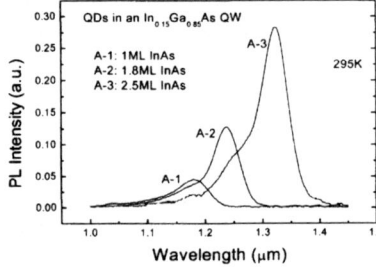

Fig. 1. Room temperature photoluminescence spectra of InAs quantum dots embedded in an 8 nm In$_x$Ga$_{1-x}$As quantum well with various InAs deposition.

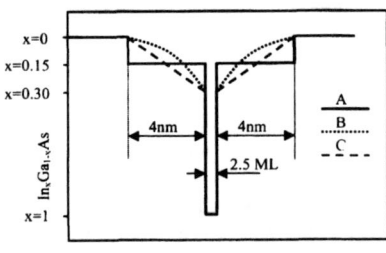 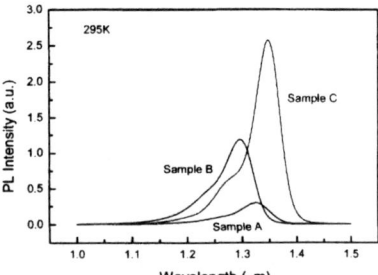

Fig. 2. Schematic structure of three different sets of quantum dot samples.

Fig. 3. Room-temperature PL spectra of QD samples at an optical excitation level of ~30 W/cm^2.

To further study the influence of Indium composition profile in the In$_x$Ga$_{1-x}$As QW, We modify the QW growth. A schematic of the generic structures is shown in Fig. 2. Structure A consists of 2.5 monolayers of InAs QDs sandwiched at the center of an 8 nm In$_{0.15}$Ga$_{0.85}$As layer. Structure B consists of 2.5 monolayers of InAs QDs sandwiched at the center of an 8 nm In$_x$Ga$_{1-}$.

$_x$As layer; where the bottom half of the In$_x$Ga$_{1-x}$As layer is graded from x = 0.03 to x = 0.3 and the top half is graded from x = 0.3 to x = 0.03. The change of Indium composition in the graded QW was achieved by gradually increasing or decreasing the Indium cell's temperature, which resulted in a curved shape of Indium distribution. The structure C is similar to structure B except that InAs/GaAs superlattice layers are used to form a quasi-linear indium distribution along the growth direction.

Room temperature photoluminescence spectra of three typical samples A, B, and C are shown in Fig. 3. The emission from the ground state transition for all three samples is at, or beyond 1.3 μm, with sample C extending to 1.35 μm. The spectral linewidths at half-maximum intensity for these three samples are Δv_A = 48 meV, Δv_B = 53 meV, and Δv_C = 39 meV, which shows the inhomogeneous broadening in all the samples. The grading of the In$_x$Ga$_{1-x}$As QW does not have apparent effect on the uniformity of the dots. What the grading appears to do is to improve the efficiency of emission. The peak intensity of sample C, for example, is about 8.7 times higher than that of sample A. Since the only difference in the three samples is the structure of the In$_x$Ga$_{1-x}$As layer into which the quantum dots are embedded, it is reasonable to speculate that the enhancement of PL intensity is due to the grading. There may be two possible reasons that account for this enhancement. One is that, in a graded QW, the energy band is also graded; this helps to drive the carriers into the quantum well, and hence into the quantum dots. Because of the three-dimensional nature of quantum dots, the graded potential also exists in the lateral direction; this enhances spatially carrier capture within the dots. The second reason for the enhancement is the smoothing effect of the InAs/GaAs superlattice. The use of superlattice causes smooth transition of the lattice parameter from the GaAs to the InAs, thus may help to reduce the defects formation. The defects, if present, would form loss channels in the electron-hole recombination process.

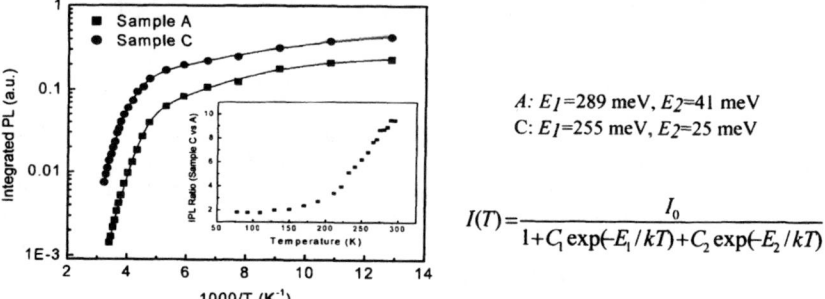

$A: E_1$=289 meV, E_2=41 meV
$C: E_1$=255 meV, E_2=25 meV

$$I(T) = \frac{I_0}{1 + C_1 \exp(-E_1/kT) + C_2 \exp(-E_2/kT)}$$

Fig. 4. The Arrhenius plots of the IPL intensity of sample A and C at an excitation level of ~10 W/cm^2. The solid lines are fitted by the equation shown on the figure. The inset is the ratio of the IPL intensity of sample C to that of sample A as a function of temperature.

We have further studied the temperature-dependent integrated photoluminescence (IPL) intensity of samples A and C. The Arrhenius plots for these two samples are shown in Fig. 4. As seen from the figure, photoluminescence enhancement of sample C is more dramatic at higher temperatures (see the inset in Fig. 4). We fit the variation of the IPL intensity data with temperature by a generic empirical relationship [11] shown on the figure, where E_1 and E_2 are the thermal activation energies for loss mechanisms active at certain temperature ranges, k is the

Boltzmann constant, T is the temperature, and C_1, C_2 and I_0 are fitting constants. In a physical model, these constants would take into account the recombination rates and the geometric dimensions of the dots. In Fig. 4, the solid lines are the fitting curves; the filled circles and squares represent experimental data. The extracted thermal activation energies are E_1=289 meV, and E_2=41 meV for sample A; and E_1=255 meV and E_2=25 meV for sample C. A calculation (which we will discuss later) shows that the energy difference between the ground state in the conduction band of a dot and the band-edge of the surrounding (In,Ga)As quantum well at 300 K is about 317 meV for structure A and 267 meV for sample C. These energy differences are quite close to the measured activation energies E_1 (which dominates the quenching at high temperature) for sample A and C. Note that the IPL intensity has a quenching threshold temperature ~190 K for sample C, and ~170 K for sample A. Either a fast trapping of carriers to the QDs or a reduced loss channel around QW can result in a higher quenching threshold. Our results here are similar to those of Ru *et al.* who used hydrogen passivation (in order to reduce the defect concentrations in their samples) to enhance the photoluminescence emission [12].

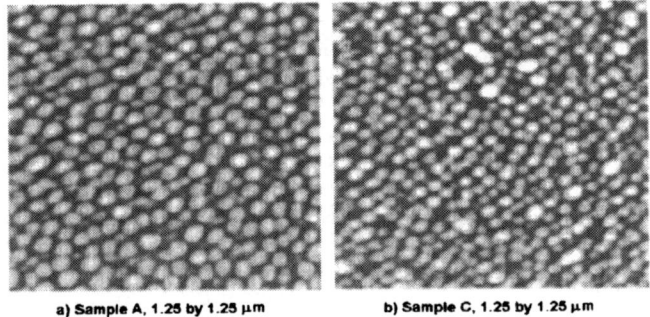

a) Sample A, 1.25 by 1.25 μm   b) Sample C, 1.25 by 1.25 μm

Fig.5. AFM photo-micrographs of the morphology of structure A and C.

We have also examined the possible effects of the grading on the dot density and uniformity. Two structures similar to sample A and C were grown for AFM observation. Figure 5 shows that there are no major differences in the two samples. Sample A, for example, has a dot density of around 2.5×10^{10} cm^{-2}, while sample C has a dot density about 2.8×10^{10} cm^{-2}. The true size of the dots cannot be determined from AFM measurement; reasonable estimates, however, can be obtained from cross-sectional TEM studies.

To show that the graded quantum well structure does not substantially modify the basic band structures, we have studied both the intensity dependent photoluminescence and photocurrent at LN$_2$ (78 K) temperature. In Fig. 6, we show the PL emission from sample A and sample C at high pump levels, and in Fig. 7, we show the photocurrent spectra of sample A and C excited by a broad band incident light (filtered by GaAs). There are clearly four transition energy peaks for each sample; the peaks occur at almost identical locations for both samples. To confirm these transition levels are from QDs, we have carried out some analytical calculations to determine the interband transition energies. These calculations require information on the strain distribution inside and around the dots. The strain generally depends on the shape and size of the dots, which are difficult to determine accurately in most experimental situations. For our dots, we estimate their lateral base to be around 15 nm; the base to height ratio for sample A is about 7; and it is

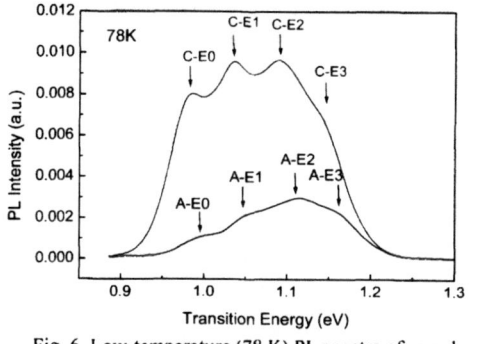

Fig. 6. Low-temperature (78 K) PL spectra of sample A and C at high optical pumping level.

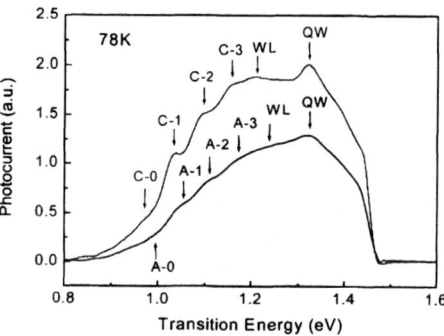

Fig. 7. Low-temperature (78 K) Photocurrent spectra of sample A and C.

5 for sample C. These estimates are based on our earlier work on buried InAs/GaAs and InGaAs/GaAs quantum dots [13], and on TEM and SEM observations. The estimate on height is further corroborated with *in-situ* RHEED observations where the thickness of an overlayer needed to completely bury an array of dots can give a rough measure of the dot height. For the calculations, we have further assumed that the dots are truncated pyramids.

Structure	Data Source	Transition Energy (eV)				
		Low Temperature (77K)				RT
		E0	E1	E2	E3	E0
A	Experimental	1.019	1.056	1.102	1.154	0.934
	Calculated	1.0162	1.0572	1.0951	1.1500	0.9405
C	Experimental	0.987	1.036	1.083	1.139	0.919
	Calculated	0.9830	1.0379	1.0809	1.1315	0.9138

Table 1. Experimental and calculated transition Energies for sample A and C

The electronic spectra here are calculated in the envelope function approximation using an eight-band, strain-dependent Hamiltonian based on the $\mathbf{k \cdot p}$ method [14]. The method takes into account the coupling among the light- and heavy-hole bands, and the split-off valence bands. It also includes the linear coupling between the conduction and valence bands. The lack of inversion symmetry in the zinc-blende structure, however, is neglected. The effect of strain is included via deformation potential theory [15], and the Luttinger-Kohn parameters are calculated according to the method by Pollak [16], and Bir and Pikus [17]. Because the quantum dots are in the strong confinement regime, additional binding energy from the Coulomb interaction is neglected. The bound states of QDs are found by numerically solving the Schrödinger equation, by invoking periodic boundary conditions, expanding the wavefunctions in terms of normalized plane-wave states, and diagonalizing the obtained matrix. The calculated transition energies for sample A and C are shown in Table I. We have also shown the experimentally determined values from the photoluminescence data by multi-peaks Gaussian fitting. The experimental and the calculated transition energies are in reasonably good agreement.

CONCLUSIONS AND ACKNOWLEDGEMENT

In summary, we have shown that the photoluminescence efficiency of InAs quantum dots can be enhanced by embedding them in a graded (In,Ga)As quantum well structures. Furthermore, our proposed approach has the virtue of promoting carrier capture in potential laser structures in addition to allowing emission at the desirable telecommunications wavelength of 1.3 μm.

This work is supported under grant number DAAD19-00-1-0442 of the US Army Research Office, Research Triangle Park, North Carolina, and by the US Army Research Laboratory, Adelphi, Maryland.

REFENENCES

[1] N. Kirkstaedter, N. Ledentsov, M. Grundmann, D. Bimberg, V. Ustinov, S. Ruvimov, M. Maximov, P. Kop'ev, and Zh. Alferov, Electron. Lett. **30**, 1416 (1994).

[2] G. Park, O.B. Shchekin, D.L. Huffaker, and D.G. Deppe, IEEE Photon. Tech. Lett. **12**, 230 (2000).

[3] D.L. Huffaker, G. Park, Z. Zou, O.B. Shchekin, and D.G. Deppe, Appl. Phys. Lett. **73**, 2564 (1998).

[4] J.A. Lott, N. Ledentsov, V. Ustinov, N. A. Maleev, A. E. Zhukov, A. R. Kovsh, M. V. Maximov, B. V. Volovik, Zh. I. Alferov, and D. Bimberg, Electron. Lett. **36**, 1384 (2000).

[5] D.L. Huffaker and D.G. Deppe, Appl. Phys. Lett. **73**, 520 (1998).

[6] R. Murray, D. Childs, S. Malik, P. Siverns, C. Roberts, J.M. Hartmann, and P. Stavrinou, Jpn. J. Appl. Phys. 38, 528 (1999).

[7] K. Nishi, H. Saito, and S. Sugou, Appl. Phys. Lett. **74**, 1111 (1999).

[8] A. Stintz, G. T. Liu, H. Li, L. F. Lester, and K. J. Malloy, IEEE Photon. Tech. Lett. **12**, 591 (2000).

[9] G. T. Liu, A. Stintz, H. Li, T.C. Newell, A.L. Gray, P.M. Varangis, K. J. Malloy, and L. F. Lester, IEEE Journal of Quan. Electron. **36**, 1273 (2000).

[10] N. Hatori, M. Sugawara, K. Mukai, Y. Nakata, and H. Ishikawa, Appl. Phys. Lett. **77**, 773 (2000).

[11] Y. Wu, K. Arai, and T. Yao, Phys. Rev. B **53**, 10485 (1996).

[12] E. C. Le Ru, P. D. Siverns, and R. Murray, Appl. Phys. Lett. **77**, 2446 (2000).

[13] V. G. Stoleru, D. Pal, and E. Towe, *Mater. Res. Soc. Proc.* Vol. **642**, J.1.7.1-6, Boston, MA (2000).

[14] T.B. Bahder, Phys. Rev. B **41**, 11992 (1992).

[15] O. Stier, M. Grundmann, and D. Bimberg, Phys. Rev. B **59**, 5688 (1999).

[16] F. H. Pollak, *Semiconductors and Semimetals*, Academic Press, Inc., New York, Volume **32**, pp. 17 (1990).

[17] G. E. Pikus and G. L. Bir, Sov. Phys. Solid State **1**, 1502 (1960).

Mat. Res. Soc. Symp. Proc. Vol. 707 © 2002 Materials Research Society

A Theoretical Study of Structural Disorder and Photoluminescence Linewidth in InGaAs/GaAs Self Assembled Quantum Dots

Yih-Yin Lin, Hongtao Jiang,[1] and Jasprit Singh
Department of Electrical Engineering and Computer Science,
University of Michigan, Ann Arbor, MI 48109-2122
[1]Broadcom Corporation, Irvine, CA 92618

ABSTRACT

The past few years have seen considerable efforts in growth and device application of self-assembled quantum dots. However, the photoluminescence (PL) linewidth, which represents structural fluctuations in dot sizes, is still in the range of 30-50 meV. This large linewidth has deleterious effects on devices such as lasers based on self-assembled dots. In this paper we will examine the configuration-energy diagram of self-assembled dots. Our formalism is based on: (1) an atomistic Monte Carlo method which allows us to find the minimum energy configuration and strain tensors as well as intermediate configurations of dots; (2) an 8-band $k \cdot p$ method to calculate the electronic spectra. We present results on the strain energy per unit cell for various distributions of InAs/GaAs quantum dots and relate them to published experimental results. In particular we examine uncovered InAs/GaAs dots and show that in the uncovered state a well-defined minimum exists in the configuration energy plot. The minimum corresponds to the size that agrees well with experiments.

INTRODUCTION

The use of strained epitaxy to create quasi-0 dimensional quantum dots has been widely studied over the last decade. Self-assembled quantum dots have now been fabricated using InGaAs/GaAs, SiGe/Ge, and many other strained systems [1-2]. In addition to the study of electronic and optical properties of these dots, devices lower as quantum dot interband lasers [3] and intersubband detectors [4] have been demonstrated. However, in spite of much progress in the area of self assembly, there is still a nagging issue - dot size nonuniformity. The self assembled dots vary in sizes and shapes - a variation that reflects in the electronic spectrum, photoluminescence linewidth, and gain spectrum etc. For many potential device applications, dot size nonuniformity creates deleterious effects. To understand the reason for nonuniformity and its extent, it is important to study the configuration energy of self-assembled dots.

There have been several studies on the strain tensor and energy in self-assembled dots. A useful model has emerged the valence force field (VFF) method [5,6], which allows us to calculate the strain tensor at an atomic level. It is found that the strain tensor in dots plays a key role in determining the electronic spectrum [7]. While the strain tensor of quantum dots in the covered state (i.e. with the dot buried in the large bandgap material matrix) has been examined, to our knowledge no work has been done on why certain mean dot sizes are chosen and why there is a distribution in the dot size. In this paper we use an extension of the VFF model to examine the strain tensor and energy of "covered" and "uncovered" InAs/GaAs dots. Then we apply the 8-band $k \cdot p$ model to obtain the electronic spectrum.

THEORETICAL MODEL

It is known that when a large strain exists between an overlayer and a substrate ($\varepsilon > 3\%$), the growth of the overlayer is described by the Stranski-Krastanow mode. In this mode a thin "wetting layer" is followed by an island growth. For InAs/GaAs system, the wetting layer is ~ 1 monolayer, and the islands are approximately 60 Å high with a more or less pyramidal shape. The base of the pyramid is about twice the height [8, 9]. This description of the self-assembled dots is; however, very qualitative since there is a considerable variation in the size and shape. There is substantial theoretical work on why the growth mode is island growth for high strain epitaxy. However, there is no quantitative work on why a particular size dominates the self-assembled system. To shed light on this issue, we calculate the strain tensor and energy for various sizes/shapes of InAs dots on a GaAs substrate. We examine the "uncovered" dots where the GaAs overlayer is not present and the "covered" dots where dots are embedded in a GaAs matrix.

The strained InAs/GaAs quantum dot system is shown in figure 1 for growth along the [001] direction. The InAs island is pyramidal shaped with a small square base, lying on a 1-monolayer InAs wetting layer. The InAs QDs are embedded in a GaAs matrix.

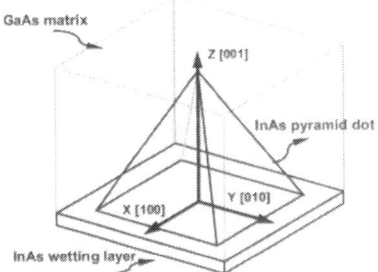

Figure 1. Schematic illustration of the Stranski-Krastanow growth mode for InAs/GaAs system.

Valence force field (VFF) model is a microscopic elastic theory, which includes the bond-stretching and bond-bending parts. The total VFF energy is taken as

$$U = \frac{1}{4}\sum_{ij}\frac{3}{4}\alpha_{ij}\left(d_{ij}^{\,2} - d_{0,ij}^{2}\right)^{2}/d_{0,ij}^{2} + \frac{1}{2}\sum_{i}\sum_{j\neq k}\frac{3}{4}\beta_{ijk}\left(\mathbf{d}_{ij}\bullet\mathbf{d}_{ik} + d_{0,ij}d_{0,ik}/3\right)^{2}/d_{0,ij}d_{0,ik} \qquad (1)$$

where i represents all atomic sites, and j, k are the nearest neighbor sites. $\mathbf{d}_{ij}$ is the vector joining sites i and j, d_{ij} is the bond's length, and $d_{0,ij}$ is the corresponding equilibrium bond length. α and β are the bond-stretching and bending constants. To fully understand the formation of self assembled InAs QDs, we simulate the uncovered InAs dots by replacing the GaAs cap with a virtual material by artificially reducing α and β to very small values (factor of 10 smaller). As a result, we can observe the behavior of InAs islands before the covering GaAs is deposited.

The strain tensors are solved by minimization of the total energy within the framework of the VFF model. We use the approach taken by several authors [10-12]. In the beginning all atoms are placed on the GaAs lattice. The atoms are allowed to deviate from this starting position and periodic boundary conditions are assumed in the plane perpendicular to the growth condition. In each process, only one atom is displaced while other atoms are held fixed. All atoms are displaced in sequence. The whole sequence is repeated until the maximum distance moved is so small that there is essentially no change in the system energy. The strain energy E_{str} can then be obtained once the strain tensors can be obtained by

$$E_{str} = \frac{a^3}{4}\left[\frac{1}{2}c_{11}\left(\varepsilon_{xx}^2 + \varepsilon_{yy}^2 + \varepsilon_{zz}^2\right) + c_{12}\left(\varepsilon_{yy}\varepsilon_{zz} + \varepsilon_{zz}\varepsilon_{xx} + \varepsilon_{xx}\varepsilon_{yy}\right) + \frac{1}{2}\left(\varepsilon_{zx}^2 + \varepsilon_{yz}^2 + \varepsilon_{xy}^2\right)\right]$$ (2)

where a is the lattice constant, ε_{ij} is the strain tensor component, and c_{ij} is the elastic constant. The 8-band $k \cdot p$ method [13] is used to calculate the electronic spectra.

RESULTS

The calculated results for strain energy per unit cell as a function of dot size in a covered InAs/GaAs system are shown in figure 2. The size variation is done when the base to height ratio is maintained at 2:1. No clear preference appears in the form of a well defined energy minimum. The strain energy of InAs dots for large sizes even show higher values than forming a flat thin layer (~ 0.17 eV), which is unexpected. It is clear that the InAs dot size/shape is determined by *the energies of the problem before the GaAs overlayer is deposited.*

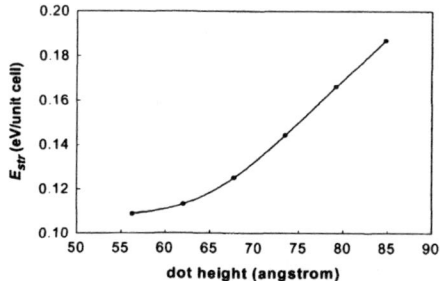

Figure 2. Strain energy per unit cell for the covered InAs dot pyramid at various dot sizes.

Figure 3 addresses the question of why certain dot sizes are preferred in high strain epitaxy. The strain energy per unit cell is illustrated as a function of the dot size. We show results for the dot alone and also the results when the wetting layer is included. Inclusion of the wetting layer is a little arbitrary since the results depend on the dot density. Nevertheless a clear minimum arises in the configuration energy plots in both cases at 62 Å dot height. This value is remarkably close to empirical observation [9]. However, due to entropy consideration we do not expect all dots to have this minimum energy size. Experimentally we know that there are variations in the dot size, causing inhomogeneous broadening of the photoluminescence line. Figure 3 also lists the values of effective bandgaps, showing how the dot size can alter the effect bandgap. We find that the effective gaps range from 0.954 to 1.2 eV as the height changes from 85 to 55 Å. The preferred size has a peak energy of ~ 1.04 eV, which agrees well with the experimental PL results [14].

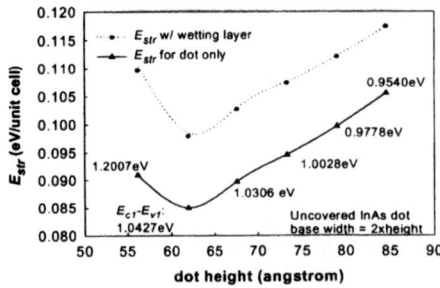

Figure 3. Strain energy per unit cell for dot only and for dot including wetting layer and effective bandgaps for the uncovered InAs dot pyramid at various dot sizes.

The electronic spectrum for different dot sizes is shown in figure 4. Here we show the ground states and the excited states of dots with heights of 56.5 Å and 62.3 Å. In each ease the base is twice the height. As can be seen, the size variation will contribute to more fluctuation for higher level transitions.

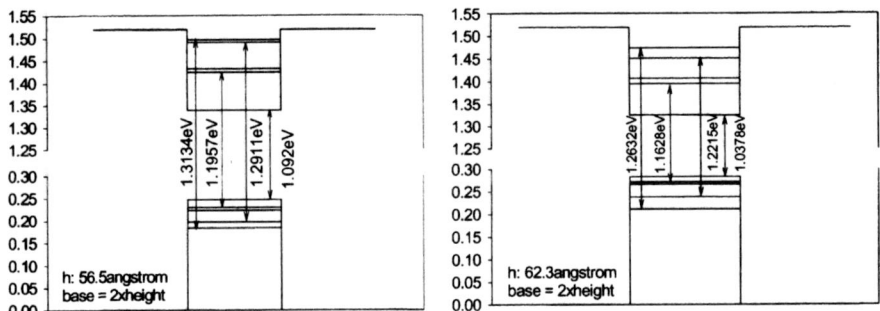

Figure 4. Electronic spectrum for InAs/GaAs dots with various heights of 56.5 Å and 62.3 Å.

Figure 5 shows how the strain energy change as a function of the shape of the dot. Here the height is fixed at the lowest configuration energy found in figure 3 while the width is adjusted. A larger base width corresponds to a wider dot pyramid and vice versa. It can be seen that when the base equals twice the island height the energy is a minimum. This result is consistent with experimental observation [9]. It is important to note that the total energy of the system would include the wetting layer strain energy and one has to then account for the dot density as well.

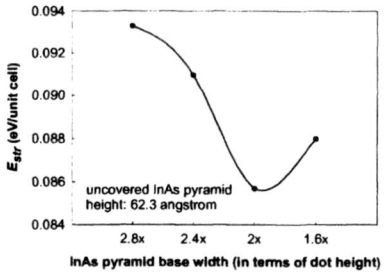

Figure 5. Strain energy of dot formation per unit cell for different dot shapes (various base).

CONCLUSION

In this paper we have examined the configuration energy profile for InAs/GaAs self-assembled dots. While it is not possible to examine the infinite possible sizes and shapes a dot can take, we find that there is a well defined size and shape as which the stain energy for the uncovered dot becomes a minimum. The shape and size calculated by us using the VFF model agree quite well with experimental findings on InAs/GaAs dots. We have also examined the configuration energy plot for covered dots, which do not show any energy minimum at the dot size observed empirically. This suggests that the dot size/shape is determined before the covering GaAs layer is deposited. Although the configuration energy is estimated assuming a pyramidal shaped dot, the calculation is done by a microscopically atomistic model. Thus this approach can be exploited to the truncated pyramidal quantum dots or even in the microlenses structure.

Another outcome of our finding is that the strain energy difference between the minimum energy configuration and other nearby dot sizes is not very large. As a result, it is expected that there will be a distribution of dots in actual growth due to entropy considerations. We have also shown how the size of dots alters the effective band-edge using the 8-band $k \cdot p$ method.

The good agreement of the minimum energy dot size and shape with published experimental results gives us confidence that the VFF model can be explored to understand how the dot uniformity can be controlled. This could be done by examining the role of pre-existing strains in the GaAs substrate. Further work will address these issues as well as the issue of dot size distributions.

ACKNOWLEDGEMENT

This work has been funded by US Army MURI (contract number F004658).

REFERENCES

1. J.-Y. Marzin, J.-M. Gérard, A. Izraël, D. Barrier, and G. Bastard, Phys. Rev. Lett., **73**, 716 (1994).

2. C. Teichert, M. G. Lagally, L. J. Peticolas, J. C. Bean, and J. Tersoff, Phys. Rev. B, **53**, 16334 (1996).

3. S. Krishna, O. Quasaimeh, P. Bhattacharya, P. J. MaCann, and K. Namjou, Appl. Phys. Lett., **76**, 3355 (2000).

4. H. Jiang and J. Singh, Physica E, **2**, 720 (1998).

5. P. N. Keating, Phys. Rev., **145**, 637 (1966).

6. R. M. Martin, Phys. Rev. B, **1**, 4005 (1969).

7. M. Grundmann, N. N. Ledentsov, O. Stier, D. Bimberg, V. M. Ustinov, P. S. Kop'ev, and Zh. I. Alferov, Appl. Phys. Lett., **68**, 979 (1996).

8. M. A. Cusack, P. R. Briddon, and M. Jaros, Phys. Rev. B, **54**, R2300 (1996).

9. S. Krishna, J. Sabarinathan, P. Bhattacharya, B. Lita, and R. S. Goldman, J. Vac. Sci. Technol. B, **18**, 1502 (2000).

10. J. Bernard and A. Zunger, Appl. Phys. Lett., **65**, 165 (1994).

11. F. Glas, J. Appl. Phys., **66**, 1667 (1989).

12. M. R. Weidmann and K. E. Newman, Phys. Rev. B, **45**, 2763 (1996).

13. H. Jiang and J. Singh, Appl. Phys. Lett., **71**, 3239 (1997).

14. F. Alder, M. Geiger, A. Bauknecht, F. Scholz, H. Schweizer, M. H. Pikuhn, B. Ohnesorge, and A. Forchel, J. Appl. Phys., **80**, 4019 (1996).

Mat. Res. Soc. Symp. Proc. Vol. 707 © 2002 Materials Research Society

InAs quantum dots in AlAs/GaAs short period superlattices: structure, optical characteristics and laser diodes

Vadim Tokranov, M. Yakimov, A. Katsnelson, K. Dovidenko, R. Todt, and S. Oktyabrsky,
UAlbany Institute for Materials, University at Albany–SUNY, 251 Fuller Rd, Albany, NY 12203

ABSTRACT

The influence of two monolayer - thick AlAs under- and overlayers on the formation and properties of self-assembled InAs quantum dots (QDs) has been studied using transmission electron microscopy (TEM) and photoluminescence (PL). Single sheets of InAs QDs were grown inside a 2ML/8ML AlAs/GaAs short-period superlattice with various combinations of under- and overlayers. It was found that 2.4ML InAs QDs with GaAs underlayer and 2ML AlAs overlayer exhibited the lowest QD surface density of 4.2×10^{10} cm^{-2} and the largest QD lateral size of about 19 nm as compared to the other combinations of cladding layers. This InAs QD ensemble has also shown the highest room temperature PL intensity with a peak at 1210 nm and the narrowest linewidth, 34 meV. Fabricated edge-emitting lasers using triple layers of InAs QDs with AlAs overlayer demonstrated 120 A/cm^2 threshold current density and 1230 nm emission wavelength at room temperature. Excited state QD lasers have shown high thermal stability of threshold current up to 130 °C.

INTRODUCTION

Quantum dot (QD) layers were proposed as an active gain medium for semiconductor laser diodes in 1982 [1]. After the discovery of self-assembly of QDs in the InAs/GaAs system via Stranski–Krastanov growth mode, significant research efforts were directed to obtain QD ensembles with uniform size, high density, and high emission efficiency [2,3], and to fabricate QD lasers [4,5,6]. Currently, the performance of QD lasers is comparable or even better than that of quantum well (QW) lasers, e.g. room temperature threshold current density of 16 A/cm^2 for single- and 36 A/cm^2 for triple-layer QD laser [6], and long wavelength (1.3 μm) lasing on GaAs substarte [5] were obtained. However, InAs/GaAs quantum-dot lasers with the lowest threshold current density (ground level near 1.3 μm emission wavelength) have not yet achieved high-temperature stability [7].

In spite of profound studies of Stranski–Krastanov growth of In(Ga)As islands on GaAs, this self-organized formation of nanoscale islands is still not completely understood. The structural and optical properties of QD ensembles are very sensitive to the growth parameters [8,9]. This phenomenon of self-assembly is further complicated by intermixing of InAs QDs with the GaAs barrier [10], segregation of In atoms when InGaAs islands are overgrown by GaAs [11,12], complex diffusion properties of adatoms. Very recently, the top Al containing layers have been used to achieve a red shift of the InAs QDs photoluminescence band [13,14].

The primary goal of the present study is the development of the self-assembled QD active medium for laser diodes operating at elevated (>100°C) temperatures. We have investigated the influence of two monolayer - thick AlAs under- and overlayers on the formation and properties of InAs QDs in a short period superlattice (SPSL) using transmission electron microscopy (TEM) and photoluminescence (PL).

EXPERIMENTAL DELAILS

The structures were grown on GaAs(100) substrates in EPI GEN II Molecular Beam Epitaxy (MBE) system equipped with As-valved cracker source. For QD growth, InAs growth rate was maintained at 0.048±0.002 ML/s, and As_2 flux was kept constant at 1.5×10^{-6} Torr. Indium flux was calibrated by InGaAs growth using reflection high energy electron diffraction oscillations. The self-assembled InAs QDs were embedded into 2ML-AlAs/8ML-GaAs SPSL barrier. The closest to the QDs four SPSL periods were grown at the QD growth temperature.

The PL and TEM measurements were carried out on the same samples with QD layers placed 30 nm below the surface. The growth was started with 500 nm of $Al_{0.45}Ga_{0.55}As$ cladding alloy followed by 280 nm SPSL grown at 650 °C. Then, the growth was interrupted to reduce the substrate temperature to that for QD growth (450 - 500 °C). InAs QDs were grown using different combinations of 2ML-AlAs and 8ML-GaAs underlayers and overlayers. The self-assembled QDs were then capped with four periods of SPSL, and after heating up the substrate by 80 °C, 3 periods of SPSL and 10 nm of AlGaAs alloy were grown on the top.

The optical properties of the self-assembled QDs were characterized using PL. The top and bottom $Al_{0.45}Ga_{0.55}As$ cladding layers were used to prevent photogenerated carriers from spreading to the substrate or to the surface. We used an Ar^+ ion laser (514.5nm) as an excitation source. The PL spectra were detected by LN_2-cooled Ge p-i-n photodiode and recorded using lock-in techniques.

The laser diode structures were grown on n-type GaAs(100) substrates. Si and Be were used for n–type and p–type doping, respectively. n^+–GaAs buffer layer and 1 μm - thick bottom n-$Al_{0.7}Ga_{0.3}As$ cladding layer were grown at 610°C. They were followed by an undoped 900 nm - thick SPSL waveguide structure with the QD active layers in the center. The bottom half of the SPSL waveguide and the QD layers were grown at 650 °C and 475 °C, respectively, following a similar temperature ramping procedure as for PL/TEM samples. Triple layers of InAs QDs with GaAs underlayer and AlAs overlayer, and 30 nm SPSL spacing were used as the laser active medium. Top half of the SPSL waveguide, 1 μm – thick p-$Al_{0.7}Ga_{0.3}As$ cladding layer and the top 250 nm - thick p^+-GaAs contact layer were all grown at 570°C to prevent thermal evolution of the QD structure. Gain-guided lasers with stripe width from 10 to 200 μm and cavity lengths from 1.5 to 10 mm were fabricated from these structures. Laser diodes were fabricated by cleaving without any coating of the facets. Stripe-up laser crystals were mounted on a heatsink using In solder.

RESULTS AND DISCUSSIONS

Transmission Electron Microscopy (TEM): QDs Imbedded into Short Period SL

To optimize the properties of QD structures for the laser gain medium, we have started with the investigation of the influence of GaAs and AlAs under- and overlayers and growth temperature on size, size distribution, surface density and PL properties of QDs. We have grown a set of samples at the substrate temperature in the range of 450 - 500 °C with 2.4ML of InAs imbedded into 8ML-GaAs/2ML-AlAs SPSL. Some of the advantages of SPSL are that this is an effective wide-bandgap material grown at relatively low temperature and another variable to control the emission wavelength. We have investigated four basic QD designs: (i) with 8ML GaAs as both under- and overlayer; (ii) the same but with 2ML AlAs; (iii) with 8ML GaAs

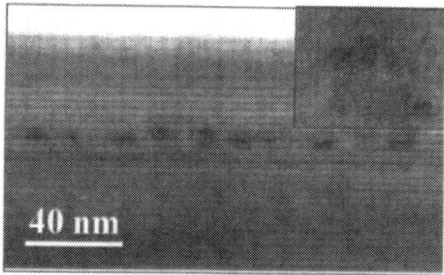

Fig. 1. Cross-sectional TEM image of QD array (2.4ML InAs with AlAs under- and overlayer). Inset: cross-sectional high-resolution TEM image of a single QD.

Fig. 2. TEM plan-view micrograph of QDs grown at 475 °C on AlAs with 2ML AlAs overlayer (small QDs ~14 nm).

underlayer and 2ML AlAs overlayer; and (iv) with 2ML AlAs underlayer and 8ML GaAs overlayer.

200 keV TEM (JEOL 2010 FEG) was used for plan-view and cross-sectional studies. Fig. 1 shows a cross-sectional TEM image of a single layer 2.4ML InAs QDs with 2ML AlAs on both sides imbedded into SPSL. The strain contrast originating from the QDs is clearly visible. One can also observe that the SPSL is an effective way for structure smoothening after the QD growth. The inset in the Fig. 1 shows a cross-sectional high-resolution TEM image of a single QD demonstrating its typical pyramidal shape.

Plan-view image of a QD ensemble grown at 475 °C with 2ML AlAs under- and overlayer is presented in Fig. 2. Statistical analysis of plan-view micrographs allowed us to evaluate the surface density and average size of quantum dots. In Fig. 3, we plot the QD surface number density as a function of the growth temperature. All the studied QD designs exhibit a reduction of surface density at higher substrate temperatures, but the samples with AlAs overlayer are less sensitive to this parameter. The highest QD density of about 2×10^{11} cm^{-2} was obtained in the structures with AlAs underlayer and GaAs overlayer. This results from the lower diffusion rate of In adatoms on the AlAs surface as we have observed recently [15].

On the contrary, QDs with GaAs underlayer and AlAs overlayer had the lowest surface density of about 3×10^{10} cm^{-2} at 500 °C. The average QD size dependance on the growth temperature is shown in Fig. 4 for various QD designs. Generally, QD sizes increase with the growth temperature, and QDs on GaAs with 2ML AlAs overlayer exhibit the largest sizes. In the case of AlAs overlayer (as compared to GaAs), we have observed an increase of the average size and reduction of the surface density likely because of dissolving of small QDs and wetting layer by AlAs overlayer [14]. GaAs underlayer instead of AlAs also leads to increased size and reduced density of QDs. In this case, it is due to the higher surface diffusion of In adatoms on GaAs surface [15]. The main result from these TEM studies is that InAs QDs grown at 475 °C on GaAs underlayer and capped with AlAs overlayer exhibit the lowest QD surface density of 4.2×10^{10} cm^{-2} and the largest QD lateral size of about 19 nm as compared to the other combinations of the cladding layers. This QD ensemble is expected to be very promising for manufacturing of the QD medium with low ground state energy near 1.3 μm and narrow size distribution. It should be noted, that the laser applications require high density and high optical

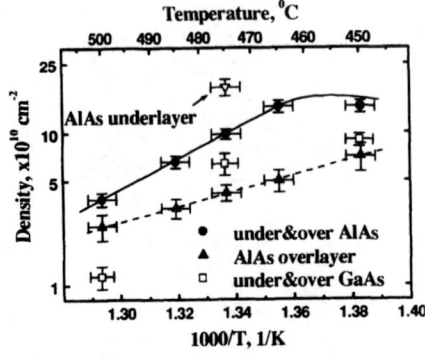

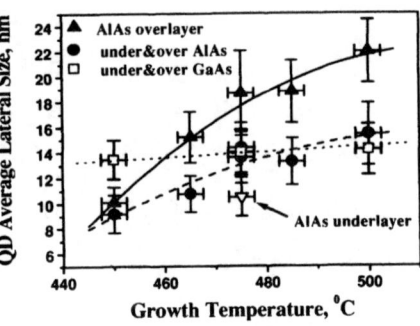

Fig. 3. Surface density of 2.4ML InAs QDs with different under- and overlayers as a function of growth temperature.

Fig. 4. Average sizes of 2.4ML InAs QDs with different under- and overlayers vs. growth temperature. Vertical error bars correspond to the QD size dispersion.

quality of the QD ensemble, which are usually quite conflicting demands. Next section is devoted to the optimization of the PL properties of the QDs.

Photoluminescence: optical properties of QDs imbedded into SPSL

Fig. 5 shows PL peak energy and FWHM of the luminescence band as a function of the average lateral size of QDs grown with different under- and overlayers at 475 °C . These data were obtained at room temperature with the excitation intensity of 10 W/cm². The PL band is found to shift towards the lower energies with increasing of QD sizes as expected from a simple quantum size effect. On the contrary, the effect of barrier bandgap is obscured by the size effect. For example, QDs capped with AlAs high-bandgap overlayers exhibit noticeable PL redshift in comparison with those overgrown by low-bandgap GaAs. Significant reduction of FWHM of the luminescence band indicates that the homogeneity of the QD sizes is also improved for large QDs. Therefore, our growth conditions with the increased In adatom diffusion length lead not only to the increase of the QD sizes, but also to the improved size distribution.

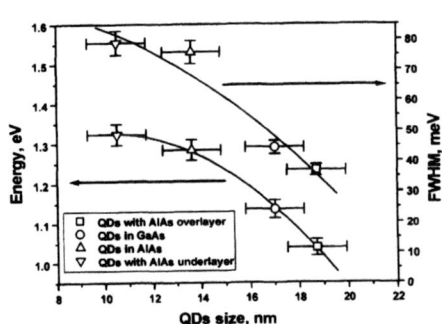

QD ensemble with 2ML AlAs overlayer has demonstrated the highest room temperature PL intensity and the lowest band FWHM of 34 meV (at low excitation level, 0.1 W/cm²) in comparison with the other QD designs. The highest room temperature PL intensity indicates the reduction of density of nonradiative recombination centers in the sample with the lowest QD surface concentration grown with AlAs overlayer.

Fig. 5. Room temperature PL peak energy and band FWHM of 2.4ML InAs QDs as a function of average sizes of QD layers with different design.

Though the AlAs layers are usually expected to have higher defect density than GaAs, we have not observed any degradation of the QD radiative recombination associated with the presence of AlAs.

Electroluminescence of triple QD layer edge-emitting lasers

All electroluminescent (EL) measurements were carried out under pulsed excitation in a temperature range 77 - 430 K. The excitation pulse width was 1 μs with a duty cycle factor of about 0.5%. The EL spectra of an edge-emitting laser with a triple 2.2ML InAs QD layers operating on the ground state transitions are shown in Fig. 6. Light-current characteristic is plotted in the inset. Ground state lasing was obtained at a threshold current density, J_{th} = 120 A/cm^2, and wavelength, 1.23 μm. Minimum cavity length for the ground state lasing was found to be 6 mm. We have estimated the maximum modal gain of our structure, 3.5 cm^{-1}, assuming similar intrinsic losses (α_i=1.8 cm^{-1}) as in the case of quantum well heterolaser with the same waveguide width (0.9 μm). We expect that these results can be significantly improved by optimization of the laser heterostructure, especially by increasing of a potentially low optical confinement factor of the QD active medium.

The EL spectra of an edge-emitting laser with a triple 2.0ML InAs QD layers operating on excited state transitions are shown in Fig. 7, with a corresponding light-current characteristic plotted in the inset. In this case, the lasing was achieved at a higher current density, J_{th} = 360 A/cm^2, and shorter wavelength, 1.14 μm. Minimum cavity length for ground state lasing was about 3 mm, corresponding to the maximum modal gain of the excited state of about 5 cm^{-1} in our structure. These results can be also improved by at least 2-3 times by optimization of the laser heterostructure, such as decreasing the waveguide width down to 0.25 m, decreasing intrinsic losses by 20-30% down to the best reported values, 1.3 - 1.5 cm^{-1} [6].

Thermal quenching of the laser threshold current is plotted in Fig.8. The maximum working temperature for the ground state laser was found to be 75 °C that corresponds to the currently reported values. Excited state lasers exhibited weaker temperature dependence of the threshold current with the maximum operating temperature exceeding 130 °C. The primary reason for the higher thermal stability of excited state QD lasers is the higher saturated modal gain [6]. An additional reason for poor thermal stability of the ground state QD lasers is the stress in a very long laser crystal mounted on a copper heatsink because of the difference in

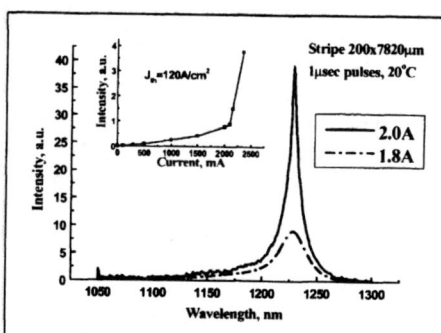

Fig. 6. EL spectra of a ground state QD laser. Inset: Ground state QD laser light-current characteristic.

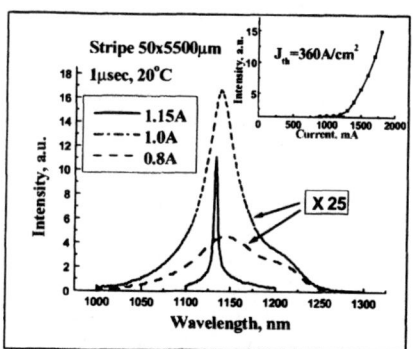

Fig. 7. EL spectra of an excited state QD laser. Inset: QD laser light-current characteristic.

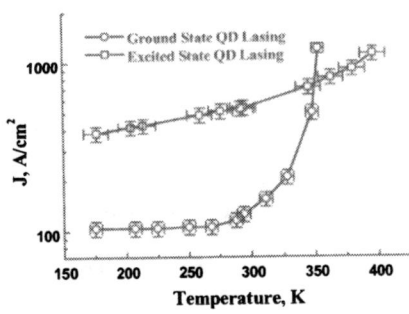

Fig. 8. Threshold current density dependence on temperature for ground state and excited state QD lasers.

thermal expansion coefficients of GaAs and copper. Therefore, synthesis of multilayered QD structures with high QD number density and low nonradiative recombination probability is still a challenging goal for low threshold QD lasers operating at high temperatures.

CONCLUSIONS

To accelerate optimization of QD growth technology and understanding of the growth mechanisms we have used QD structures suitable for both photoluminescent characterization and TEM plan-view measurements of QD density and average size. By employing of this combined method, we have found that 2.4ML InAs QDs with GaAs underlayer and 2ML AlAs overlayer exhibited the highest PL efficiency at room temperature, the narrowest band FWHM (34 meV), the lowest QD surface density of $4.2 \cdot 10^{10}$ cm^{-2} and the largest QD lateral size of about 19 nm, as compared to the other combinations of cladding layers. Edge-emitting laser with triple 2.2ML InAs QD layers operating at the ground state transitions demonstrated low room temperature threshold current density of 120 A/cm^2 and working temperature up to 75 °C. Excited state triple 2ML InAs laser showed threshold current density of 360A/cm^2, and high operation temperature up to 130 °C.

ACNOWLEDGMENTS

The work was supported by MARCO and DARPA under the National Focus Center for Interconnects for Gigascale Integration. This support is greatly appreciated.

REFERENCES

1. Y. Arakawa, and H. Sakaki, Appl. Phys. Lett., **40**, 939 (1982).
2. D. Leonard, M. Kishnamurthy, C. M. Reaves, et al., Appl. Phys. Lett., **63**, 3203 (1993).
3. N. N. Ledentsov, V. M. Ustinov, A. Yu. Egorov, et al., Semicond., **28**, 832 (1994).
4. D. Bimberg, N. N. Ledentsov, M. Grundmann, et al., Physica E., **3**, 129 (1998)
5. G. Park, D L. Huffaker, Z. Zou, et al., IEEE Photon. Technol. Lett., **11**, 301 (1999).
6. P. G. Eliseev, H. Li, A. Stintz, T. C. Newell, et al., Appl. Phys. Lett., **77**, 262 (2000).
7. X. Huang, A. Stintz, C. P. Hains, et al., IEEE Photon. Technol. Lett., **12**, 227 (2000).
8. L. Chu, M. Arzberger, G. Böhm, and G. Abstreiter, J. Appl. Phys., **85**, 2355 (1999).
9. S. Fafard, Z. R. Wasilewski, C. Ni. Allen, D. Picard, et al., Phys. Rev. B, **59**, 15368 (1999).
10. P. B. Joyce, T. J. Krzyzewski, G. R. Bell, et al., Phys. Rev. B., **58**, R15981 (1998).
11. O. Brandt, L. Tapfer, K. Ploog, R. Bierwolf, et al., Appl. Phys. Lett., **61**, 2814 (1992).
12. U. Woggon, W. Langbein, J. M. Hvam, et al., Appl. Phys. Lett., **71**, 377 (1997).
13. M. Arzberger, U. Käsberger, G. Böhm, et al., Appl. Phys. Lett., **75**, 3968 (1999).
14. A. F. Tsatsul'nikov, A. R. Kovsh, A. E. Zhukov, et al., J. Appl. Phys., **88**, 6272 (2000).
15. M. Yakimov, V. Tokranov, and S. Oktyabrsky, MRS Symp. Proc., **648**, P2.6.1 (2001).

Mat. Res. Soc. Symp. Proc. Vol. 707 © 2002 Materials Research Society   H11.6

InP Self Assembled Quantum Dot Lasers Grown on GaAs Substrates by Metalorganic Chemical Vapor Deposition

R. D. Dupuis[1], J. H. Ryou*[1], R. D. Heller[1], G. Walter[2], D. A. Kellogg[2], N. Holonyak, Jr[2], C. V. Reddy[3], V. Narayanamurti[3], D. T. Mathes[4], and R. Hull[4]
[1] Microelectronics Research Center, The University of Texas at Austin
10100 Burnet Road, Building 160, Austin, TX 78758 USA
    Phone: +1-512-471-0537, Fax: +1-512-471-0957, e-mail: dupuis@mail.utexas.edu
[2] Center for Compound Semiconductor Microelectronics, The University of Illinois at Urbana-
    Champaign, Urbana, IL
[3] Gordon McKay Laboratory of Applied Science, Harvard University, Cambridge, MA 02138
[4] Department of Materials Science and Engineering, The University of Virginia, Charlottesville,
    VA
*Now with Honeywell VCSEL Products Division, Plymouth MN 55441

ABSTRACT

We describe the operation of lasers having active regions composed of InP self-assembled quantum dots embedded in $In_{0.5}Al_{0.3}Ga_{0.2}P$ grown on GaAs (100) substrates by MOCVD. InP quantum dots grown on $In_{0.5}Al_{0.3}Ga_{0.2}P$ have a high density on the order of about $1-2 \times 10$ cm^{-2} with a dominant size of about 10-15 nm for 7.5 ML growth.[1] These $In_{0.5}Al_{0.3}Ga_{0.2}P/InP$ quantum dots have previously been characterized by atomic-force microscopy, high-resolution transmission electron microscopy, and photoluminescence.[2] We report here the 300K operation of optically pumped red-emitting quantum dots using both double quantum-dot active regions and quantum-dot coupled with InGaP quantum-well active regions. Optically and electrically pumped 300K lasers have been obtained using this active region design; these lasers show improved operation compared to the lasers having QD-based active regions with threshold current densities as low as $J_{th} \sim 0.5$ KA/cm^2.

INTRODUCTION

III-phosphide self-assembled quantum-dot (SAQD or simply QD) structures having delta-functional behavior of the density of states and the discrete energy levels of carriers induced by three-dimensional quantum confinement offer the potential to realize injection lasers operating in the visible spectral region with improved performance characteristics, such as low threshold current density, high characteristic temperature, and high differential gain [3,4,5]. The direct growth of coherently strained defect-free self-assembled quantum dots on planar substrates using the coherent Stranski-Krastanow (SK) growth mode [6,7] offers the potential to develop QD visible laser devices with the theoretically predicted and experimentally realized improved performance. Also, the SAQD growth process can overcome the limitation of lattice matching between the substrate and epitaxial active region due to the intrinsic strain-compliant nature of the SK growth mode.

III-As quantum dot-related structures for infrared optoelectronic applications at ?~1.3?m have been extensively researched for growth condition optimization, material property characterization, and device applications including lasers in the infrared spectral region. Since the first demonstration of InAs quantum dot lasers [8], a great deal of research on III-As based quantum dot structures have led to quantum-dot lasers with improved threshold current densities and characteristic temperatures [9] as compared to more conventional quantum-well lasers. However, in comparison to the III-As material system, less research has been performed in the development of III-phosphide SAQD structures for applications in the visible-light spectral region. Initially, it was reported that the properties of III-P SAQDs had a reduced QD density and a broader size distribution than those of the III-As SAQDs. Due to the "blue-shift" effect of the emission from the SAQDs induced from multiple orders of quantum confinement and compressive strain on QDs, there is the potential to extend the wavelength of light emitters to the yellow or green spectral regions using binary and ternary III-Phosphide SAQD structures.

In the past, structures with active regions containing InP SAQDs on GaAs substrates, the QDs have generally been grown on $In_{0.49}Ga_{0.51}P$ matrix layers by MBE [10,11], MOCVD [12,13], or by hydride vapor phase epitaxy (VPE) [14]. InP SAQDs embedded in $In_{0.49}Ga_{0.51}P$ grown on GaAs (100) substrates grown by MOCVD have been reported to yield relatively low densities of SAQDs ($\sim 10^7$-10^9 cm^{-2}), compared to III-As SAQDs, and a bimodal size distribution of coherent islands has been reported. InP SAQDs have also been grown on GaP substrates [15] to modify the strain applied to InP QDs; however, no PL emission from the QDs was observed, possibly due to the expected Type II conduction band alignments in this system. Moreover, the growth of InP SAQDs is not as well developed as the growth of III-As SAQDs and further investigation is required.

To improve the QD density, the effect of different matrix layers lattice-matched to GaAs substrates having larger bandgap, such as $In_{0.49}(Al_xGa_{1-x})_{0.51}P$ (x=0.3, 0.6, 1.0), needs to be studied to improve the carrier confinement. Furthermore, a different matrix material system is expected to change the morphology and growth characteristics of the InP SAQDs. Additionally, the matrix material affects the optical properties of SAQD structures. Recently, we studied the growth of InP SAQDs on InAlGaP matrix layers and the effect of coupling quantum dot states with the electronic states of the InGaP quantum well [20,21,22,23].

EXPERIMENT

In the present study, InP SAQD growth conditions are employed which have been optimized with various matrix layers with regards to QD size, uniformity, and density as well as optical quality to fabricate InP QD based laser operating in visible light spectral region with improved characteristics. InP quantum dots coupled to InGaP quantum wells have been studied. These lasers have lased CW optically and pulsed electrically pumped at 77K and 300K. QD material and optical properties are characterized by atomic force microscopy (AFM), photoluminescence (PL), and high-resolution transmission electron microscopy (TEM).
The InAlGaP/InP QD and laser structures in this work were grown by low-pressure metalorganic chemical vapor deposition (MOCVD) in an EMCORE Model GS3200 UTM rotating disk reactor at ~60 Torr using a H_2 ambient. The group III precursors used are trimethylindium (TMIn), trimethylaluminum (TMAl), and triethylgallium (TEGa); the group V hydride sources are arsine (AsH_3) and phosphine (PH_3). Disilane (Si_2H_6) is used as a source of Si for n-type doping and bis-cyclopentadienyl magnesium (Cp_2Mg) is used for p-type (Mg) doping. The growth was preformed simultaneously on two-inch diameter GaAs:Si (100) on-axis, (100) -10? <111>A, and (100) -15? <111>A substrates.

The surface morphology of SAQDs is characterized by tapping-mode AFM on the exposed QD samples to determine the QD average (or dominant) size, density, and uniformity. The QD size is expressed in terms of the height as measured by AFM. The density is taken from the multiple sets of relatively large area AFM scans. To evaluate the morphology, the $In_{0.5}Al_{0.5}P/InP$ QD structure used consists of an $In_{0.49}(Al_xGa_{1-x})_{0.51}P$ lower matrix layer (250 nm thick) followed by an InP QD active layer all grown on a GaAs buffer layer/GaAs:Si (100) substrate. Following the QD deposition, the growth is terminated and the wafer is cooled down under a PH_3 overpressure to prevent desorption. The lattice parameters of the $In_{0.49}(Al_xGa_{1-x})_{0.51}P$ (x=0.3, 0.6, 1.0) matrix layers are analyzed using a Rigaku high-resolution five double-crystal X-ray diffractometry (using a four-crystal monochromator) to determine the degree of lattice matching to the substrate. For lattice-matched matrix layers, the epitaxial layers are calibrated to have less than -200 arc-seconds of angle separation relative to the GaAs substrate measured from an (004) X-ray rocking curve. This is done in order to minimize the strain effect from the matrix layer on the growth of active QD layer. The optical properties of InP SAQDs embedded in various matrices are characterized by room-temperature and low-temperature PL. For the PL measurements, a ?~488 nm excitation source from an Ar^+ laser operating at a constant power density (50 – 200 mW) and a GaAs photomultiplier with a GaAs photocathode are used. For these quantum-dot heterostructures (QDHs), the $In_{0.49}(Al_{1-x}Ga_x)_{0.51}P$ upper matrix layer is grown on the InP SAQD layer with a certain post-purge time after the completion of SAQD layer deposition. TEM is also used to study the microscopic morphology and material quality of individual QDs with and without upper matrix layer, as shown in

The InP SAQD growth conditions are optimized by altering the growth temperature, growth time, and V/III ratio. The growth temperature of the QDs is optimized at 650°C for 1 minute (7.5 ML) with a V/III ratio ~2,100. The estimated nominal growth rate of InP has been reduced to ~0.125 monolayer/sec (ML/s), as calibrated by measuring the thickness of a thin InP planar layer using glancing grazing incidence X-ray reflectivity.

METHODS USED

In this study, we report some of the characteristics of the InP SAQDs. We also report the device results of optically and electrically pumped lasers based on these InP SAQDs. The InP QD growth studies are performed by altering growth temperatures and times and using various $In_{0.49}(Al_xGa_{1-x})_{0.51}P$ matrices (x=0.0, 0.3, 0.6, and 1.0). The morphology changes of the exposed SAQDs depend on the growth time and the matrix material, and are characterized by atomic force microscopy (AFM). Photoluminescence (PL) spectra were taken at 4K and 300K to determine the light-emitting characteristics of the $InP/In_{0.49}(Al_xGa_{1-x})_{0.51}P$ quantum-dot heterostructures (QDHs). 4K PL spectra from the InP SAQDs embedded in $In_{0.49}(Al_xGa_{1-x})_{0.51}P$ cladding layers exhibit PL emission in the visible orange and red spectral regions, as shown below. Also, transmission electron microscopy is used to characterize the microscopic material quality and morphology of the individual QD and the interfaces between SAQD and cladding layers.

RESULTS OBTAINED

AFM studies of InP SAQDs grown on InAlGaP matrix layers have shown that under some conditions, a relatively high density of small quantum dots is produced, as described in Figure 1. As the QD growth proceeds with longer deposition times, the average QD height (and base width) increases and the average density correspondingly decreases. Our work has concentrated on the 7.5ML effective QD thickness where the size and density appear to be optimal for light emitting properties determined by PL.

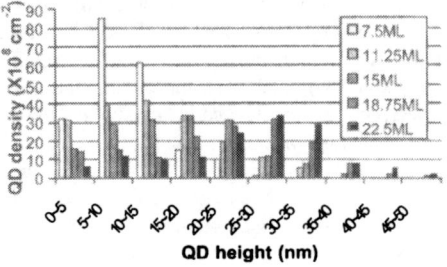

Figure 1: Variation of height and density vs. nominal layer thickness for InP SAQDs.

The 4K PL intensity vs. wavelength spectra for InP/In$_{0.49}$(Al$_{0.6}$Ga$_{0.4}$)$_{0.51}$P QDHs grown for different deposition times is shown in Figure 2. The PL spectral peak position changes from 591 nm (2.10 eV for 3.75 MLs), to 653 nm (1.90 eV for 7.5 MLs) and 681 nm (1.82 eV for 15 ML), as the deposition time increases. This PL emission is at a lower energy than that observed from comparable-sized InP QDs embedded in In$_{0.49}$Al$_{0.51}$P.

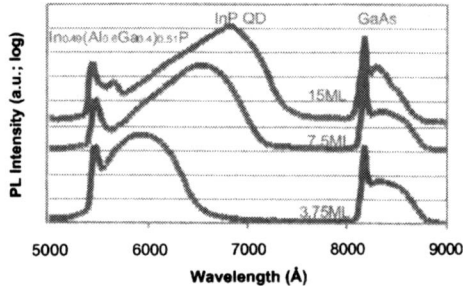

Figure 2: 4K PL spectra of InP SAQDs embedded in InAlGaP cladding layers grown at 650 °C for various deposition times.

However, InP/In$_{0.49}$(Al$_{0.6}$Ga$_{0.4}$)$_{0.51}$P QDHs exhibit more efficient luminescence than InP/In$_{0.49}$Al$_{0.51}$P QDHs at room temperature, possibly due to better electron confinement. TEM studies of the InP SAQDs have shown the dislocation-free structure of single- and multiple-layer QDHs and the vertical alignment achieved for multiple QD layers, as shown in Figure 3. These InP QD active regions have been incorporated into various separate-confinement laser active regions. In the present work, we have grown three types of quantum-confined "red-emitting" active region laser structures on GaAs substrates: (1) single- and multiple-QDHs; (2) single-QW active-region QDHs; and couples QD+QW structures and compared the 300K and 77K PL emission, as shown in Figure 4. Recently, we have described the effects of coupling of the InP QD states to the electronic states of InGaP quantum wells grown below and above the quantum dots, resulting in the coupling of the quantum dots through electronic states in the quantum wells [24]. Optically and electrically pumped 300K lasers using QD+QW active regions have been obtained using this unique design; these lasers show improved operation compared to lasers having QD-based active regions with pulsed 300K threshold current densities J_{th} ~1.5 KA/cm^2, as shown in Figure 5. By optimizing the coupling between the quantum well and the quantum-dots, other QD+QW laser devices have been grown that exhibit pulsed J_{th} values at 300K as low as ~0.5 KA/cm^2.

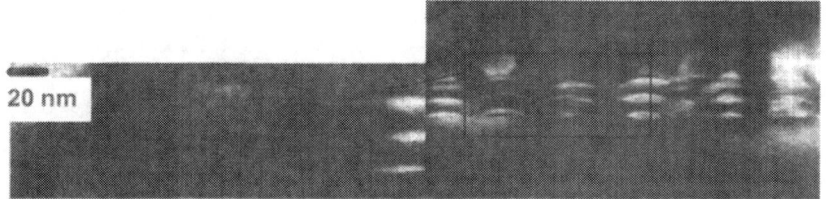

Figure 3: TEM micrograph of stacked InP SAQDs.

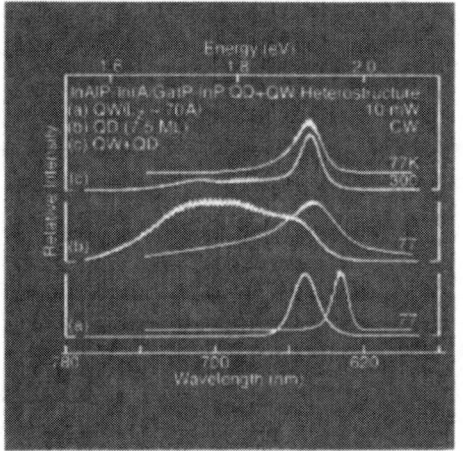

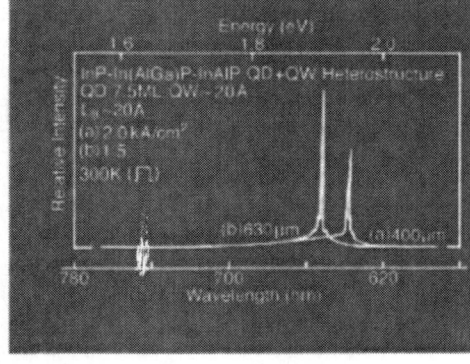

Figure 4: PL spectra for InAlP-InAlGaP-InP QW, QD, QD+QW samples.

Figure 5 : InAlP-InAlGaP-InP QD+QW 300K electrically pumped optical spectra.

CONCLUSIONS

In summary, we have grown and characterized InP SAQDs embedded in $In_{0.49}(Al_xGa_{1-x})_{0.51}P$ matrices (x=0.0, 0.3, 0.6, and 1.0). The InP SAQDs grown at 650C exhibit uniform morphology with a 5-10 nm dominant height and ~2x10^{10} cm^{-2} density for 7.5 ML. As the deposition time increases, the QD size increases, while the density decreases slightly, and the PL peak position shifts to lower energy. For $InP/In_{0.5}Ga_{0.5}P/In_{0.49}(Al_{0.6}Ga_{0.4})_{0.51}P/In_{0.49}Al_{0.51}P$ SQD+SQW injection lasers, we have achieved electrically pumped lasing at 681 nm at 300 K with reduced threshold current densities as low as 0.5 KA/cm^2. We believe this approach is applicable to other QD laser materials systems.

REFERENCES

[1] J. H. Ryou, R. D. Dupuis, G. Walter, N. Holonyak, Jr., D. T. Mathes, R. Hull, C. V. Reddy, and V. Narayanamurti, J. Appl. Phys., to be published.

[2] J. H. Ryou, PhD. Dissertation, The University of Texas at Austin (2001).

[3] Y. Arakawa and H. Sakaki, Appl. Phys. Lett. **40**, 939 (1982).

[4] M. Asada, Q. Miyamoto, and Y. Suematsu, IEEE J. Quantum Electron. **QE-22**, 1915 (1986).

[5] N. N. Ledentsov, M. Grundmann, F. Heinrichsdorff, D. Bimberg, V. M. Ustinov, A. E. Zhukov, M. V. Maximov, Zh. I. Alferov, and J. A. Lott, IEEE J. Select. Topic. Quantum Electron. **6**, 439 (2000).

[6] I. N. Stranski and L. Krastanow, Akad. Wiss. Wien, Math-Naturwess. Klasse **146**, 797 (1937).

[7] V. A. Shchukin and D. Bimberg, Rev. Mod. Phys. **71**, 1125 (1999).

[8] N. N. Ledentsov, V. M. Ustinov, A. Y. Egorov, A. E. Zhukov, M. V. Maksimov, I. F. Tavatadze, and P. S. Kop'ev, Semiconductors **28**, 832 (1994).

[9] G. Park, O. B. Shchekin, S. Csutak, D. L. Huffaker, and D. G. Deppe, Appl. Phys. Lett. **75**, 3267 (1999).

[10] A. Kurtenbach, K. Eberl, and T. Shitara, Appl. Phys. Lett. **66**, 361 (1995).

[11] A. Kurtenbach, C. Ulrich, N. Y. Jin-Phillipp, F. Noll, K. Eberl, K. Syassen, and F. Phillipp, J. Electron. Mater. **25**, 3 (1996).

[12] S. P. DenBaars, C. M. Reaves, V. Bressler-Hill, S. Varma, W. H. Weinberg, and P. M. Petroff, J. Cryst. Growth **145**, 721 (1994).

[13] N. Carlsson, W. Seifert, A. Petersson, P. Castrillo, M. E. Pistol, and L. Samuelson, Appl. Phys. Lett. **65**, 3093 (1994).

[14] J. Ahopelto, A. A. Yamguchi, K. Nishi, A. Usui, and H. Sakaki, Jpn. J. Appl. Phys. Part 2 **32**, L32 (1993).

[15] Y. Nabetani, K. Sawada, Y. Fukukawa, A. Wakahara, S. Noda, A. Sasaki, J. Cryst. Growth **193**, 470 (1998).

[16] K. Eberl, A. Kurtenbach, M. Zundel, N. Y. Jin-Phillipp, F. Phillipp, A. Moritz, R. Wirth, and A. Hangleiter, J. Cryst. Growth **175/176**, 702 (1997).

[17] T. Riedl, E. Fehrenbacher, A. Hangleiter, M. K. Zundel, and K. Eberl, Appl. Phys Lett. **73**, 3730 (1998).

[18] Y. M. Manz, O. G. Schmidt, and K. Eberl, Appl. Phys. Lett. **76**, 3343 (2000).

[19] J. Porsche, M. Ost, F. Scholz, A. Fantini, F. Philipp, T. Riedl, and A. Hangleiter, IEEE J. Select. Topic Quantum Electron. **6**, 482 (2000).

[20] G. Walter, N. Holonyak, Jr., J. H. Ryou and R. D. Dupuis, Appl. Phys. Lett. **78**, 26, 4091 (2001).

[21] J. H. Ryou, R. D. Dupuis, G. Walter, D. A. Kellogg, N. Holonyak, Jr., D. T. Mathes, R. Hull, C. V. Reddy, and V. Narayanamurti, Appl. Phys. Lett., **79**, 4091 (2001).

[22] J. H. Ryou, R. D. Dupuis, D. T. Mathes, R. Hull, C. V. Reddy, and V. Narayanamurti, Appl. Phys. Lett. **78**, 3526 (2001).

[23] J. H. Ryou, R. D. Dupuis, C. V. Reddy, V. Narayanamurti, D. T. Mathes, R. Hull, A. Mintairov, and J. L. Merz, J. Electron. Mat. **30**, 471 (2001).

[24] G. Walter, N. Holonyak, Jr., J. H. Ryou and R. D. Dupuis, Appl. Phys. Lett. **79**, 3215 (2001).

Mat. Res. Soc. Symp. Proc. Vol. 707 © 2002 Materials Research Society

Formation of GaN Self-Organized Nanotips by Nanomasking Effect

Harumasa Yoshida, Tatsuhiro Urushido, Hideto Miyake and Kazumasa Hiramtsu
Department of Electrical and Electronic Engineering, Faculty of Engineering, Mie University,
1515 Kamihama, Tsu, Mie 514-8507, Japan

ABSTRACT

We have successfully fabricated self-organized GaN nanotips by reactive ion etching using chlorine plasma, and have revealed the formation mechanism. Nanotips with a high density and a high aspect ratio have been formed after the etching. We deduce from X-ray photoelectron spectroscopy (XPS) analysis that the nanotip formation is attributed to nanometer-scale masks of SiO_2 on GaN. The structures calculated by Monte Carlo simulation of our formation mechanism are very similar to the experimental nanotip structures.

INTRODUCTION

GaN based III-nitride semiconductors are attractive wide direct band gap semiconductors for applications in optoelectronic and electronic devices such as ultraviolet-blue-green light-emitting diodes, laser diodes, field-effect transistors and high-temperature transistors. To fabricate nanometer-scale fine structure of GaN is one of the key technologies for further development of group-III nitride semiconductor devices. The formation of the fine tip and pillar structure of GaN becomes important in the advanced applications of stress-free epitaxy, cold-cathode field emitters and quantum dots. The stress-free epitaxy on GaN nanocolumn structure was carried out by molecular beam epitaxy (MBE) [1]. The GaN pyramid structure formed by selective area growth is expected to be the application of a cold-cathode field emitter [2]. Self-assembled GaN quantum dots grown by plasma-assisted MBE were demonstrated [3]. The GaN whiskers formed by photoelectrochemical (PEC) wet etching [4] and metalorganic vapor phase epitaxy (MOVPE) were reported [5]. It was reported that the former whiskers were related to threading dislocations in the GaN layer.

However, there are no reports on the formation and the mechanism of GaN fine tips by the dry etching technique. It is difficult for conventional lithographic methods to meet this requirement due to the limitation of lithographic technique. In this paper, we demonstrate a novel method and the mechanism of forming the self-organized nanotips of GaN by reactive ion etching (RIE).

EXPERIMENTAL PROCEDURE

RIE of GaN was carried out using chlorine plasma. A GaN layer was grown by MOVPE on buffer layer deposited at low temperature on (0001) sapphire substrate. The thickness of the GaN was ~2 μm. After the deposition of a SiO_2 film (300 nm thickness) on the GaN by RF sputtering, stripe windows of 5 μm width with a periodicity of 10 μm were developed by conventional photolithography technique in the <1-100> directions of the GaN layer, in order to examine the etch rates and to form banks for the protection of fine tips.

The dry etching apparatus used in this study was a RIE system with a 6 inch sample stage of the cathode electrode (Samco RIE-5). A quartz (SiO_2) plate was put on the cathode electrode, then the GaN samples were placed in the center of the quartz plate in two configurations as follows. One sample was placed such that the surface of GaN and the electrode became parallel. Another sample was placed by tilting it approximately 20 degrees from the electrode. Chlorine plasma was generated by applying 13.56 MHz RF power of 200 W to the electrodes at a total Cl_2 flow of 30 sccm and a total pressure of 6 Pa. The self-bias voltage of the cathode was measured at -700 V under the condition. The apparatus was operated at room temperature, and the cathode electrode was water-cooled. The surfaces of the samples were observed by scanning electron microscopy (SEM), and were surveyed by XPS with a monochromatic Mg Kα source. The optical emission from the plasma was measured via the quartz view port on the side of the RIE chamber, and spectrally analyzed using an optical spectrometer (Hamamatsu C7460).

RESULTS AND DISCUSSION

Figure 1 shows the SEM image of GaN nanotip structures formed by RIE using Cl_2 plasma. Cone-shaped GaN islands were formed as the initial stage of nanotips after 1 min of etching, as shown in Fig. 1(a). Tip-shaped GaN pillars with a density of approximately 8×10^9 cm^{-2} were formed after 4 min of etching, as shown in Fig. 1(b). The nanotips exhibit a length of about 0.7 μm and a high aspect ratio. The average diameter of the middle part of tips and the standard deviation are 19.0 nm and 4.5 nm, respectively. This nanotip structure is similar to the structures of whiskers formed by PEC wet etching [4] and MOVPE [5]. In the former report, it was suggested that the PEC whiskers had close correlation with the threading dislocations in the GaN layer. The whiskers are perpendicularly formed to the substrate due to the dislocations along the <0001> direction of the GaN layer.

However, the nanotip feature depends on the direction of ion injection from the plasma in our experiment using a tilted sample. The nanotip structure of the tilted sample is shown in the SEM image in Fig. 2(a). The tips were formed at an equal gradient to the tilted angle of the sample as illustrated in Fig. 2(b), in which the angle between the <11-20> direction of GaN and the electrodes was approximately 20 degrees. The ions of the plasma perpendicularly descend to the sample surface in the case of Fig. 1. In the meantime, the ions descend with declined angle

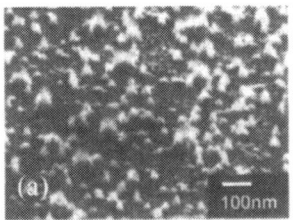 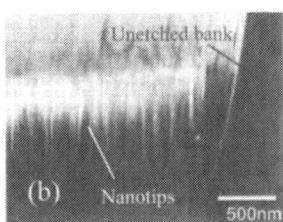

Figure 1. SEM images of GaN nanotip structures.
(a) After 1 min of etching.
(b) After 4 min of etching.

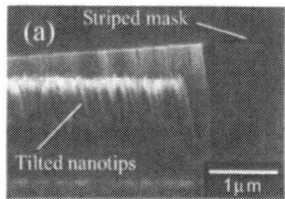 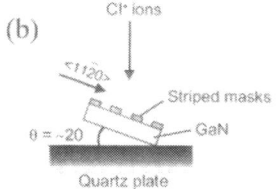

Figure 2. Nanotip structure formed in the tilted GaN sample. (a) SEM image of the tilted nanotips. (b) Schematic illustration of etching the tilted sample.

to the surface in the case of the configuration illustrated in Fig. 2(b). The anisotropic etch profile is achieved by RIE due to the acceleration of energetic ions from the plasma to the sample surface. Therefore, this phenomenon suggests that the nanotip formation has no correlation with threading dislocations in the GaN layer.

Figure 3 shows the XPS spectra of Si_{2p} and Al_{2s} on the surface of a sapphire (Al_2O_3) sample which was treated by RIE for 2 min under the same condition. The peak of Si_{2p} and the saturated peak of Al_{2s} are located at 104 eV and 122 eV, respectively. The drastic reduction in the Si_{2p} intensity can be seen with etching by Ar ion-gun installed in the XPS. The XPS spectrum after 20 min etching by Ar is as same as that of an untreated sapphire sample. None of interesting XPS spectrum on the nanotips of the GaN sample was observed. The total top-area of tips is too small to detect Si_{2p} signal, because the area is estimated to be 1/1000 of the whole area by assuming 4 nm top-diameter of the tip. We deduce from these results that the nanotip formation is attributed to masks of SiO_2 on the GaN.

Figure 4 shows the optical emission spectrum of the Cl_2 plasma with the quartz plate on the cathode without a GaN sample. The lines over 20 peaks of Cl radicals that react with GaN are observed over the range from 720 nm to 950 nm. Several broad lines from 195 nm to 315 nm of Cl^+ ions and a continuous spectrum from 400 nm to 650 nm of both molecular Cl_2 and molecular Cl_2^+ ions are also observed. The Cl radicals and Cl^+ ions, respectively, seem to be the main species of chemical and physical reactions in the Cl_2 plasma.

We suggest a masking effect of nanometer-scale SiO_2 masks, and discuss the mechanism of forming the nanotip structure below. The model of nanotip formation is illustrated in Fig. 5. The quartz plate is sputtered by Cl^+ ions accelerated by a cathode bias voltage of –700 V, and then

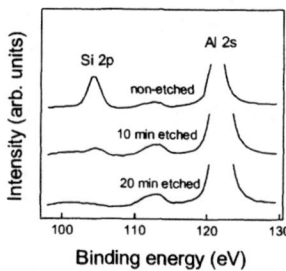 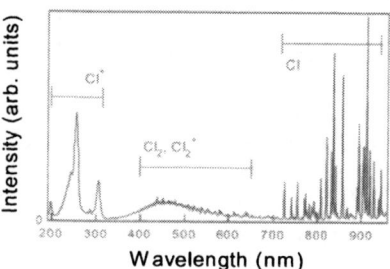

Figure 3. XPS spectra of Si_{2p} and Al_{2s} on the surface of a sapphire sample treated under the same condition. Si_{2p} intensity decreases with etching by Ar ion-gun for 10 min and 20 min.

Figure 4. Optical emission spectrum from the Cl_2 plasma.

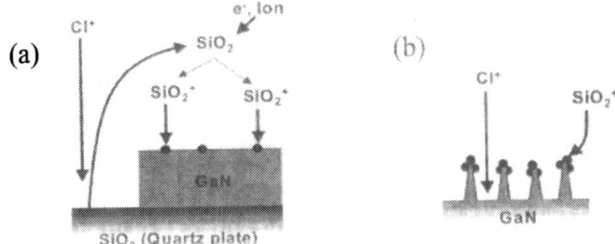

Figure 5. Schematic illustration of the formation mechanism of nanotips. (a) SiO_2 particles are sputtered by Cl^+ ions. Low-energy SiO_2 particle is ionized by electron and ion impacts. Positively ionized SiO_2 particles deposit on the surface of GaN. (b) The ionized SiO_2 particles concentrate at the upper part of the tip. Etching proceeds faster at the nonmasked area.

neutral and ionized SiO_2 particles are generated as shown in Fig. 5(a). Low-energy SiO_2 particle within the scattered SiO_2 particles from the quartz is easily ionized by electron and ion impacts in the plasma because of its low speed, i.e., long residence time in the plasma. Next, the positively ionized SiO_2 particles deposit on the surface of the negatively biased GaN.

An etch selectivity of ~10:1 for GaN : SiO_2 is derived from both the etch-rate ratio of the striped SiO_2 mask, which has been intentionally deposited, and the bottom part of the nanotip structure in the GaN layer. This high selectivity shows that SiO_2 is sufficiently effective as an etching mask for GaN.

The tip-shaped structure indicates that the SiO_2 particles cannot uniformly deposit on the surface of GaN, but can self-selectively deposit at a specific site. This fact suggests the following formation mechanism. In the initial stage of etching, small level differences are formed as initial tiny tips between the SiO_2 masked site and the nonmasked site at the surface. The positively ionized SiO_2 particles are drawn toward the proximity of the tiny tips because the electric field concentrates at the upper part of the tip. Once the tiny tips are formed as observed in Fig. 1(a), the ionized SiO_2 particles concentrate at the upper part of the tip due to a higher electric field, and then the etching of the nonmasked area proceeds more quickly than the masked area, as shown in Fig. 5(b). According to this mechanism, we can explain that the nanotips are formed along the incident direction of Cl^+ ions in both the tilted and untilted GaN samples.

We also investigated the effects of the Si plate for comparison with the effects of the quartz plate. A 6 inch Si wafer was set between the GaN sample and the quartz plate, and then etching of GaN was carried out using Cl_2 plasma. A smooth etched surface without a nanotip was obtained. Si is so reactive with Cl_2 plasma that the masking effect was not observed. We report detailed investigations concerning this effect elsewhere.

Numerical simulations of the nanotip formation model are performed under the condition summarized in Table I. The 'mesh' denotes minimum unit of area. The length and width of etching area are divided into 100 by 100 meshes. The unit of thickness is arbitrary. The etch rate for 1 min is made to be a half of that for 4 min due to the change in the sample temperature [6]. The generation and the landing sites of SiO_2 particles are calculated by Monte Carlo simulation.

Table I. Conditions for simulation.

Etching time	1 min	4 min
Etch rate	200 unit/min	400 unit/min
Nano-mask thickness	1 unit	1 unit
Masking probability	5×10^{-2} sec^{-1} mesh^{-1}	$5 \times 10^{-4} \sim 5 \times 10^{-1}$ sec^{-1} mesh^{-1}
Selectivity (GaN/SiO$_2$)	10	10

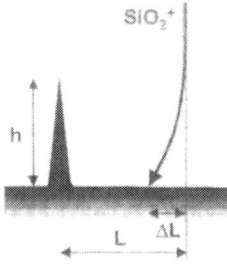

Figure 6. Schematic illustration of the ionized particles landing near the tip on the GaN surface.

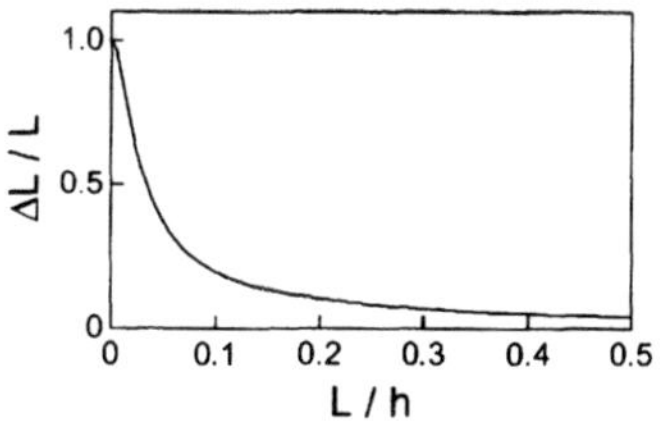

Figure 7. Ratio $\Delta L/L$ as a function of ratio L/h.

The mass and energy distributions of the particles are not taken into consideration. A positively ionized SiO$_2$ particles, landing on the surface of GaN, is drawn toward the direction of the nearest tip due to the slope of the electric field, as illustrated in Fig. 6. We assume, for simplicity, that the drawn distance ΔL of the particles depend on both the distance L from the nearest tip and the height h of the tip, and the ratio $\Delta L/L$ depends on the ratio L/h as the function shown in Fig. 7. According to the function, some particles which are far form the tip perpendicularly descend to the surface, or other particles which are close to the tip drawn toward the upper part of the tip.

Typical calculated structures are shown in Fig. 8. The calculated structures with masking probabilities (MP) of 5×10^{-2} sec^{-1} mesh^{-1} and 3×10^{-2} sec^{-1} mesh^{-1} for 1 min and 4 min etch-time are, respectively, almost similar to the experimental nanotip structures in Fig. 1(a) and 1(b). The tip-shaped structures are formed with an MP range from 10^{-2} sec^{-1} mesh^{-1} to 0.2 sec^{-1} mesh^{-1}.

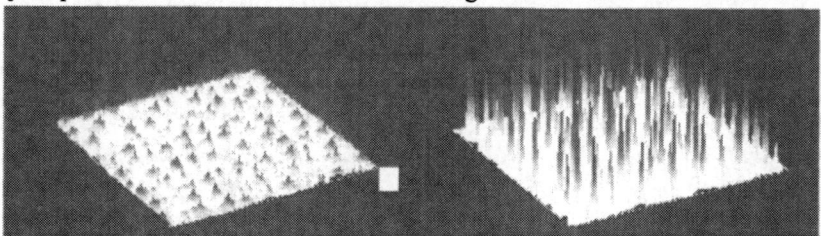

Figure 8. Typical structures calculated by numerical simulation. (a) Tiny tip structure resembling the structure in Fig.1(a), at MP of 5×10^{-2} sec^{-1} mesh^{-1} for 1 min of etching. (b) Nanotip structure resembling the structure in Fig.1(b), at MP of 3×10^{-2} sec^{-1} mesh^{-1} for 4 min of etching.

A smooth surface is formed by lower masking effect, such as the case using a Si wafer, with a lower MP of $<5 \times 10^{-3}$ sec^{-1} $mesh^{-1}$. The simulation results can reasonably explain the experimental results. A mesa structure is formed with a higher MP of >0.5 sec^{-1} $mesh^{-1}$ because of etch-stop effect by excessive masking. Actual MP values and actual tip densities are estimated by setting the calculated tip density at MP of 3×10^{-2} sec^{-1} $mesh^{-1}$ corresponding to the experimental tip density of 8×10^9 cm^{-2}. This estimation shows that the appropriate process window depends on the actual MP range from 1×10^2 to 2×10^3 sec^{-1} μm^{-2}, and the actual tip density range from 5×10^8 cm^{-2} to 1×10^{10} cm^{-2} are expected.

CONCLUSION

We have found that the GaN self-organized nanotip structure is formed by RIE using a quartz cathode plate and Cl_2 plasma. The nanotip structure depended on the direction of ion injection. XPS analysis implies the deposition of SiO_2 on the surface of GaN. These phenomena suggest that the nanotip formation is attributed to nanometer-scale masks of SiO_2 with a slower etch rate than GaN. Numerical simulations were carried out under an assumption that the positively ionized SiO_2 in the plasma concentrates at the upper part of the tip due to higher electric field. The structures calculated by this model closely resembled to the experimental nanotip structures. These results prove the validity of the formation mechanism by the nanomasking effect of ionized SiO_2 particles sputtered by Cl^+ ions.

ACKNOWLEDGMENTS

The authors would like to thank Prof. Y. Takeda and Mr. T. Ichikawa for the XPS measurements.

REFERENCES

1. K. Kusakabe, A. Kikuchi and K. Kishino, Jpn. J. Appl. Phys. **40**, L192 (2001).
2. S. Kitamura, K. Hiramatsu and N. Sawaki, Jpn. J. Appl. Phys. **34**, L1184 (1995).
3. B. Daudin, G. Feuillet, H. Mariette, G. Mula, N. Pelekanos, E. Molva, J. L. Rouvière, C. Adelmann, E. Martinez-Guerreo, J. Barjon, F. Chabuel, B. Bataillou and J. Simon, Jpn. J. Appl. Phys. **40**, 1892 (2001).
4. C. Youtsey, L. T. Romano and I. Adesida, Appl. Phys. Lett. **73**, 797 (1998).
5. M. Haino, M. Yamaguchi, H. Miyake, A. Motogaito and K. Hiramatsu, Jpn. J. Appl. Phys. **39**, L449 (2000).
6. H. Yoshida, T. Urushido, H. Miyake and K. Hiramatsu, Jpn. J. Appl. Phys. **40**, L313 (2001).

Shape Transitions of Self-Assembled Ge Islands on Si (001)

Armando Rastelli[1], Matthias Kummer[2], Hans von Känel[2]
[1]INFM - Università di Pavia, Via Bassi 6, I-27100 Pavia, Italy
[2]Laboratorium für Festkörperphysik, ETH Zürich, CH-8093 Zürich, Switzerland

ABSTRACT

Coherently strained Ge islands were grown at a substrate temperature of 550°C by magnetron sputter epitaxy on Si (001) and studied by scanning tunnelling microscopy (STM). The shape changes induced by exposure to a Si-flux at 450°C were investigated as a function of the Si-coverage. During Si-capping, multifaceted domes were found to flatten and to transform into {105}-faceted pyramids and subsequently into stepped mounds through intermediate shapes. The observed sequence of morphological changes is induced by Si-Ge intermixing and is shown to be the inverse of that occurring during Ge or $Si_{1-x}Ge_x$ growth on Si (001). The results are interpreted with a model in which the stable shape of an island mainly depends on its volume and composition.

INTRODUCTION

In a wide range of growth parameters, the epitaxial deposition of Ge on Si (001) leads to the spontaneous formation of three-dimensional islands which allow to partially relieve the strain due to the ~4.2% lattice-mismatch. By embedding these islands in a Si matrix, they can be converted into self-assembled quantum dots [1]. The knowledge of possible shape and composition changes taking place during the Si-capping is essential to produce structures with predictable properties.

Several previous investigations [1-3] on Ge and $Si_{1-x}Ge_x$ islands overgrown with Si pointed out that the island shape is affected during the capping process. Buried islands are generally flatter than the as-grown ones, but, particularly for dome-shaped islands, the details of the shape changes have been only recently reported [4].

We present here an STM investigation of the shape evolution of Ge/Si (001) islands during Si-capping and a comparison with that occurring during Ge and SiGe deposition on Si (001).

EXPERIMENTAL DETAILS

The specimens used in this study were prepared by ultra high vacuum (UHV) magnetron sputter epitaxy. p^+-Si (001) substrates were outgassed in UHV and then flash-cleaned by direct current heating to remove the native oxide. A 60 nm-thick Si buffer was grown to get a reproducibly clean surface. Ge islands were obtained by depositing Ge at a substrate temperature T_s=550°C and a rate of 0.3 ML/s. Samples were cooled to 450°C before depositing the Si cap layer and eventually to room-temperature (RT) for the STM characterization. We chose the capping temperature such as to avoid appreciable island ripening and possible intermixing with the Si substrate. In fact, while samples annealed for several minutes at T_s=450°C show no significant change of island-size-distribution with respect to those cooled to RT immediately after growth, the volume of islands annealed at 550°C clearly increases. Thus, at 450°C, islands can be

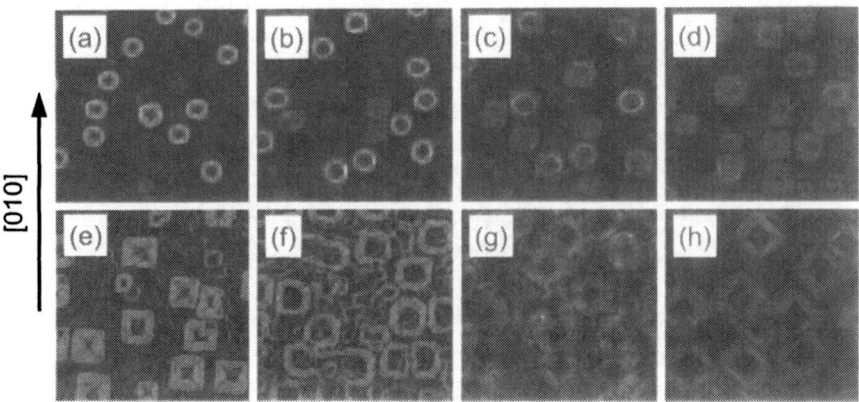

Figure 1. Shape evolution of Ge islands (obtained by depositing 7 ML of Ge on Si (001) at T_s=550°C and a rate of 0.3 ML/s) during overgrowth with Si. 450×450 nm^2 STM topographs of an uncapped sample (a) and samples overgrown with 0.5, 1, 2, 4, 8, 16, 32 ML of Si (b)-(h). The grey-scale corresponds to the local surface slope (normalized for each image): steeper facets appear darker.

considered as almost closed systems and their evolution during capping could provide information on the stability of individual islands.

RESULTS AND DISCUSSION

The details of the shape evolution of the Ge islands during exposure to a Si flux were investigated by producing eight samples with different thickness of the Si cap layer while keeping the other growth parameters fixed. Several uncapped specimens were grown as well. They served as reference and as a means to evaluate the reproducibility of sample features obtained under the same nominal conditions. Figure 1(a) displays an STM topograph of a typical uncapped sample obtained by depositing 7 ML of Ge on Si (001). Islands are square based pyramids and domes. The former are bounded by four {105} facets while the latter exhibit also steeper facets ({15 3 23} and {113}), and have larger volume and height-to-base-diameter ratio (aspect ratio). {105} facets are observed at the apex and at the foot of the domes.

Figures 1(b)-(h) illustrate the effect of the Si capping on the shape of the Ge islands. For domes, appreciable shape changes are noticeable at a Si coverage as low as Q=0.5 ML [Fig. 1(b)], with the {105} facets expanding and the steep facets shrinking. This trend is even more pronounced at Q=1 ML [Fig. 1(c)] where domes have been transformed to islands with a shape approaching that of {105} pyramids. In the following we will refer to islands with such a shape as "T-domes". The steep facets disappear almost completely at Q=2 ML [Fig. 1(d)]. Subsequently, the pyramid apices are gradually replaced by (001) facets at Q=4 ML [Fig. 1(e)]. The top (001) facets expand and the edges between adjacent {105} facets are replaced by steps parallel to the 110 directions at Q=8 ML [Fig. 1(f)]. We will call the islands with this shape "T-pyramids". The {105} facets disappear completely at Q=16 ML [Fig. 1(g)].

The shape changes for pyramids and domes are accompanied by a continuous flattening (Fig. 2) due to height reduction and base expansion [4]. For Q 4 ML, the average dome height

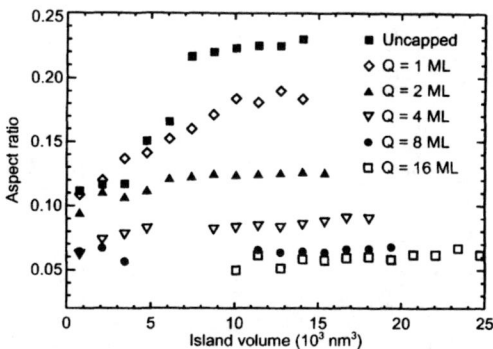

Figure 2. Aspect ratio as a function of island volume for Ge/Si (001) islands overgrown with the indicated amount Q of Si. Each point represents an average over several islands.

decreases by about 8 Å per each Å of deposited Si, clearly indicating the occurrence of atom rearrangement during exposure to a Si flux. The process mainly responsible for this behaviour is probably Ge surface segregation [5], allowing Si to penetrate the sub-surface layers of the islands. In turn, Si incorporation leads to a local strain relief destabilizing the facets and making Ge atoms diffuse from the top towards the base of the island. A similar mechanism has been proposed by Joyce *et al.* [6] to explain the observed shape changes of InAs islands during capping with GaAs. From Fig. 2 a significant volume increase is noticeable as the capping proceeds due to Si incorporation. Kinetic limitations prevent the complete alloying of the islands and their consequent dissolution, allowing to encapsulate the intermixed dots under a Si layer, as shown in Fig. 1(h) for Q=32 ML.

Interestingly, the shape evolution described above appears as the exact inverse of that previously reported for Ge and SiGe islands during growth or post-growth annealing [7-11]. In fact, both Ge and SiGe islands on Si (001) are known to undergo several shape transitions as their volume exceeds certain critical values V_c, although the mechanisms responsible for such transitions are still under debate [12]. As illustrated in Fig. 3, they evolve from shallow stepped mounds [Fig. 3(a), also called prepyramids [11]] to "T-pyramids" [Fig. 3(b)], truncated pyramids [Fig. 3(c)], {105}-pyramids [Fig. 3(d)], "T-domes" [Fig. 3(e)] and finally to domes [Fig. 3(f)].

Prepyramids, pyramids and domes are likely to be thermodynamically stable shapes [15, 16] in three different volume ranges. During growth, or post-growth annealing, islands increase their volume undergoing shape transitions involving intermediate (and often asymmetric [10]) shapes.

Islands shown in Figs. 3(a)-(c) and in Figs. 3(d)-(f) were obtained by depositing on Si (001) 14 ML of $Si_{0.5}Ge_{0.5}$/Si (001) at a substrate temperature T_s=600°C, and 7 ML of Ge at 550°C, respectively. In the $Si_{1-x}Ge_x$/Si system the lattice mismatch is proportional to x making the linear dimensions (height and base side) of the islands approximatively proportional to x^{-2} [13]. Hence we assume the critical volumes to scale as $V_c(x) \sim V_c(1)/x^6$, where $V_c(1)$ are the measured critical volumes for pure Ge islands on Si. We estimate $V_c(1)$ to be about 100 and 5000 nm³ for the prepyramid-pyramid [11] and for the pyramid-dome [4] transitions. The predicted $V_c(x)$ are compatible with the critical size reported in [10] for the pyramid-dome transition (x=0.3) and with the V_c we found for the prepyramid-pyramid transition for x=0.5.

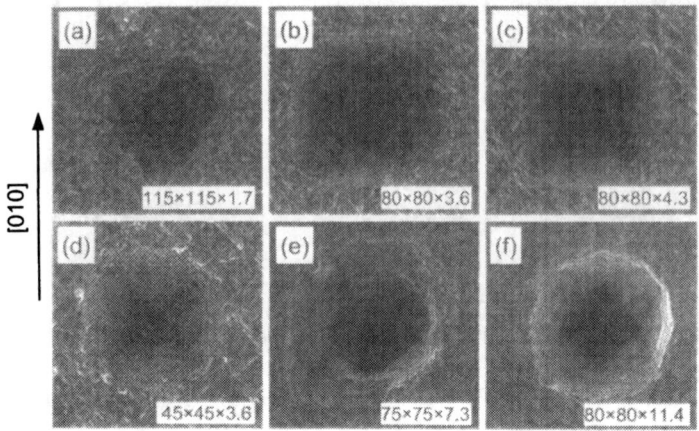

$[010]$

Figure 3. Island evolution during growth. (a) Prepyramid, (b) "T-pyramid" (c) truncated pyramid, (d) pyramid, (e) "T-dome", (f) Dome. (a)-(c): 14 ML of $Si_{0.5}Ge_{0.5}$/Si (001); (d)-(f) 7 ML of Ge/Si (001). Size of the STM topographs is indicated in nanometers.

We explain the observed *reverse evolution* of Ge domes during exposure to a Si flux by the following simple model: Si-Ge alloying occurs during Si overgrowth of Ge islands at T_s=450°C; hence the average Ge fraction x inside a dome decreases making the critical volume V_c for the pyramid-dome transition increase proportionally to x^{-6}; at the same time the island volume increases, but at a much lower rate, because of the incorporated Si; thus $V_c(x)$ will sooner or later exceed the dome volume rendering the pyramid shape more favourable. In this way all the domes can be converted to pyramids by providing the suitable amount of Si (slightly exceeding 2 ML). A similar argument was first used by Kamins et al. [14] to explain the dome-to-pyramid transition observed during annealing at 650°C. In that case, the high temperature allowed the islands to intermix with the Si substrate. As the Si overgrowth proceeds pyramids undergo a second shape change as their volume drops below the composition-dependent critical volume for the prepyramid-pyramid transition. In order to compare our data with this model, we estimated the island composition during Si capping by simply assuming that just the Si impinging on the island is incorporated. This is supported by the observation that a dense array of Ge islands

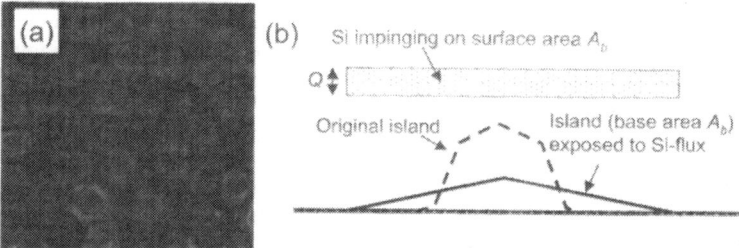

Figure 4. (a) Dense array of Ge islands (produced by depositing 8 ML of Ge at a rate of 1ML/s at a substrate temperature T_s= 550°C) overgrown with 2 ML of Si at T_s= 450°C. Other data as in Fig. 1. (b) Sketch illustrating how the volume of the incorporated Si was estimated.

capped with 2 ML of Si [Fig. 4(a)] qualitative by behaves as the "dilute" ensemble shown in [Fig. 1(d)]. With reference to Fig. 4(b) the volume of Si incorporated can be estimated as the product of the equivalent thickness Q (e.g. 2 ML) times the measured base area of the island. By assuming that the uncapped islands are composed of pure Ge (for a discussion on this point see Ref. [4]), the Ge fraction x of an island capped with an amount Q of Si is simply given by $x \sim (1 - A_b \times Q/V)$, where A_b and V are the measured base area and volume of the island.

Figure 5 shows the calculated average composition x as a function of the island volume for uncapped (x=1) and capped (x<1) samples. The $V_c(x)$ curves divide the V-x plane in three regions, in which prepyramids, pyramids and domes are separately stable. According to this picture, transitions from one shape to another gradually occur when island volume and/or composition are allowed to change. Our data are in qualitative agreement with the simple model presented above, in spite of the uncertainties in the composition determination. Besides, the island composition during Si capping is likely to be inhomogeneous, with a Ge-rich core and Si-rich surface layers.

CONCLUSIONS

We have presented a detailed analysis of the shape evolution of Ge/Si (001) islands during exposure to a Si flux. The comparison between this evolution and that previously reported for growth of Ge islands reveals that the former is the inverse of the latter.

ACKNOWLEDGMENTS

The authors would like to thank P. Raiteri, L. Miglio, J. Tersoff, A. Madhukar, W. Seifert and E. Müller for fruitful discussions and suggestions and M. Döbeli for Rutherford backscattering measurements.

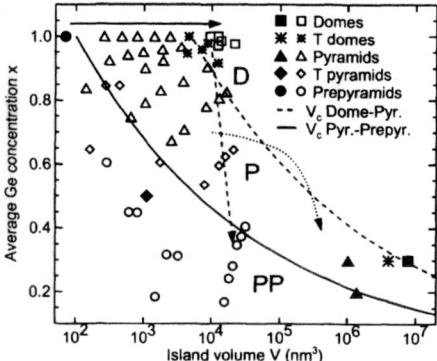

Figure 5. Island shape as a function of volume and estimated average composition. Filled symbols refer to typical islands observed in previous experiments [7-11]. Curves represent the critical volumes $V_c(x)$ for the prepyramid-pyramid (solid) and pyramid-dome (dashed) transitions and separate the regions in which the stable shape is the dome (D), pyramid (P) and prepyramid (PP). Arrows show schematic pathways followed during growth (solid), Si overgrowth (dashed) and post-growth annealing (dotted).

REFERENCES

1. O.G. Schmidt *et al., Appl. Phys. Lett.* **77,** 2509 (2000).
2. P. Sutter and M.G. Lagally, *Phys. Rev. Lett.* **81,** 3471 (1998).
3. M. Kummer, B. Vögeli, and H. von Känel, *Mat. Sci. and Eng.* **B69,** 247 (2000).
4. A. Rastelli, M. Kummer, H. von Känel, *Phys. Rev. Lett.* (in press).
5. K. Nakagawa and M. Miyao, *J. Appl. Phys.* **69,** 3058 (1990).
6. P.B. Joyce, T.J. Krzyzewski, P.H. Steans, G.R. Bell, J.H. Neave, and T.S. Jones, *Surf. Sci.* **492,** 345 (2001).
7. J.A. Floro *et al., Phys. Rev.* **B59,** 1990 (1999).
8. K.M. Chen *et al., Phys. Rev.* **B56,** 1700 (1997).
9. G. Medeiros-Ribeiro *et al., Science* **279,** 353 (1998).
10. F.M. Ross, R.M. Tromp, and M.C. Reuter, *Science* **286,** 1931(1999).
11. A. Vailionis *et al., Phys. Rev. Lett.* **85,** 3672 (2000).
12. I. Daruka, J. Tersoff, and A.L. Barabási, *Phys. Rev. Lett.* **82,** 2753 (1999).
13. W. Dorsch *et al., Appl. Phys. Lett.* **72,** 179 (1998).
14. T.I. Kamins *et. al., Appl. Phys.* **A67,** 727 (1998).
15. F.M. Ross, J. Tersoff, and R.M. Tromp, *Phys. Rev. Lett.* **80,** 984 (1998).
16. P. Raiteri *et al.* (in press).

Mat. Res. Soc. Symp. Proc. Vol. 707 © 2002 Materials Research Society

Shape Transition in Self-Organized InAs/InP Nanostructures

H.R. Gutiérrez, M.A. Cotta, M.M.G. de Carvalho.
Instituto de Física Gleb Wataghin, DFA/LPD, UNICAMP, CP 6165, 13081-970 Campinas-SP, Brazil.

ABSTRACT

In this letter we report the transition from self-assembled InAs quantum-wires to quantum-dots grown on (100) InP substrates. This transition is obtained when the wires are annealed at the growth temperature. Our results suggest that the quantum-wires are a metastable shape originated from the anisotropic diffusion over the InP buffer layer during the formation of the first InAs monolayer. The wires evolve to a more stable shape (dot) during the annealing. The driving force for the transition is associated with variations in the elastic energy and hence in the chemical potential produced by height fluctuations along the wire. The regions along the wires with no height variations are more stable allowing the formation of complex, self-assembled nanostructures such as dots interconnected by wires.

INTRODUCTION

InAs nanostructures in an InP matrix (mismatch ~ 3.2%) have received much attention in the last years. Their potential application in light-emitting devices in the 1.3-1.55 μm wavelength range makes it necessary to understand and control the formation of these structures. Both InAs wires and dots grown on InP substrates with random distribution have been reported[1-3]. Recent results suggest that the surface morphology[2] and/or the chemical composition[4] of the buffer layer (InP, InAlAs, or InGaAs) determine the final shape of the InAs nanostructures. Nevertheless, the conditions and mechanisms that originate each kind of nanostructure are not clear. Dots have been shown to form only on smooth surfaces, with no anisotropies in the buffer layer morphology. When this was not the case, InAs wires were obtained[2]. However, in other work InAs dots were obtained for both smooth (2D) as well as highly anisotropic in shape, rippled buffer layers[1].

In this letter we report the transition from self-assembled InAs quantum-wires (QWr) to quantum-dots (QD) grown on (001) InP substrates. We observe that this shape transition is obtained when the QWr's are annealed at the growth temperature, for both smooth and rippled InP buffer layer surfaces. Our results suggest that the QWr is a metastable shape resulting from anisotropic diffusion over the InP buffer layer during the formation of the first InAs monolayer. The wires evolve to a more stable shape (dot) during annealing.

EXPERIMENTAL DETAILS

The samples were grown by Chemical Beam Epitaxy (CBE) using trimethylindium (TMI) diluted with H_2 carrier gas as the group III source and thermally decomposed phosphine (PH_3) and arsine (AsH_3) as group-V sources. The substrate native oxides were removed by heating the substrate for 10 minutes at 535°C in the growth chamber under P_2 overpressure. Two set of samples were prepared, the first consisting of 3 ML of InAs deposited on an InP buffer layer at

490°C and 0.25 Å/s. The second set was prepared under the same conditions but, after growth, the samples were annealed at the growth temperature for 1 minute in As_2 overpressure. The effect of surface configuration (terraces and steps) on the nanostructure shape was investigated by growing the InAs films simultaneously on (001) InP substrates both nominally-oriented and misoriented by 2° towards [011] (A surface) and [101] (C surface). Reflection High-Energy Electron Diffraction (RHEED) was carried out *in situ* on the growing surfaces. After removal from the growth chamber, in-air Atomic Force Microscopy (AFM) was used to analyze the surface morphologies of the samples. A high-resolution C-nanotube tip was used in non-contact mode. Using a JEM 3010 URP 300 kV Transmission Electron Microscope (HRTEM) cross section images of the nanostructures in different crystalline directions were obtained.

RESULTS AND DISCUSSION

Figures 1a, b and c show the AFM images corresponding to the first set of samples grown over nominal, A-type and C-type InP substrates, respectively. The InAs films were grown on top of an InP buffer layer with smooth morphology. In all three cases the InAs self-assembled nanostructures present a wire shape with axis oriented along the [1 $\bar{1}$ 0] direction. The wire period ranges between 150-400Å with an average value of 250Å. The average height is about 10Å and 30Å for QWr's grown on nominal and vicinal substrates, respectively. These results indicate that the step profile and orientation can influence parameters such as height and homogeneity of the InAs QWr's; however, the wire-like form of the nanostructures is not affected.

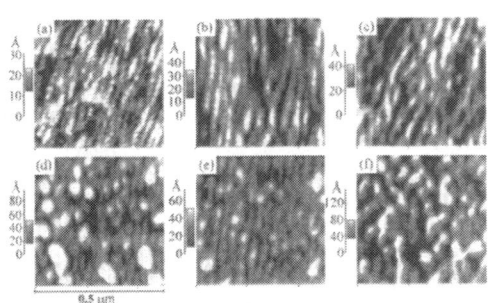

Figure 1. AFM images of 3ML InAs films grown over nominal (001) (a and d), A-type (b and e) and C-type (d and f) InP substrates. a, b and c are unannealed samples. d, e and f are samples annealed for 1 min. at 490°C.

Figure 2. Histograms (upper set) and Radius vs Height distributions (lower set) of the quantum dots formed after annealing for nominal (001), A and C-type InP substrates.

The morphology observed for the second set of samples (Fig.1 d-f) was completely different. Dots with densities around 3 - 4 x10^{10} cm^{-2} can be observed on the surface. A shape transition occurs during annealing, from QWr to QD nanostructures. The radius distribution and the height vs radius plots for the self-assembled QD's are shown in Figure 2. The higher dispersion in the height distribution for the C-type sample may be associated with the larger roughness of the buffer layer frequently obtained in this kind of substrate. Note that for all the samples the maximum of the radius distribution remains closer to 250Å, the same values of the QWr average period. On the other hand, for A-type substrates the QD's of the second set of samples seem to be

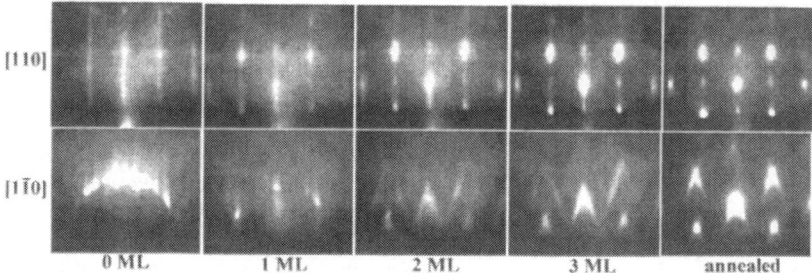

[110]

[1 1̄ 0]

0 ML   1 ML   2 ML   3 ML   annealed

Figure 3. RHEED patterns for electron beam incidence parallel to [110] and [1 1̄ 0] directions for different InAs growth times; in this case the InP substrate is vicinal A-type.

aligned in rows along the [1 1̄ 0] direction (see Figure 1e). The rows of dots have the same density along the [110] direction as the wires on the unannealed samples. These results suggest that there is no significant exchange of mass between neighboring wires during the transition. That is, the reassembling of the nanostructure shape occurs due to mass transport along the wire. For longer annealing times the islands continue to grow.

We monitored in real time the RHEED patterns during the formation of the InAs nanowires. Figure 3 shows the RHEED patterns along [110] and [1 1̄ 0] azimuths for different InAs growth times. During QWr's formation the [110] azimuth RHEED pattern gradually changes from streaks along the [001] direction, corresponding to a flat surface, to elongated spots on the streaks corresponding to the transmission-diffraction from the wires. However, for the [1 1̄ 0] azimuth both 3D transmission and chevron features are observed. After annealing (island formation) the RHEED streaks in [110] azimuth disappear, with only the symmetrical spots remaining. For the [1 1̄ 0] azimuth the chevron edge becomes brighter and streaks shorten and lose intensity.

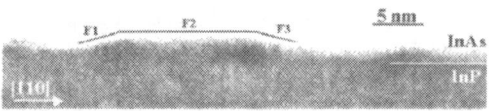

Figure 4. HRTEM image of the InAs nanowire cross-section showing three different facets F1, F2 and F3 corresponding to (1̄ 1̄ 4), (001) and (114) facets, respectively. The [1 1̄ 0] azimuth is perpendicular to the Figure plane.

The striking coincidence between the chevron streaks orientation angles (reciprocal space) and the facets angles (direct space) is an experimental fact commonly accepted as a method to identify the presence of facets in three-dimensional nanostructures[6,7]. In our experiment the angle between chevron streaks and the [001] direction is approximately $20°$, corresponding to [114] and [1̄ 1̄ 4] directions perpendicular to (114) and (1̄ 1̄ 4) facets. Figure 4 shows a HRTEM cross-section of the InAs nanowires along the [110] direction. F1, F2 and F3 are the (114), (001) and (1̄ 1̄ 4) facets, respectively. HRTEM of the islands (not shown) reveals the same (114) and (1̄ 1̄ 4) facets along the [110] direction.

The form of the chevron streaks and the angle between them change noticeably during the formation of the InAs QWr's. Figure 5 shows the time dependence of the angle with respect to the [001] direction for the three streaks forming the chevron. Insets at the top of the Figure show the real shape of the chevron structure for different growth times.

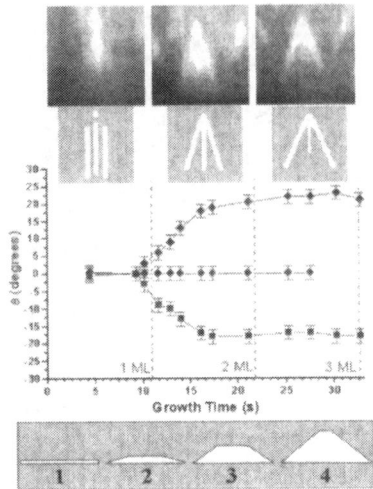

Figure 5. Temporal evolution of the chevron streaks angles with respect to the [001] direction. This graphic corresponds to the RHEED shown in Figure 3. Insets at the top of the Figure show the actual RHEED chevron structure for different growth times and the schematic chevron shape. The schematic at the bottom of the Figure shows the evolution of the InAs wires cross section inferred from the graphic. Both growth time and InAs volume increase from 1 to 4.

The schematic chevron shape is also shown. For early growth times the chevron is composed of three streaks with little or no difference in the orientation. When 0.8 ML were deposited the angle of the left and right streaks begin to increase, saturating at approximately -20° and 20°, respectively, for 1.5 ML. The central streak does not change in orientation. Above 1.5 ML and up to 3 ML the intensity of the chevron central streak, corresponding to the (001) facet, begins to decrease and disappears when approximately 2.6 ML were deposited, suggesting that the area of the top (001) facet decreases during growth. This shape change occurs slowly with no modification of the facet angles. These results suggest that facets begin to form at 0.8 ML and the gradual increase in facets angles indicates that the growth rate at the center of the nanostructure is higher than those at the edges. This will occur if a flux of adatoms (along [110]) from the edges to the center of the initial InAs structure exists. This reasoning is in agreement with the self-assembling mechanism proposed by Barabási[8] for the initial states of island formation. Strain energy on the initial islands produces a chemical potential gradient that generates a net current of adatoms to the top of the nanostructure[8], as described above. The wire shape of the initial island (Figure 6) and the variation of facets oriented parallel to the wires make a one-dimensional treatment of the problem possible. The height of these initial islands is 1 - 2 ML. The schematics at the bottom of Figure 5 represent the evolution of the InAs wire cross section as expected from the chevron streaks behavior. From 1 to 3 a gradual increase of the lateral facet angle occurs; when the facet angle stabilizes (3) the area of the (001) facet begins to decrease (4).

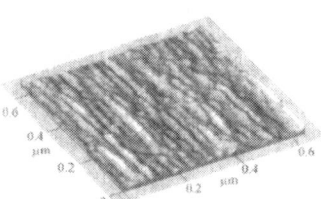

Figure 6. AFM images of 1 ML of InAs deposited on a flat InP surfaces (A-type substrate) showing the initial elongated islands. These islands act as a template for the InAs QWr's formation.

Earlier works have reported that homoepitaxial InP rippled structures with periods about 100-200 nm can be grown at particular grown conditions[9]. InAs films were grown on such InP rippled surfaces using the same growth conditions as for samples shown in Figure 1. In particular, for C-type substrates the InP ripples in the buffer layer form at ~45° off the [1 1̄ 0] direction. Even in this case, and for all rippled InP surfaces, the InAs wires formed aligned in the [1 1̄ 0] direction (Figure 7 a, c) showing that surface morphology does not affect their formation. If InAs is obtained by As/P exchange during the exposure of the InP rippled surfaces to an As_2 flux, then only dots (and no wires) are obtained (Figure 7 b, d). This

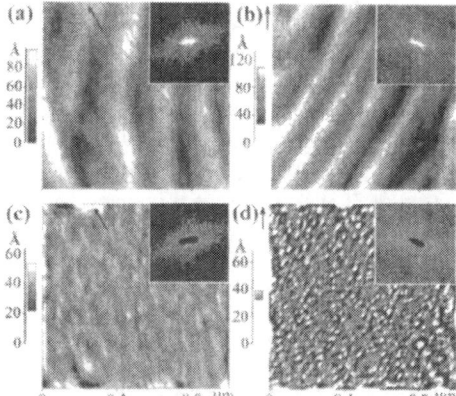

Figure 7. AFM images of: (a) 3 ML of InAs deposited on a rippled InP surfaces (C-type substrate); (b) InAs islands formed by the As/P exchange during the exposure of InP ripples to an As$_2$ flux. The FFT filtering of the above images eliminate the lower frequencies (white disk at the center of the FFT spectrum in the inset of Figures (a) and (b)) corresponding to the InP ripples morphology. (c) and (d) are the FFT filtered images of (a) and (b), respectively. The arrow indicates the [1 1̄ 0] direction.

suggests that the kinetics of adatoms strongly influences the InAs QWr's formation.

In our experiments the 2x4 reconstruction was observed by RHEED on the InP buffer layer surface before the InAs growth. A similar 2x4 reconstruction is observed for As-stabilized GaAs (001) surfaces[10]. Shiraishi[11] calculated, using first principles (pseudopotential) methods, that the activation barrier for adatom diffusion through the GaAs missing dimer rows is three times lower than in the perpendicular direction. For the InP surface the missing dimer rows are parallel to the [1 1̄ 0] direction[12]. Therefore, a higher diffusion along this direction may be expected to produce the nucleation of elongated InAs islands during deposition of the first InAs monolayer, leading to the QWr formation.

The InAs QWr's in Figure 1 show height variations of ± 1 ML along the [1 1̄ 0] direction (Figure 8a). However, in the perpendicular direction these height variations are at least three times higher. A schematic diagram for the chemical potential in both directions is shown in Figure 8b. During annealing, mobile surface atoms can move over the surface biased by the chemical potential gradients. There are minima of potential along the [110] direction over each wire. These minima may explain the low mass exchange between wires, since atoms moving on the surface remain confined in the low potential

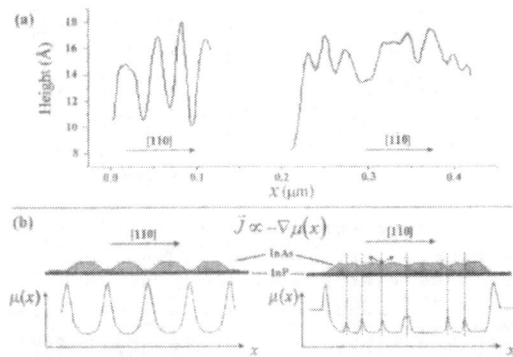

Figure 8. (a) Actual profiles of the InAs wires measured by AFM in the [110] and [1 1̄ 0] directions with a high-resolution C-nanotube tip; (b) Schematic showing the chemical potential profiles after the wire formation.

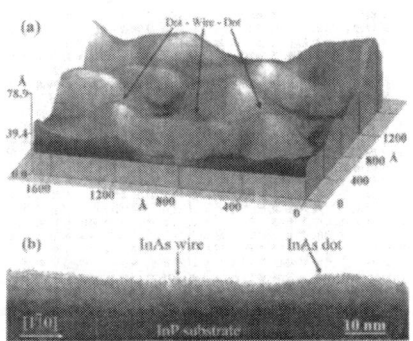

Figure 9. (a) AFM image of an InAs DWD nanostructure formed along the [1 1̄ 0] direction after annealing (1 min.). (b) TEM image of the same sample.

regions (over the wires). On the other hand, for the [1 $\overline{1}$ 0] direction (along the wire), height fluctuations produce local potential maxima that represent an unstable position for any mobile atom on the surface. These maxima correspond to sites with lower heights along the wire. Consequently a net current is generated to the higher sites, further increasing its height. The wire thus becomes more discontinuous along the [1 $\overline{1}$ 0] direction during annealing, leading to the QD's formation. Consequently, we propose that the driving force inducing the wire-dot shape transition is related to the height variations along the wire. If our description is correct only the wires with height fluctuations will be unstable; regions with a constant height will remain stable after annealing. The process results in a more complex nanostructure formed by dots interconnected by a wire. This kind of self-assembled nanostructure was indeed observed in our experiment as shown in Figure 9. For higher growth temperatures we have obtained InAs QWr's more uniform in height and period, making the wire-dot transition more controllable. This will be the subject of another work.

CONCLUSIONS

In summary, we have shown a shape transition from InAs QWr's to QD's for the InAs/InP system. Our results suggest that QWr's are a metastable shape that can evolve to a more stable shape (QD's) during annealing. Neither InP substrate miscut nor the InP surface morphology affected the transition. Based on non-equilibrium mechanisms we conclude that the driving force for this transition is the chemical potential gradient produced by height fluctuations along the wire.

ACKNOWLEDGMENTS

One of the authors (H.R.Gutiérrez) acknowledges financial support from CNPq and FAPESP. This work was supported by FAPESP, CNPq and FINEP. The HRTEM measurements were made at the LME, National Laboratory of Synchrotron Light (LNLS)(Brazil).

REFERENCES

1. C.A.C. Mendonça, M.A. Cotta, E.A. Meneses and M.M.G Carvalho. *Phys.Rev.* B **57**, 12501 (1998).
2. L.Gonzalez, J.M. Garcia, R. Garcia, F. Briones, J. Martinez-Pastor and C. Ballesteros. *Appl. Phys. Lett.* **76**, 1104 (2000).
3. C. Walther, W. Hoerstel, H. Niehus, J.Erxmeyer and W.T. Masselink. *J. Cryst. Growth* **209**, 572 (2000).
4. J. Wu, Y.P. Zeng, Z.Z. Sun, F. Lin, B. Xu and Z.G. Wang. *J. Cryst. Growth* **219**, 180 (2000).
5. J. Brault, M. Gendry, G. Grenet and G. Hollinger. *Appl. Phys. Lett.* **73**, 2932 (1998).
6. Y. Nabetani, T. Ishikawa, S. Noda and A. Sasaki. *J. Appl. Phys.* **76**, 347 (1994).
7. D.W. Pashley, J.H. Neave and B.A. Joyce. *Surf. Sci.* **476**, 35 (2001).
8. A.L. Barabási. *Appl. Phys. Lett.* **70**, 2565 (1997).
9. M.A. Cotta, R.A. Hamm, T.W. Staley, S.N.G. Chu, L.R. Harriott, M.B. Panish and H. Temkin. *Phys.Rev.Lett.* **70**, 4106 (1993).
10. M.D. Pashley, K.W.Haberern and J.M. Gaines. *J. Vac. Sci. Technol.* B **9**, 938 (1991).
11. K. Shiraishi. *Appl. Phys. Lett.* **60**, 1363 (1992).
12. Q. Guo, M.E. Pemble, E.M. Williams. *Surf. Sci.* **433**, 410 (1999).

Mat. Res. Soc. Symp. Proc. Vol. 707 © 2002 Materials Research Society   N5.10

Morphologies of Self-Assembled Quantum Dots: A Variational Approach

R. Arief Budiman and Harry E. Ruda

University of Toronto, Toronto, Ontario M5S 3E3, Canada

Abstract

We construct a 3D model for coherent island formation by (i) using a novel 3D strain tensor to account for bulk strains and (ii) representing adatom diffusion as an external field that perturbs an otherwise flat strained layer. Equilibrium shapes of coherent islands and wetting layer thickness are obtained. Coherently compressed layers are typically unstable, but become stable in tension. Comparisons with $Si_{1-x}Ge_x/Si(001)$ and $Si_{0.5}Ge_{0.5}/Si_{1-x}Ge_x(001)$ layers are discussed.

Introduction

Nanometer-size 3D coherent islands form spontaneously during strained layer deposition of several monolayers (ML) thick. Applications of these self-assembled quantum dots hinge on our ability to predict and obtain a uniform island size distribution [1]. For this reason, several fundamental issues on the nature of coherent islanding transition need to be clarified. Two such issues revolve around the physics of wetting layer thickness. What governs its thickness? Why does it coexist with coherent islands? In this paper, a simple answer to the first issue is offered by our 3D model: in the limit of zero shear strains, flat wetting layer thickness is determined by the force balance between surface and interface tension, and the surface stress from (vertical) tetragonal strain.

The basic idea of our single-component model is that shape transition that gives rise to the 3D islands must be accompanied by the emergence of bulk shear strains. Several experiments [2, 3, 4, 5] have consistently verified this notion. Our formulation allows for the tetragonal strain to change so that plane strain condition ($u_{zz} = 0$, or $h_z = 0$ as later defined) is no longer needed. The removal of plane strain assumption is crucial, since the resulting force equilibrium that includes the tetragonal strain yields the wetting layer thickness. In addition, shapes of the 3D islands are predicted from the accumulation of shear and tetragonal strains throughout the layer thickness.

The model also addresses three additional issues. First, the coexistence of a wetting layer and coherent islands is explained by coupled nonlocal shear and longitudinal surface force equilibria along the growth direction. The nonlocality arises from the accumulation of bulk stresses with layer thickness, which defines the elastic surface stresses. Second, the layer stability against shear strains, thus islanding, depends on the sign of biaxial misfit strain. Coherently compressed layers are typically unstable above their wetting layer thickness, but become stable in tension. Third, the nanometer-size property is shown to arise from the exponential decay length $\approx (\gamma_s/\mu)$ of bulk shear strains with thickness, where γ_s and μ are surface tension and shear modulus, respectively. The exponential decay of shear strains also explains shape metastability of coherent islands.

3D Model

For sufficiently thin strained layers, a 3D deformation map: $(x, y, z) \rightarrow (x+u_x(x,y), y+u_y(x,y), z+h(x,y,z))$, can be defined to obtain a 3D nonlinear strain tensor u_{ij}. The deformation map demands that only the deformation component along the z-coordinate, h, depend on z, since for such layers, the lateral strains throughout the layer thickness are pinned by the biaxial misfit strain, ε_f. For this reason, the lateral deformations, u_x and u_y, depend only on the lateral coordinates (x, y). With an unstrained layer as the reference state, we obtain u_{ij} by computing the distance change due to the deformation map, $d^2r = d^2r_o + 2u_{ij}\, dx\, dy$ [6]:

$$
u_{ij} = \frac{1}{2}
\begin{bmatrix}
2u_{x,x} + (h_x)^2 & 2u_{x,y} + h_x h_y & h_x \\
2u_{x,y} + h_x h_y & 2u_{y,y} + (h_y)^2 & h_y \\
h_x & h_y & 2h_z
\end{bmatrix},
\tag{1}
$$

where the subscript after each comma denotes differentiation. The lateral strain pinning is done by equating $u_{x,x}$ and $u_{y,y}$ to the biaxial misfit strain: $u_{x,x} = u_{y,y} = \varepsilon_f$, with $u_{x,y} = u_{y,x} = 0$.

The elastic free energy density is $f_E = \frac{\lambda}{2}u_{ii}{}^2 + \mu u_{ij}{}^2$, where the repeated index summation convention is used, so that the isotropic elastic free energy is $F_E = \int f_E\, dV$:

$$
F_E = \int dx\, dy \int_0^\theta dz \left[f_1 + f_2\{h_x{}^2 + h_y{}^2\} + f_3\{h_x{}^2 + h_y{}^2\}^2 \right],
\tag{2}
$$

where λ and μ are strained layer's Lamé constants; $f_1 = 2(\lambda + \mu)\varepsilon_f{}^2 + 2\lambda\varepsilon_f h_z + \frac{1}{2}(\lambda + 2\mu)h_z{}^2$; $f_2 = \frac{1}{2}\mu + (\lambda + \mu)\varepsilon_f + \frac{1}{2}\lambda h_z$; $f_3 = \frac{1}{8}(\lambda + 2\mu)$; and θ is layer thickness. The substrate occupies $-\infty < z < 0$, so that the lower limit is set at the substrate-layer interface $(z = 0)$.

Although a coherent, flat interface is imposed, the substrate is not rigid as it absorbs a constant fraction, ε_s, of the total misfit strain, $\varepsilon_m = (a_s - a_f)/a_f$, using the compatibility condition at the coherent interface: $\varepsilon_f - \varepsilon_s = \varepsilon_m$ [7], where a_f and a_s are lattice constants of layer and substrate, respectively. The compatibility condition produces $\varepsilon_f = \varepsilon_m M_s a_s/[M a_f + M_s a_s]$, and $\varepsilon_s = -\varepsilon_m M a_f/[M a_f + M_s a_s]$, where ν and $M = 2\mu(1+\nu)/(1-\nu)$ denote layer's Poisson's ratio and biaxial modulus, while the subscript s is of substrate. For rigid substrates: $M_s \to \infty$, so that $\varepsilon_f \to \varepsilon_m$. The nature of whether the substrate is rigid or elastic cannot be determined from our model; it is a model's parameter and must be determined from microscopic considerations. We shall assume an elastic substrate throughout this paper.

Isotropic surface and interface tension are defined with respect to flat, low-index surface and substrate-layer interface, respectively. While the surface tension γ_s is proportional to the number of dangling bonds, the interface tension γ_i is the energy cost to deposit material per unit interface area: $\gamma_i = \gamma_{sub} - 2[E_{bond} - E_{coh}]/A$, where E_{bond} is the bond strength across the interface and A is the unit interface area of a particular orientation. The cohesive energy E_{coh} of the layer serves as a reference for the bond energy. An unstrained unit volume dV changes into $dV' = (1 + u_{ii})dV$ during deformations; thus, the strained surface area is $dA' = \int (1 + u_{ii})\delta(z - \theta)\, dV$, where the Dirac-$\delta$ function collapses the unit volume to a strained areal measure at $z = \theta$. From Eq. (1), $u_{ii} = 2\varepsilon_f + h_z + \frac{1}{2}(h_x{}^2 + h_y{}^2)$, so that the surface and interface free energies become

$$
F_S + F_I = \int dx\, dy \int_0^\theta dz\, \{1 + 2\varepsilon_f + h_z + \frac{1}{2}(h_x{}^2 + h_y{}^2)\}\left[\gamma_s\delta(z - \theta) + \gamma_i\delta(z) \right],
\tag{3}
$$

where $\delta(z)$ ensures that γ_i is defined only at the interface.

Adatoms also contribute additional free energy due to their entropy, and their chemical bonds to the layer atoms [8]. We set the bonding part of the free energy to zero, since our zero energy is that of the unstrained layer; once adatoms are incorporated, their free energy is zero unless the layer is strained. We are thus left with their entropy, whose density is proportional to their local concentration $\vec{F} \cdot \vec{n}$, where $\vec{F} = -\frac{1}{4}F(1,1,1)$ is the hemispherically averaged deposition flux vector [9], F is the average deposition flux, and $\vec{n} = (-h_x, -h_y, 1)[1 + h_x{}^2 + h_y{}^2]^{-1/2}$ is the unit normal vector. The entropic contribution of the adatoms is thus given by

$$F_A \approx k_B T F \tau_d \int dx \, dy \int_0^\theta \frac{1 - (h_x{}^2 + h_y{}^2)^{1/2}}{\left[1 + h_x{}^2 + h_y{}^2\right]^{1/2}} \delta(z - \theta) \, dz, \qquad (4)$$

where $\delta(z - \theta)$ accounts for the fact that the adatoms only live on the layer surface. The mean diffusion time for adatoms, $\tau_d \approx 10^{-13} \exp(E_d/k_B T)$, measures their average lifetime before incorporated into the layer and appears as we consider only their steady-state density. E_d and T are adatom diffusion barrier and substrate temperature, respectively.

Order Parameter and Equilibrium Morphologies

Shape transition that produces coherent islands requires the shear strains along the growth direction, $\frac{1}{2}h_x$ and $\frac{1}{2}h_y$, to become nonzero. We thus conclude that $\vec{\phi} = \frac{1}{2}(h_x, h_y)$ is the vector order parameter for the islanding transition. To get analytical results, the average shear strain: $\phi \equiv \frac{1}{2}(h_x{}^2 + h_y{}^2)^{1/2}$ is used and the total free energy $F[\phi] = F_E + F_S + F_I + F_A$ can be written as a functional of ϕ. The variation in $F[\phi]$ with respect to ϕ yields the shear force equilibrium:

$$\frac{\delta F}{\delta \phi} \approx \int_0^\theta \left[\{2\mu + 4(\lambda + \mu)\varepsilon_f + 2\lambda h_z\} \phi + 4(\lambda + 2\mu)\phi^3 \right] dz + 2\phi(\theta)\gamma_s + k_B T F \tau_d = 0, \qquad (5)$$

where only the leading order term of $\delta F_A/\delta \phi$ is included, and the flat, coherent interface assumption: $\phi(z = 0) = 0$, is used. Eq. (5) describes a *nonlocal* shear force equilibrium as the surface shear force arises from the accumulation of shear strains throughout the layer thickness.

The other degree of freedom in the free energy is the tetragonal strain h_z. The variation in $F[\phi]$ with respect to h_z yields the force equilibrium due to the longitudinal strain h_z:

$$\frac{\delta F}{\delta h_z} = \int_0^\theta \left[2\lambda\varepsilon_f + (\lambda + 2\mu)h_z + 2\lambda\phi^2 \right] dz + \gamma_s + \gamma_i = 0, \qquad (6)$$

by substituting the expressions for f_1 and f_2. An equilibrium coexistence of a stable wetting layer and coherent islands will be obtained by solving the two coupled integral equations simultaneously. It was not possible, however, to do this. Nonetheless, as explained below, we may decouple these equations by imposing a mechanical stress equilibrium for Eq. (5) and a flat wetting layer assumption for Eq. (6). In doing so, we have restricted the solutions for Eqs. (5) and (6) to be valid only for coherent islands and a flat wetting layer, respectively.

The flat wetting layer assumption demands that $\phi = 0$ inside the wetting layer, so that Eq. (6) becomes

$$\int_0^{\theta_t} [2\lambda\varepsilon_f + (\lambda + 2\mu)h_z] \, dz = -(\gamma_s + \gamma_i), \qquad (7)$$

where θ_t is defined as the *transition thickness*, which is the flat layer thickness needed to balance forces from surface and interface tension. The total deformation from the tetragonal strain is proportional to the Poisson's ratio ν: $\int_0^{\theta_t} h_z \, dz \approx -2\nu\varepsilon_f\theta_t$, so that Eq. (7) gives:

$$\theta_t = -\frac{\gamma_s + \gamma_i}{2\varepsilon_f[\lambda - \nu(\lambda + 2\nu)]}. \qquad (8)$$

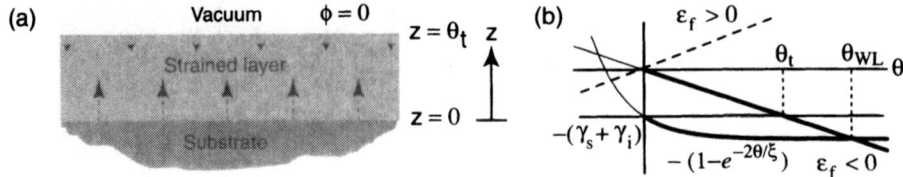

Figure 1: *Layers in compression.* (a) Tensile tetragonal strain from compressive misfit strain produces a force directed in the positive-z direction, which balances the compressive force from surface and interface tension. (b) Typically, we get $\gamma_s + \gamma_i > 0$. The exponential curve intersects the ε_f-line at larger $\theta > 0$ than the $(\gamma_s + \gamma_i)$-line does. For layers under compression, surface shear strains will increase wetting layer thickness from transition thickness θ_t value to θ_{WL}. *Layers in tension.* Forces from compressive tetragonal strain [shown as dashed lines in (a)] and from surface and interface tension are now directed in the same negative z-direction. The ε_f-line now has a positive slope, shown as a dashed line in (b); thus, it no longer intersects with the exponential curve, making $\theta_{WL} \to \infty$.

Thus, $\theta_t \propto -\varepsilon_f^{-1}$ and is *sensitive* to the sign of ε_f. $\theta_t > 0$ for compressive strains ($\varepsilon_f < 0$). The vertically compressive force from the surface and interface tension is thus balanced by the tensile force from tetragonal strain as shown in Fig. 1. In contrast, $\theta_t < 0$ for $\varepsilon_f > 0$, indicating that both surface-and-interface-tension and tetragonal stresses are directed in the negative z-direction. Layers under compression are thus unstable against shear deformations when $\theta > \theta_t$, while layers under tension are stable as there is no free surface below the interface. Homoepitaxial layers are stable since for $\varepsilon_f = 0$, $\theta_t \to \infty$.

The mechanical stress equilibrium along the growth direction dictates that the stress $\sigma_{zz} \approx 0$ due to the typical ultrahigh vacuum deposition condition. Thus, $\sigma_{zz} = \partial f_E / \partial u_{zz} = 2\lambda(\varepsilon_f + \phi^2) + (\lambda + 2\mu)h_z \approx 0$. Substituting this for h_z in Eq. (5) gives

$$\phi^2(z = \theta) \approx C^2 \exp[-2A(\theta - \theta_t)], \quad \theta \geq \theta_t, \tag{9}$$

where $C = -\frac{1}{2}k_B T F \tau_d / \gamma_s$, and $A = (\mu/\gamma_s)[1 + 2\varepsilon_f(3\lambda + 2\mu)/(\lambda + 2\mu)] \approx (\mu/\gamma_s)$, which gives the decay length for shear strains with layer thickness. Eq. (9) is valid only for shear strains inside coherent islands.

A more realistic situation to get the wetting layer thickness can be obtained by allowing for shear deformations to perturb a flat wetting layer, thus promoting an adatom exchange between wetting layer and coherent islands. This is done by a first-order iteration in solving the coupled integral equations, by using the solution of Eq. (7) to connect Eq. (5) to Eq. (6). The iteration yields a surface force equilibrium, which may be solved for the wetting layer thickness θ_{WL}:

$$2\varepsilon_f \theta_{WL} \{\lambda - \nu(\lambda + 2\mu)\} = -(\gamma_s + \gamma_i) - \lambda c^2 \xi \{1 - \exp[-2\theta_{WL}/\xi]\}, \quad \theta \geq 0, \tag{10}$$

where $\xi = (2\mu\gamma_s - \lambda\gamma_i)/[2\mu^2(1 + 2\varepsilon_f) + \lambda\mu(1 + 6\varepsilon_f)]$, and $c = -\frac{1}{2}k_B T F \tau_d(\lambda + 2\mu)/(2\mu\gamma_s - \lambda\gamma_i)$. Graphical solutions presented in Fig. 1 illustrate two typical cases for layers under compression and tension. It may be rare for $\gamma_s + \gamma_i < 0$. For this situation, layers under compression may be stable, which can be ascertained from Fig. 1 by shifting the $-(\gamma_s + \gamma_i)$-line sufficiently above the θ-axis, while layers under tension will be unstable.

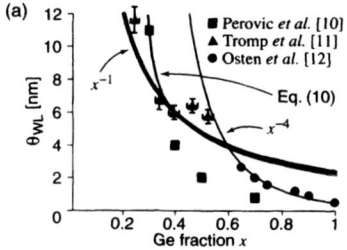 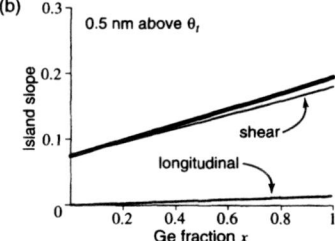

Figure 2: (a) Wetting layer thickness θ_{WL} as a function of Ge fraction x. Increased x produces higher compressive misfit strain. (b) SiGe island slope as a function of x at 5 Å above $\theta_t(x)$. Note that $\theta_t \to \infty$ for $x = 0$; thus, an infinitely thick Si layer is needed to have a Si/Si(001) island slope of ≈ 0.08.

SiGe Layers

Fig. 2(a) shows that θ_{WL} increases with decreasing x for $Si_{1-x}Ge_x/Si(001)$ layers. All parameters are assumed to follow Vegard's law. The x^{-4} curve [10] overestimates for low Ge fractions despite setting its Ge/Si(001) value at 3 ML. The x^{-1} curve from Eq. (8) gives a reasonable agreement for layers grown between 650-680°C [11], but it cannot produce a sudden increase of θ_{WL} at around $x = 0.4$ for layers grown at 525°C [10]. An even lower growth temperature of 500°C [12] also yields a poor agreement. These facts are in accord with the more equilibrated epitaxial growth becomes with increased temperature.

One possible explanation for the sudden jump in θ_{WL} for the 525°C data is the higher Si adatom diffusion barrier than that of a Ge adatom. For relatively high growth temperatures, the barrier difference yields a negligible correction to the x^{-1} curve, when plotting θ_{WL} using Eq. (10). Using $E_{d,Si} = 1.20$ eV, $E_{d,Ge} = 0.64$ eV, and $T = 37°C$, we were able to simulate the sudden jump in θ_{WL} with Eq. (10), as shown in Fig. 2. Our diffusion barrier parameters are consistent with the literature values [13], but the temperature is much too low. However, the qualitative result of our model in explaining the sudden jump should remain. A higher adatom concentration due to a higher diffusion barrier stabilizes the strained layer, since increased adatom concentration would demand a larger wetting layer thickness in order for the tetragonal stress to balance both surface-and-interface-tension and adatom forces.

Fig. 2(b) shows the SiGe island slope as a function of x, which is obtained by dividing the sum of shear and longitudinal deformations with the lateral unit length scale given by the lattice constant of the layer, a_f:

$$\text{slope} \equiv \frac{1}{a_f} \left[\left\{ \int_0^\theta |\phi|^2 \, dz \right\}^{1/2} + \int_0^\theta h_z \, dz \right]. \tag{11}$$

The shear component dominates for $\theta - \theta_t = 5$ Å, but is virtually unchanged with increased thickness due to its exponential decay. The longitudinal component steadily increases with θ and becomes roughly equal to the shear component at $\theta - \theta_t = 30$ Å (not shown). Our 3D model predicts an island slope of 0.2 for $x = 1$, which agrees well with the observed {105} facets for Ge/Si(001) islands [11]. The monotonic increase of longitudinal component will eliminate the shape stability from the shear component; It is expected that the increase would eventually nucleate dislocations.

We also consider the stability of $Si_{0.5}Ge_{0.5}$ layers deposited on $Si_{1-x}Ge_x(001)$. For $x > 0.5$, the layers are in tension, and we find $\theta_t < 0$ under this condition. Using the first-order correction, the wetting layer thickness becomes infinite; thus, the tensile-strained layers are morphologically stable, as were found experimentally [3]. In contrast, we find that $Si_{0.5}Ge_{0.5}$ layers in compression (with $x < 0.5$) are unstable.

Conclusions

We have presented a new continuum approach in predicting morphologies of self-assembled quantum dots using the 3D strain tensor that allows for bulk strains to freely change to minimize the free energy. Equilibrium shapes of 3D coherent islands are determined by a nonlocal surface force equilibrium. The presence of adatoms is included by calculating their entropy. Coherent layers under tension in typical conditions are stable, but they are unstable when compressed. Good agreement with $SiGe/Si(001)$ and $Si_{0.5}Ge_{0.5}/SiGe(001)$ data is obtained.

References

[1] F. M. Ross, IBM J. Res. Develop. **44**, 489 (2000).

[2] A. A. Williams, J. M. C. Thornton, J. E. Macdonald, R. G. van Silfhout, J. F. van der Veen, M. S. Finney, A. D. Johnson and C. Norris, Phys. Rev. B **43**, 5001 (1991).

[3] Y. H. Xie, G. H. Gilmer, C. Roland, P. J. Silverman, S. K. Buratto, J. Y. Cheng, E. A. Fitzgerald, A. R. Kortan, S. Schuppler, M. A. Marcus and P. H. Citrin, Phys. Rev. Lett. **73**, 3006 (1994).

[4] J. E. Guyer, S. A. Barnett and P. W. Voorhees, J. Cryst. Growth **217**, 1 (2000).

[5] I. Kegel, T. H. Metzger, A. Lorke, J. Peisl, J. Stangl, G. Bauer, J. M. Garcia and P. M. Petroff, Phys. Rev. Lett. **85**, 1694 (2000).

[6] R. A. Budiman and H. E. Ruda, J. Appl. Phys. **88**, 4586 (2000).

[7] L. B. Freund and W. D. Nix, Appl. Phys. Lett. **69**, 173 (1996).

[8] J. Y. Tsao, *Materials Fundamentals of Molecular Beam Epitaxy* (Academic Press, Boston, 1993), Ch. 6.

[9] D. L. Smith, *Thin-Film Deposition: Principles and Practice* (McGraw-Hill, New York, 1995), Ch. 1.

[10] D. D. Perović, B. Bahierathan, H. Lafontaine, D. C. Houghton and D. W. McComb, Physica A **239**, 11 (1997).

[11] R. M. Tromp, F. M. Ross and M. C. Reuter, Phys. Rev. Lett. **84**, 4641 (2000).

[12] H. J. Osten, H. P. Zeindl and E. Bugiel, J. Cryst. Growth **143**, 195 (1994).

[13] V. Milman, D. E. Jesson, S. J. Pennycook, M. C. Payne, M. H. Lee and I. Stich, Phys. Rev. B **50**, 2663 (1994).

Mat. Res. Soc. Symp. Proc. Vol. 707 © 2002 Materials Research Society

Morphology of Self-Assembled InAs Quantum Dots on GaAs(001).

F.Arciprete, F.Patella, M.Fanfoni, S.Nufris, E.Placidi, D.Schiumarini, and A.Balzarotti

Dipartimento di Fisica, Università di Roma "Tor Vergata", and Istituto Nazionale per la Fisica della Materia, Via della Ricerca Scientifica 1, 00133 Roma, Italy

ABSTRACT

We have followed by Atomic Force Microscopy (AFM) the epitaxial growth of InAs on GaAs(001) starting from the initial formation of a strained two-dimensional wetting layer up to the self-assembled nucleation and growth of 3D nanoparticles. In this work we underline many aspects of the morphology of this system, which substantiate the role either of kinetics on thermodynamics in the process of growth as well as the role of surface instabilities in controlling lateral ordering of the nanoaggregates.

INTRODUCTION

Self-organization of nanostructures is a promising way to produce Quantum Dots (QDs) with a narrow distribution of size in presence of lattice mismatch. Epitaxial Stranski-Krastanov (SK) growth on a substrate starts usually with the formation of a pseudomorphic strained layer, referred to as the wetting layer (WL), and proceeds with the spontaneous formation of dots above a critical thickness, θ_c. The strain energy accumulated in the thicker layers is partially relieved by misfit dislocations (MD) or by the formation of coherent, dislocation-free QDs. The morphology (size and shape) and density of these dots depend on the interplay of thermodynamic and kinetic effects, which, in turn, vary with the growth parameters, and the details of the heteroepitaxy (substrate temperature, growth rate, flux ratio, etc.). To simplify the matter, the structural properties of the nanostructures are often discussed in terms of an equilibrium phase diagram based on the minimization of the free-energy in a restricted space of parameters [1].

In this paper we focus on the InAs/GaAs(001) system having a lattice mismatch as high as 7%. It is clearly important to identify and understand the kinetic pathways to island formation, particularly at the atomic scale. The growth of InAs on GaAs(001) does not proceed following a pure SK mode since, for equal InAs depositions on substrates at increasing temperature, larger fractions of WL and substrate participate to the formation of the dots [2-3,11]. We have grown samples of InAs on GaAs(001) at different coverage below and above the critical thickness by molecular beam epitaxy (MBE) either in modified migration enhanced (MEG) and conventional continuous (CG) growth modes in order to vary the migration length of In atoms keeping the other growth parameters fixed. The AFM analysis evidenced how different growth procedures that effectively alter relevant kinetic factors - in the present case the migration length of the element III - strongly affect the final structure and composition of the film. This result substantiates the overwhelming role of kinetics on thermodynamics in the non-equilibrium MBE growth. Large scale morphology of the epitaxial GaAs(001) has also been studied.

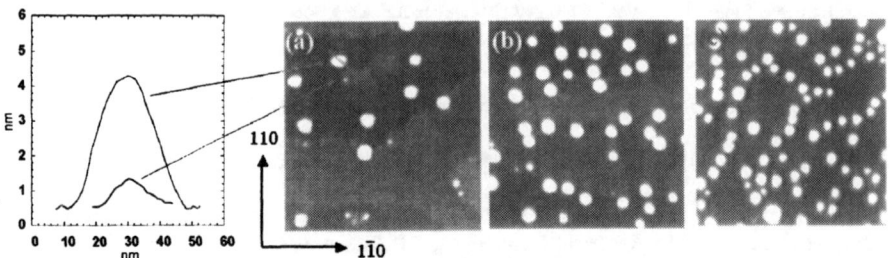

Figure 1. AFM images of InAs QDs on the GaAs(001) 400x400 nm^2 for InAs depositions of 1.5 ML (a), 1.7 ML (b), and 1.9 ML (c). On the left side the line profiles of the small and large 3D islands for the lower coverage are shown.

The singular GaAs(001) surface grows layer by layer through step flow and nucleation of 2D islands on terraces and is unstable to step bunching and step edge meandering giving rise to the formation of large mounded structures on the surface that is non planar on a microscopic scale, independently of the initial miscut [4-7]. The observed morphology shows that these instabilities influence the Stranski-Krastanov self-assembling of coherent islands.

EXPERIMENTAL

The GaAs(001) wafer was deoxidized at about 640 ^{O}C until a weak 2x4 RHEED pattern appeared. Afterwards, the substrate temperature was lowered to 590 ^{O}C and an epitaxial GaAs buffer layer of approximately 0.75 μm was grown at a rate of ~1 μm/h in As$_4$ overflow (As$_4$ BEP=8x10^{-6} Torr). After 10 min annealing, the temperature was further lowered to 500 ^{O}C for the InAs growth which has been performed at a rate of 0.028 ML/s ±5% with an In/As$_4$ flux ratio of about 1/10 (As$_4$ BEP=4x10^{-7} Torr). The InAs deposition has been performed in two ways. While keeping the As beam incident on the sample, in one case the In flux was delivered continuously until to reach the given thickness (CG), in the second case the In evaporation time was partitioned in 5 s of growth followed by 25 s of growth interruption (MEG). The overall effect is an increased migration length of the element III, which flattens the surface and produces higher quality layers than those obtained by CG [3,8]. MEG technique has been used for all the samples studied, for coverage ranging from 1.0 to 3.0 ML, to investigate the role of a kinetic parameter, like the In migration length, in QDs formation. A set of CG samples has also been grown for comparison for coverage higher than θ_c.

The 2D-3D transition was monitored by RHEED. A sudden change of the diffraction pattern from streaky to spotty and a rapid increase of the RHEED intensity signals the morphology transition. It starts after deposition of 1.5 ML of InAs (as evidenced by the AFM images – see Fig.1) and is centered at about 1.7 ML (in the following referred to as the critical thickness), independently of the growth mode.

A Variable Temperature STM/AFM (Omicron) was used to characterize the samples at mesoscopic and atomic scales, This instrument has a broad range of accessible temperatures (25K -1500 K) and a large scanning area (15x15 μm^2) allowing to perform, e.g., accurate statistical distributions of the shape, density and dimensions of the QDs.

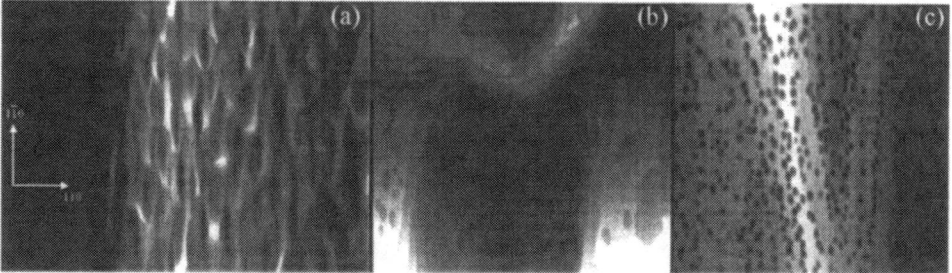

Figure 2. AFM images of InGaAs wetting layer for 1.4 ML of InAs depositions: (a) 10x10 μm^2, (b) 1.5x1.5 μm^2. (c) AFM image, 800x800 nm^2, of QDs for 1.9 ML of InAs deposition.

Atomic force microscope was used in non-contact mode (needle-sensor AFM) with conductive and non-conductive tips.

RESULTS AND DISCUSSION

2D-3D transition

In Fig. 1 are reported three AFM images representative of MEG samples grown with InAs coverage of 1.5, 1.7, and 1.9 ML respectively, around θ_c. The surfaces show steps and 2D large islands, which decrease in number with InAs coverage. The phase transition occurs at about 0.3 ML around the critical thickness 1.7 ML. 3D islands are in fact recognizable already at 1.5 ML of InAs coverage and can be divided approximately into two families: large and small ones. The latter are quasi-2D islands; the line profiles show that they are 2 or 3 ML high, and increase their size quite quickly with respect to the large ones. For coverage as high as 1.9 ML their number become negligible. It is also evident that the process of QDs formation, beginning with nucleation of very small islands, sets off at the upper edge of steps and 2D islands. This fact indicates the presence of a minimum in the hopping potential in proximity of the step edge caused by the readjustment of InAs atoms to lower the lattice misfit (7%) with GaAs. The step edges act controlling the lateral ordering of self-assembled QDs. From this point of view the surface morphology of GaAs and wetting layer are particularly important. The morphology of a singular GaAs(001) surface is made [4-7] of complex highly connected bunched steps with finger-like extensions perpendicular to the initial straight step edge. According to theory [4-6], the asymmetry in the rate of attachment and detachment of adatoms on the lower and higher terraces of step leads to step bunching and wandering. The Schwoebel barrier [5] as well as the presence of "generalized impurities" [6] on the terraces can account for this asymmetry. The starting surface strongly influences the Stranski-Krastanov self-assembling of coherent islands formed upon elastic-energy relaxation of the highly strained InAs layers deposited on GaAs(001). In Fig. 2a and 2b are shown large and small scale AFM images of an intermixed 2D InGaAs layer (1.3 ML of InAs deposit). The morphologies show as the wetting layer grows, like for GaAs, by step flow and nucleation of 2D islands whose distance from the step edges is of the order of the migration length. As shown in Fig. 1, marked differences with GaAs growth start being observed when elastic energy relaxation of the strained InAs layer begins nucleation of very small islands.

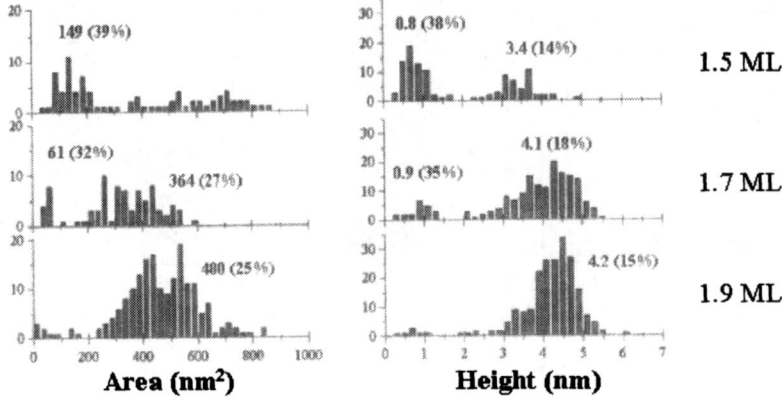

Area (nm²)   **Height (nm)**

Figure 3. Histograms of height and basal area of the dots are shown for MEG samples. The average values of the single dot area and height with the standard deviation of the distributions are indicated for each sample.

In fig. 2c is shown an AFM image, 800x800 nm², of well formed 3D islands at 1.9 ML of InAs deposition. The decoration of the step edges clearly evidences the constraint on the lateral ordering of QDs induced by the substrate morphology.

Distribution of the QD on the surface

By analyzing a large number of AFM topographies for each sample, like those shown in Fig. 1, the number density and the statistical distribution of dots size have been derived. The histograms of height and basal area of the dots for the MEG samples at the earliest stage of growth are reported in Fig.3. The plots show that the areal distribution is initially bimodal with the small dots family vanishing for coverage as high as 1.9 ML, whereas the increased dots density determines self-organization and the distribution begins to get narrow.

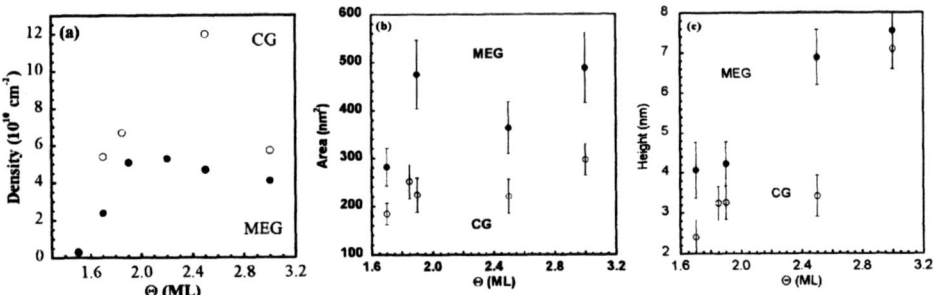

Figure 4. Density (a), mean basal area (b), and mean height (c) of dots versus InAs deposition for MEG and CG samples.

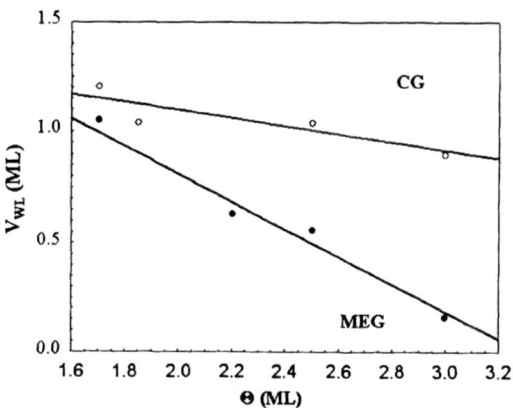

Figure 5. Volume of an effective wetting layer V_{WL} as a function of coverage for MEG and CG samples. V_{WL} is obtained as difference between the deposited volume and the total volume of non-ripened islands.

We believe that at the earliest stage of 3D growth the nucleation rate of the islands is constant and the growth rates are independent of each other. When the dots density increase, the islands interact, because of the strain field induced in the substrate, and the small dots grow more rapidly according to the kinetic model of Barabasi [12]. In fact, for higher coverages (not shown), the full-width-at-half-maximum of the distributions narrows.

For comparison, a set of CG samples has also been grown and analyzed for coverages higher than 1.7 ML, and the results are summarized in Fig. 4. For equal thicknesses the mean area values of the MEG samples are greater than those of the CG samples, as a consequence of the larger dot number density of the latter ones. This fact confirms that the In monomer diffusion length is enhanced in MEG mode; it is, in fact, an established result that the number of islands is inversely proportional to the diffusion coefficient and to the average lifetime of the monomers [13]. Furthermore, in MEG samples, the dots increase their size at almost constant density while the CG samples have a more efficient size regulation mechanism with dots that increase density linearly up to 2.5 ML. For higher coverage the density decreases rapidly as a consequence of coalescence and ripening [3,11]. Although many aspects of the equilibrium phase diagram of the mismatched heteroepitaxial growth [1] are verified, the differences observed by changing one kinetic parameter only, i.e. the migration length of In, emphasize the important role of kinetics on thermodynamics in MBE. An important aspect of the MEG growth mode is enlightened by the comparison of the behavior of the total volume of the dots as a function of coverage. In Fig. 5 the volume of an effective wetting layer, obtained by subtracting the total volume of the non-ripened islands to the deposited material, is plotted as a function of coverage. For CG samples all the evaporated material is accumulated in dots on top of an effective wetting layer of thickness ~1 ML, while in MEG sample dots of increasing size are grown at the expense of deposited material and of the wetting layer. So the substrate and/or wetting layer participate to the dots growth and, again, this effect is kinetic controlled too. We cannot rule out interdiffusion in CG sample because 1 ML is only an effective thickness, but in MEG samples this phenomenon in much more important and could be relevant in determining optical properties of the QDs.

CONCLUSIONS

In conclusion, we have studied by AFM the morphology and evolution of quantum dots of InAs on GaAs(001). We studied the 2D-3D transition, the effects of intermixing and evidencing the role of kinetics against thermodynamics in MBE. The role of large scale surface instabilities in determining nucleation and lateral ordering of 3D islands has also been discussed.

REFERENCES

1. I.Daruka and A.L.Barabasi, Phys. Rev. Lett. **79**, 3708 (1997).
2. P.B.Joyce, T.J.Krzyzewski, G.R.Bell, B.A.Joyce, T.S.Jones, Phys. Rev. B **58**, R15981 (1998).
3. F.Patella, M.Fanfoni, F.Arciprete, S.Nufris, E.Placidi, and A.Balzarotti, Appl. Phys. Lett. **78**, 320 (2001).
4. G.S.Bales and A.Zangwill, Phys. Rev. B **41**, 5500 (1990),
5. M.D.Johnson, C.Orme, A.W.Hunt, D.Graff, J.Sudijono, L.M.Sander, and B.G.Orr, Phys. Rev. Lett. **72**, 116 (1994); M.Rost, P.Smilauer, J.Krug, Surf. Sci. **369**, 393 (1996).
6. D.Kandel and J.Dweeks, Phys. Reb. B **49**, 5554 (1994),
7. F.Patella, F.Arciprete, E.Placidi S.Nufris, M.Fanfoni, and A.Balzarotti, to be published.
8. Y.Horikoshi, M.Kawashima, 1989, J. of Crystal Growth **95**, (1989) 17; Y.Horikoshi, Semicond. Sci. Technol. **8**, 1032 (1993).
9. J.G.Belk, J.L.Sudijono, X.M.Zhang, J.H.Neave, T.S.Jones, B.A.Joyce, Phys. Rev. Lett. **78**, 475 (1997).
10. J.G.Belk, C.F.McConville, J.L.Sudijono, T.S.Jones, B.A.Joyce, Surf. Sci. **387**, 213 (1997).
11. F.Arciprete, A.Balzarotti, M.Fanfoni, N.Motta, F.Patella, and A.Sgarlata, in *Recent Res. Devel. Vacuum Sci. & Tech.*, edited by Transworld Research Network, India, 2001, vol 3, p. 71.
12. A.L.Barabasi, Appl. Phys. Lett. **70**, 2565 (1997).
13. J.A.Venables, G.D.T.Spiller, M.Hanbücken, Rep. Prog. Phys. **47**, 399 (1984).

Mat. Res. Soc. Symp. Proc. Vol. 707 © 2002 Materials Research Society   N7.5

Optical Characterization Of Self-Assembled Ge Dots On Silicon

F.Marabelli[1], A.Rastelli[1,2], A.Valsesia[1], H.von Känel[2]

[1]INFM-Phys.Dept.``A.Volta", University of Pavia, Italy
[2]Laboratorium für Festkörperphysik, ETH Zürich, Switzerland

ABSTRACT

Self assembled quantum dots of Ge were obtained by magnetron sputter epitaxy of seven monolayers of Ge on a 33nm thick undoped Si buffer grown on top of a p-doped (100) Si substrate. The samples obtained in this manner were then capped with an increasing number of silicon layers in order to study the effect of Si deposition on the strain and the morphology of the dots. They were characterized ``ex situ" by spectroscopic ellipsometry and Raman spectroscopy. The optical experiments revealed well defined differences between the capped and uncapped samples and among samples with different cap thicknesses.

By monitoring the energy and the splitting of the E_0', E_1 and E_2 interband optical transitions of Ge and the Ge-Si vibrational mode, the optical measurements evidence strain effects as well as the formation of SiGe alloy, in agreement with the ``in situ" STM measurements.

INTRODUCTION

Structural and Morphological features of self assembled island are connected to a series of other different behaviours more directly related to electronic and vibrational properties. In a system like Si/Ge both composition and strain strongly affect the electronic structure of the resulting material. [1,2] Optical spectroscopy, and spectroscopic ellipsometry in particular, is a very sensitive tool of investigation of this kind of problems. [2-5] Actually, many data exist in the literature on Si_xGe_{1-x} bulk or relaxed samples [3,6]. The situation is more difficult for the strained samples, on which only theoretical data and investigations of only few selected stoichiometries are available [1,2,4,7].

In the present work we present some optical results obtained on a series of samples obtained by self assembling of Ge island on a silicon substrate.

The aim is to discriminate between strain and intermixing effects on such a series of samples.

EXPERIMENT AND DISCUSSION

The samples were prepared by ultra high vacuum magnetron sputtering epitaxy. 60 nm of silicon buffer were deposited on pre-cleaned Si(001) substrates; on top of the buffer 7 monolayers (ML) of Ge were epitaxially grown, followed in many cases by Si capping at 450°C. Further details on the sample preparation and structural characterization are given in ref.[8]. A series of samples were prepared by capping the Ge island formed after the Ge deposition with 1, 4, 8 and 16 ML of silicon, respectively. One samples was kept uncapped. An epitaxial film of 7

ML of germanium capped with 16 ML of silicon was prepared at low temperature in order to have as a reference a fully strained, homogeneous Ge layer.

Spectroscopic ellipsometry were measured with a Sopra spectrometer mod. ES 4G at an incidence angle of 70° and 75°. The spectral resolution was less than 5 meV over the whole spectral range (1.4-5 eV). A beam condenser used to focus the light beam on a spot of 100x300 μm^2 on the sample surface caused a small spread of the incidence angle which was taken into account in the data treatment. Analysis and inversion of the ellipsometric functions Tan(Ψ) and cos(Δ) were performed with the Woolam program [9].

Raman spectra were collected with a Dilor instrument equipped with a microscope and a Peltier cooled CCD detector. The 632 nm line of an He-Ne laser was used as an excitation source.
The ellipsometric functions obtained from the measurements for some samples are shown in Fig.1.

As one can notice from the figures, a well detectable difference can be determined from ellipsometric measurements. Actually possible effects due to oxidation and roughness must be taken into consideration. A test with several Si substrates, having different degree of roughness and contamination, showed that such features are affecting the optical response for a smaller amount than the observed differencies introduced by the Ge islands.

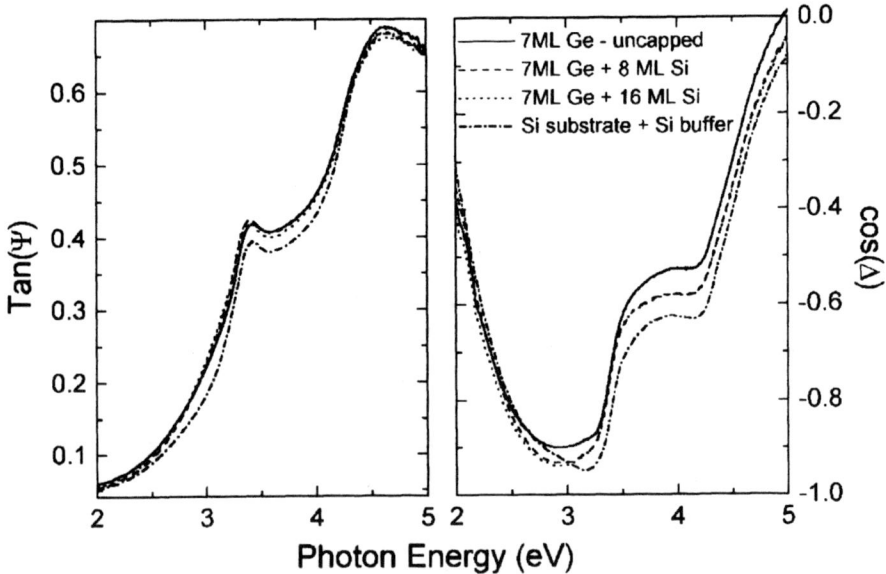

Fig.1   Ellipsometric functions Tan(Ψ) and cos(Δ) measured with an incidence angle of 75° on a set of samples containing Ge islands grown on a silicon substrate and capped with few Si overlayers. The spectrum of an uncapped sample and of the silicon substrate are shown too.

As concerning the roughness introduced by the island themselves, we used an effective medium approach modelling the sample as a series of layers: we introduced a layer equivalent to the amount of deposited Ge (~1 nm), whose optical functions have to be determined by numerical inversion of the ellipsometric functions; an overlayer (2-6 nm) with a mixture of oxide (SiO_2), voids and (eventually) a small part of silicon has been used to simulate roughness. Variations from such a simple model by changing the thicknesses and the component concentration, have been tested within the constrain of having physical meaning results. Due to the very small thickness of the layer considered, no improvement is obtained by increasing the number of layers. Since no significant changes occurs in the spectral details, apart from the absolute value, we use the model to obtain the optical functions of the samples, in order to compare the spectra and evaluate the changes introduced by the capping. The imaginary part of the dielectric function so calculated is shown in Fig.2, together with the results obtained on a strained Ge film and with the literature data of germanium [3].

One can notice that the spectral weight of the E_1 optical transitions of germanium at 2.14, 2.4 eV has been shifted towards the highest energies in all the spectra of the Ge dots; the main differences among the resulting spectra of the different samples can be observed in the region around 3–3.5 eV, where in germanium emerges the E_0' optical structure. The situation is not

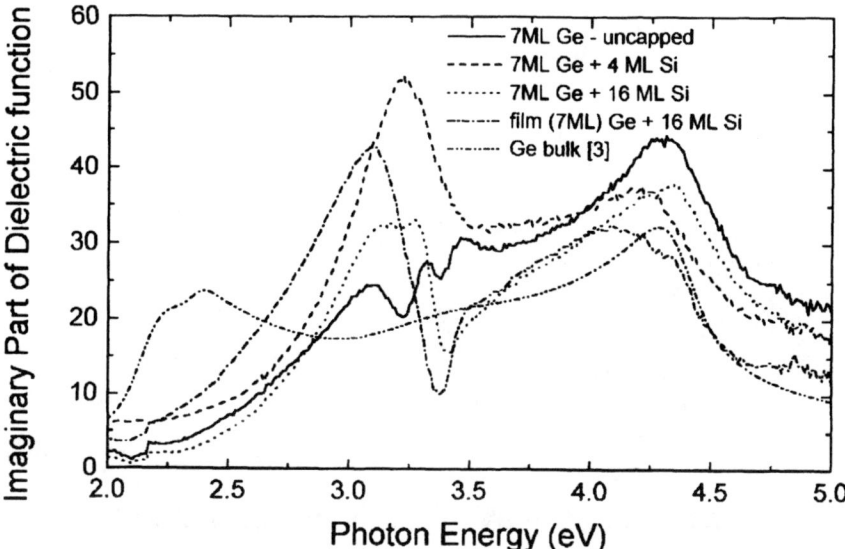

Fig.2   Imaginary part of the dielectric function obtained by numerical inversion of ellipsometric data. A layer thickness of 1 nm has been used in the case of the uncapped sample and the film. It has been slightly increased (1.3 nm) for the capped samples. The dielectric function of Ge is also shown for the sake of comparison.

clear above 4 eV, where both silicon and germanium presents the large E_2 peak and the inverted spectra become particularly noisy and dependent on the oxide overlayer used in the model; so no reliable discussion can be made in this region.

The shift of the E_1 ($E_1+\Delta_1$) peaks is a common feature, observed experimentally, due to intermixing of silicon and germanium; on the other hand, theoretical band structure calculations show that also strain is producing a blue shift of such a structure, together with a widening of the splitting Δ_1 (the intermixing effect reduces such a splitting).

Moreover, band structure calculations support the fact that the largest change in the electronic structure induced by strain affect the region around the Γ point of the Brillouin zone. Besides a widening of the direct gap E_0 and a lowering of E_0', a mixing occours of heavy hole and light hole bands with the widening of the splitting Δ_0', and changes are introduced in the conduction band when the strain is not isotropic (as it is usually the case) [1,10].

All these effects change the optical response in the region 3-3.5 eV. Since intermixing is not very effective on the E_0' structure [4,5], we start our analysis from such a spectral region.

Fig.3 shows a detail of the ε_2 structures in the spectra of Ge islands capped with an increasing number of Si layers.

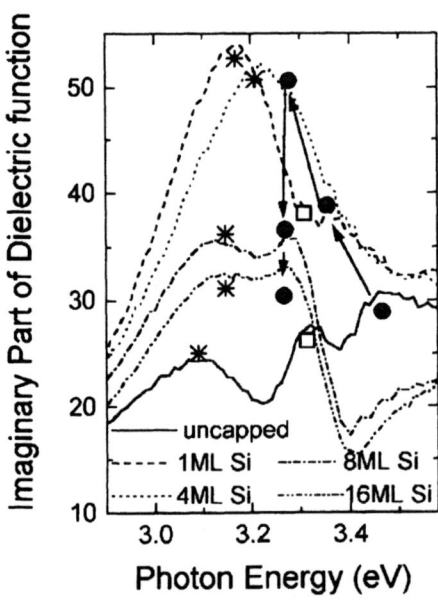

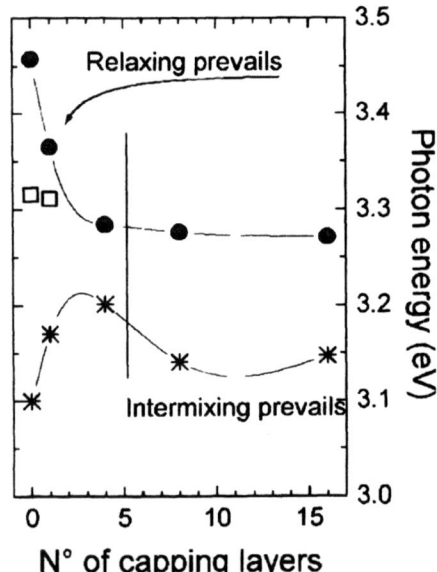

Fig.3 Imaginary Part of the dielectric function obtained from ellipsometry for the dot samples (7ML Ge on Si(001)) capped with a different number of Si monolayers.

Fig4 Evolution as a function of the Si capping of the peaks shown in Fig.3. Within a few ML of capping the main effect is related to relaxing of the strain. For a larger coverage intermixing effect prevails.

Three structure are present in the spectrum of the uncapped sample and this is consistent with the spectrum obtained on the strained film where three analogous features can be found (the two first are a little bit closer and cannot be observed directly in the spectrum of Fig.1, but they are resolved by looking at the derivative spectrum). By adding a few Si monolayers, the three structures seem to shrink into a singular structure and then, for an increasing number of capping layers, to split again into two separate structures. Fig.4 shows the evolution of the observed peaks with the number of capping layers.

The low energy peak is likely a convolution of $E_1+\Delta_1$ and E_0' structures, the first shifted at larger energies and the second at lower from the strain. The high energy peak is probably related to a splitting of E_0' or $E_0'+\Delta_0'$ and is more sensitive to strain than to intermixing. By capping the dots with silicon, intermixing begins, but the first observable effect is in relaxing the strain. Then, the E_0' components tend to the unstrained value and the $E_1(+\Delta_1)$ relax a bit toward low energies. After this step intermixing remains as the main driver of spectra evolution.

The role of E_1 structure can be better noticed by directly looking at the differences between the Tan(Ψ) spectra of the samples and that of silicon substrate. Such a difference is shown in Fig.5.

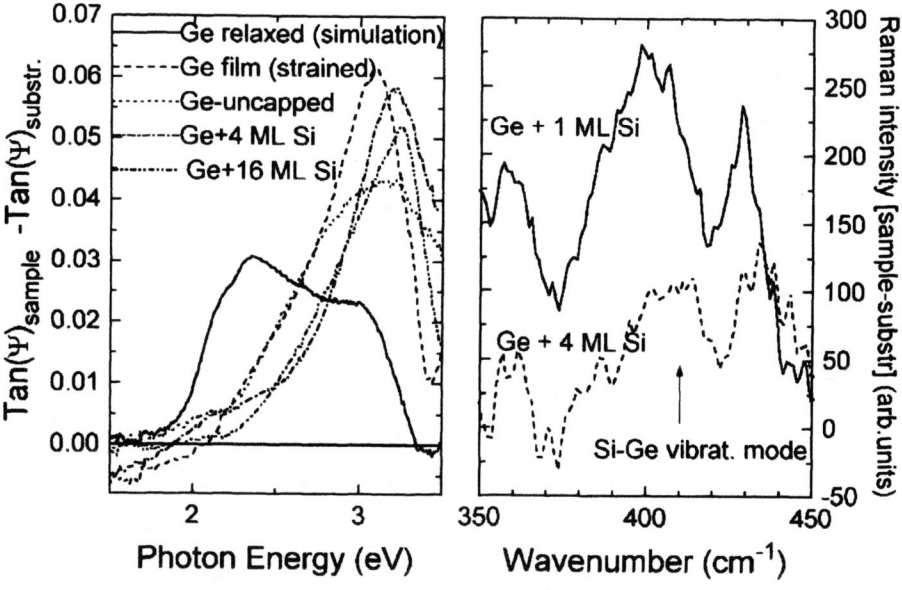

Fig.5 Difference between the Tan(Ψ) spectra of some Ge island samples capped with a different number of Si layers and the Tan(Ψ) spectrum of the silicon substrate. A simulated difference for a 1nm film of Ge relaxed is shown for comparison.

Fig.6 Difference between the Raman spectra of some Ge island samples capped with a different number of Si layers and the spectrum of the silicon substrate. A peak ascribed to Si-Ge vibrational mode emerges at about 410 cm^{-1}.

The shift of the E_1 structure in the strained film and in the dot samples with respect to the unstrained germanium is well observable in the spectra. The evolution with Si capping is a combination of strain and intermixing effects. One should take into consideration the fact that strain (as well as intermixing) is not uniformly distributed within the Ge islands, and this is probably reflected in the broad and asymmetric shape of some structures (e.g. for the uncapped sample). The upturn at about 2.5 eV for the sample capped with 4 ML could be ascribed to an inhomogeneous intermixing distribution in the (relaxed) island.

The presence of Si-Ge intermixing with formation of alloy, even in the early stages of Si coverage, is confirmed by Raman measurements. Fig.6 shows the difference between the Raman spectra of Ge island samples capped with 1 and 4 ML of silicon and the Raman spectrum of silicon substrate: a weak and broad peak appears around 400 cm^{-1}. A vibrational mode structure ascribed to Si-Ge bonds is reported in the literature at 410 cm^{-1}[5].

CONCLUSIONS

The effects of strain relax and intermixing related to the capping with silicon of (strained) germanium islands grown on Si(001) have been investigated by spectroscopic ellipsometry and Raman spectroscopy. The main results is that, whereas Si-Ge intermixing ocours even at the coverage with few ML, the first effect on the optical spectra is related to strain relaxation producing a shift of the transition energies at the critical points (E_0', E_1). After such a step intermixing effects drive the further evolution of samples.

REFERENCES

[1]   G.Theodorou, P.C.Kelires, C.Tserbak, Phys.Rev.B **50**, 18355 (1994)
[2]   R.Lange, K.E.Junge, S.Zollner, S.S.Iyer, A.P.Powell, K.Eberl, J.Appl.Phys.**80**, 4578 (1996)
[3]   J.Humliček, M.Garriga, M.I.Alonso, M.Cardona, J.Appl.Phys.**65**, 2827 (1989)
[4]   C.Pickering, R.T.Carline, D.J.Robbins, W.Y.Leong, S.J.Barnett, A.D.Pitt, A.G.Cullis, J.Appl.Phys.**73**, 239 (1993)
[5]   P.Y.Yu and M.Cardona, ''*Fundamentals of Semiconductors*'', (Springer Verlag, Berlin Heidelberg 1996) p.239 ; p.320 ; p.378.
[6]   J.Humliček, F Lukes, E.Schmidt, in *"Handbook of Optical Constants of Solids II"*, ed by E.Palik (Academic Press inc, San Diego 1991), p.607
[7]   F.Ferrieu, F.Beck, D.Dutartre, Solid State Comm. **82**, 427 (1992)
[8]   A.Rastelli, M.Kummer, H.von Känel, MRS Symposium Proc. Vol. 696 (Pittsburg, 2001).
[9]   WVASE32 (Woolam, Lincoln, NE, 2000)
[10]   L.Miglio, private communication.

Mat. Res. Soc. Symp. Proc. Vol. 707 © 2002 Materials Research Society

Atomic Self-ordering in Heteroepitaxially Grown Semiconductor Quantum Dots due to Relaxation of External Lattice Mismatch Strains

Peter Möck[1]*, Teya Topuria[1], Nigel D. Browning[1], Robin J. Nicholas[2], and Roger G. Booker[3]

[1] Department of Physics (MC 273), University of Illinois at Chicago, 845 W. Taylor Street, Chicago, IL 60607-7059, U.S.A; *pmoeck@uic.edu
[2] Department of Physics, Clarendon Laboratory, University of Oxford, Parks Road, Oxford, OX1 3PU, U.K.
[3] Department of Materials, University of Oxford, Parks Road, Oxford, OX1 3PH, U.K.

ABSTRACT

Thermodynamic arguments are presented for the formation of atomic order in heteroepitaxially grown semiconductor quantum dots. From thermodynamics several significant properties of these systems can be derived, such as an enhanced critical temperature of the disorder-order transition, the possible co-existence of differently ordered domains of varying size and orientation, the possible existence of structures that have not been observed before in semiconductors, the occurrence of atomic order over time, and the occurrence of short range order when the growth proceeds at low temperatures. Transmission electron microscopy results support these predictions. Finally, we speculate on the cause for the observed increase in life time of (In,Ga)As/GaAs quantum dot lasers [H-Y. Liu et al., Appl. Phys. Lett. 79, 2868 (2001)].

INTRODUCTION

The growth of self-assembled semiconductor quantum dots (QDs) by the heteroepitaxial Stranski-Krastanow mode and its variants has developed rapidly over recent years [1]. Regardless of the deposition parameters, alloying of the elements or compounds (e.g. the formation of (Ge,Si) [2], (In,Ga)As [3], (Cd,Zn)Se [4], ect) that constitute the QD system is known to take place during the growth. This alloying is even considered to be crucial for the Stranski-Krastanow growth mode to operate in semiconductors [5]. Since the deposition and alloying are random events, these QDs will have the structure of the prototype of the constituents of the alloy, i.e. the atoms will (initially) be randomly distributed.

A tacit assumption throughout the quantum dot community is that Stranski-Krastanow grown random alloy QDs are structurally stable. Recent experimental evidence, however, indicates that this is for several III-V and II-VI compound semiconductor systems not the case [6-8]. In these systems it appears that it is atomically ordered or phase separated QDs that are structurally stable, i.e. the initial random alloy transforms to a partially ordered superlattice structure. In hind sight, it is surprising that structural transformations by means of atomic ordering and phase separation within Stranski-Krastanow grown QDs have so far not been taken into account, as there is a large body of experimental and theoretical literature which describes such transformations in heteroepitaxial semiconductor alloy layers, e.g. [10], [11], and review [12]. As these effects lead to both a reduction of the stored lattice mismatch energy and a lower band gap, such ordering should in principle lead to QDs with enhanced properties.

In this paper, we will advance qualitative thermodynamic arguments in favor of atomic ordering in heteroepitaxially grown QDs. From these arguments we will predict a significant increase in the critical temperature for the disorder-order transition, the co-existence of differently ordered domains of varying size and orientation, and the possible existence of structures that have not been observed before in semiconductors. Furthermore, these arguments also predict that long- and short-range atomic ordering occurs over time, even at room temperature, and that short-range order may occur when the growth proceeds at low temperatures. All of these predictions are borne out by experimental observations [6-8]. Representative transmission electron microscopy (TEM) images and diffraction patterns are presented here to support these predictions. Using thermodynamic arguments, we will speculate on the cause of the observed increase in the life time of (In,Ga)As/GaAs quantum dot lasers [13].

THERMODYNAMICAL ARGUMENTS

In thermodynamic terms, any structural transformation in a crystal can be explained by the minimization of the Gibbs free energy

$$G = E - TS + pV \qquad (1)$$

where E is the internal energy, T the absolute temperature, S the entropy, p the pressure, and V the volume of the crystal. The requirement of a smaller band gap in the quantum dot than in the surrounding matrix is usually obtained for solid solutions with larger lattice constants. This means that heteroepitaxially grown QDs are typically compressively strained to a few percent and we will refer to them below as ordinarily strained quantum dots (OS-QDs) and use the subscript $_{os}$. Assuming that OS-QDs are completely embedded in a matrix with a smaller lattice constant but the same structural prototype, the product of the hydrostatic pressure on them (due to external lattice mismatch stresses) and their volume can be approximated by the product of their bulk modulus (B) and the change in their volume (ΔV),

$$p\Delta V = B_{os} \Delta V_{os} \qquad (2).$$

Now, we consider a transformation from an OS-QD with the sphalerite structure into an atomically ordered quantum dot (subscript $_{ao}$) with a different structural prototype that is negligibly strained, i.e. $\Delta V_{ao} \sim 0$, and where the contribution of the product of the atmospheric pressure and V to the Gibbs free energy is also neglected). This type of transformation is observed experimentally [6-8], (Figs. 1 and 2). Combining the energy minimization principle, (1), and (2) we obtain for this kind of structural transformation

$$E_{os} - T S_{os} + B_{os} \Delta V_{os} > E_{ao} - T S_{ao} \qquad (3).$$

Solving for the critical temperature, T_c, for the disorder-order transition where both phases are in equilibrium, we obtain

$$T_c = \frac{E_{os} - E_{ao} + B_{os} \cdot \Delta V_{os}}{S_{os} - S_{ao}} \qquad (4).$$

As T_c (and any other T) is positive, and $S_{os} - S_{ao}$ is also considered to be positive, the numerator in (4) has to be positive as well. For a sufficiently large product $B_{os} \Delta V_{os}$, this condition can be met by both possibilities $E_{os} > E_{ao}$ or $E_{os} < E_{ao}$. In either case, the larger the product $B_{os} \Delta V_{os}$ is, i.e. the effect of the external lattice mismatch stress, the larger T_c will be. The Gibbs-Helmholtz equation (1) and the energy minimization principle tell us only which structural transformations will eventually happen, but not how fast such transformations will happen and how high the obtainable degree of order will be.

Such information can be derived qualitatively for an enhanced T_c from the classical theories of long-[14] and short-range [15] atomic ordering as cooperative phenomena. These theories state that at any T below T_c the equilibrium degree of order will be the larger the higher T_c is. In order to approach the equilibrium degree of order at any T below T_c, thermal treatments can be performed at this particular temperature. If T_c is enhanced, a higher thermal treatment temperature can be chosen and the approach to equilibrium will be faster. On cooling below T_c, partial order will initially be quickly established in the vicinity of T_c but complete order will only be achieved at T = 0. Although the order parameter for long range order drops to zero at T_c [14], short range order exists above the critical temperature to an appreciable amount [15] and may act as seeds for the formation of long range order.

Finally, quenched-in vacancies should play an important role in the atomic rearrangement processes in OS-QDs. In strain gradients that exist in the surroundings of OS-QDs, vacancies will move preferentially to places of higher compressive strain [16]. Provided that the kinetic energy of the vacancies is large enough to pass the tensile strain barrier that surrounds the OS-QDs, they will end up inside these entities and facilitate local atomic rearrangements. The quotient of the number of vacancies (n_V) to the number of atoms (N) in a crystal is given by

$$\frac{n_V}{N} = \exp \frac{-F_V}{k \cdot T} \qquad (5)$$

where F_V is the formation energy of a vacancy and k is Boltzmann's constant. Assuming, for example, F_V ~ 1 eV, the effect of a fast quench from a growth temperature of 500 °C to room temperature (300 K) may be an up to ten orders of magnitude higher vacancy concentration (i.e. $\exp (8.614 \cdot 10^{-5} \cdot 300)^{-1}$ - $\exp (8.614 \cdot 10^{-5} \cdot 773.15)^{-1}$) than in the same system in thermodynamic equilibrium. Provided that ordinarily strained quantum dots are capable of attracting a sizable amount of these excess vacancies, thermodynamics driven atomic rearrangement processes will speed up significantly.

THEORETICAL IMPLICATIONS AND EXPERIMENTAL SUPPORT

To demonstrate the structural instability of heteroepitaxially grown OS-QDs, we will estimate the values of the terms in (3) and (4) for one particular example, i.e. an $In_{0.75}Ga_{0.25}P$ QD embedded in a $Ga_{0.25}In_{0.75}P$ matrix. Since the material parameters in (3) and (4) are the same order of magnitude for other semiconductor systems, other OS-QDs should follow the same trends. To extrapolate between the respective values for InP and GaP, we will employ Vegard's law. Although an order of magnitude estimation would be sufficient to support the qualitative reasoning, we will use accurate values where available. Based on the elastic constants given in ref. [17], the bulk modulus of $In_{0.75}Ga_{0.25}P$ (in the sphalerite structure) is 75.4 GPa. Also from [17], the lattice constant of an unstrained $In_{0.75}Ga_{0.25}P$ quantum dot with the sphalerite structure is 0.5764 nm and the lattice constant of the unstrained surrounding $Ga_{0.75}In_{0.25}P$ matrix is 0.5555 nm. This results in an external lattice misfit strain of ~ 3.6 %, which we assume to be evenly distributed between the QD and the immediate surrounding matrix. We will assume typical dimensions for an unstrained QD as 25 nm (length) · 25 nm (width) · 4 nm (height). Assuming that half the strain, i.e. 1.8 %, is accommodated by the QD, we obtain a volume of ~ 2367.4 nm^3 and an absolute volume change of ~ 132.6 nm^3. With the strain evenly distributed between the OS-QD and the matrix, the coherent $In_{0.75}Ga_{0.25}P$ quantum dot (with sphalerite structure) has a lattice constant of 0.56595 nm. This results in ~ 13.06 · 10^3 unit cells per quantum dot and with 8 atoms per unit cell, the QD contains ~ 104.48 · 10^3 atoms.

Calculations based on density functional perturbation theory obtained a value of 18.3 meV for the internal energy per atom of unstrained $In_{0.75}Ga_{0.25}P$ with sphalerite structure and an internal energy of 11.8 meV for unstrained $In_{0.75}Ga_{0.25}P$ with famatinite structure [10], (space group $I\overline{4}2m$ with 16 atoms per unit cell resulting from cation ordering into a superlattice [12]). From

$$\frac{\delta S}{\delta T} = \frac{C_p}{T} \qquad (6)$$

and the specific heat at constant pressure $C_p = 0.28 + 10^{-4}$ T [J g^{-1} K^{-1}] between 298 and 910 K for InP in the sphalerite structure [18], we obtain by integration

$$S(T) = 0.28 \ln T + 10^{-4} T \quad [J \ g^{-1} \ K^{-1}] \qquad (7).$$

We assume (for lack of more specific data and the fact that the entropy of mixing [19] contributes only ~ 2.4 eV K^{-1} to the total entropy when 25 % of the In atoms are replaced on random positions by Ga atoms) that this relation is also valid for the OS-QD of $In_{0.75}Ga_{0.25}P$. The entropy for the atomically ordered QD is assumed to be half that of the OS-QD. With a relative molar weight of $In_{0.75}Ga_{0.25}P$ of 134.515 g mol^{-1} and Advogadro's constant, we obtain at T = 300 K the following entropies: S_{os} ~ 237 eV K^{-1} and S_{ao} ~ 118.5 eV K^{-1}.

Equation (3) gives now for T = 300 K the structural stability determining inequality (18.3 · 104.48 − 237 · 300) eV + 6.241 · 10^4 eV > (11.8 · 104.48 − 118.5 · 300) eV. Resolved and divided by the number of atoms in the QD, we obtain the inequality: - 0.66 $^{eV}/_{atom}$ + 0.6 $^{eV}/_{atom}$ > - 0.33 $^{eV}/_{atom}$. These inequalities mean that without the external lattice mismatch strain, the sphalerite structure would possess the lower Gibbs free energy and be stable at room temperature (as observed in unstrained bulk crystals), but with the estimated contribution of the lattice mismatch strain on the OS-QD, the energy balance is reversed and the unstrained atomically ordered superlattice is the preferred structure of the system. While one may argue that our value for S_{ao} (and, thus, the right hand side of these inequalities) appears to be somewhat arbitrary, it is obvious that the external lattice mismatch strain contributes significantly to the Gibbs free energy of the OS-QD (on the left hand side of these inequalities) and, therefore, destabilizes it. Using the same values and approximations, we obtain from (4) an estimation of the critical temperature

$$T_c = (\frac{9.3234}{0.28 \cdot \ln T_c + T_c \cdot 10^{-4}} + \frac{8.5678 \cdot 10^2}{0.28 \cdot \ln T_c + T_c \cdot 10^{-4}}) \, K \qquad (8)$$

for the reversible transformation between the two possible structure of an $In_{0.75}Ga_{0.25}P$ QD a value of ~ 486 K. The second term in (8) is due to the effect of the external lattice mismatch strain and obviously enhances T_c significantly. However, using the same values as above, if all of the external lattice mismatch strain is accommodated by the OS-QDs alone and the lattice of the matrix remains undistorted, T_c ~ 852 K, i.e. a temperature within the range that is commonly used to growth (In,Ga)P quantum dots [20]. An enhancement of T_c of this kind has been observed to take place in Stranski-Krastanow grown (In,Ga)Sb quantum dots in GaSb matrix, as shown in Figure 1.

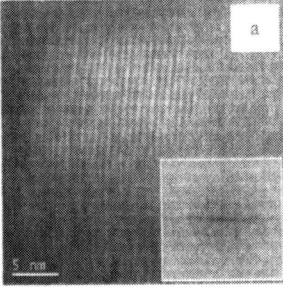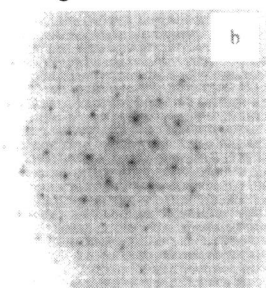

Figure 1. a) High-resolution [001] plan-view TEM image of an atomically ordered (In,Ga)Sb QD in a GaSb matrix, indicating that there is essentially no strain field associated with the atomically ordered phase. The inset Fourier transform shows double diffraction spots, i.e. the QD has a different phase to the sphalerite matrix; b) Selected area electron diffraction pattern from the same region. The images were recorded at 500 °C, after the specimen was held at this temperature for 2 hours (and previously held at 475 °C and 350 °C, for 2 hours each). This suggests that the atomically ordered QD is structurally stable and that T_c is of the order of magnitude of the growth temperature.

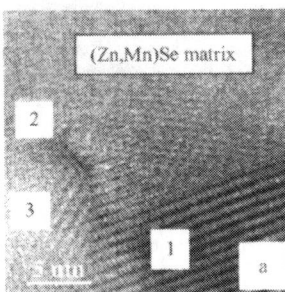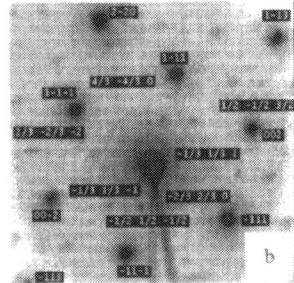

Figure 2. (a) Atomic resolution Z-contrast image, viewed in the <110> direction, showing atomic ordering in (Cd,Mn,Zn)Se agglomerates in a (Mn,Zn)Se matrix. Agglomerate 1 is actually ~ 200 nm and all the larger agglomerates in this sample posses the same orientation relationship [8,9]. Although quite large, these agglomerates are usually free of structural defects, indicating that external lattice mismatch strains are negligible; (b) Selected area

electron diffraction pattern from at least two agglomerates showing a variety of extra spots due to long range atomic ordering that indicate the existence of phases that have not been observed before in compound semiconductors [12].

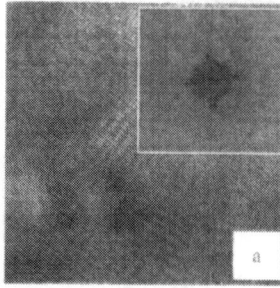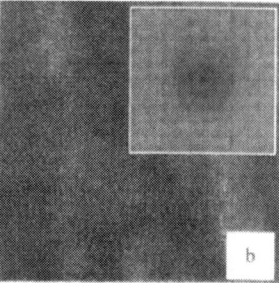

Figure 3. High-resolution TEM [001] plan-view images of atomically ordered (Cd,Zn)Se quantum dots in a ZnSe matrix showing long-range (a) and short-range (b) order in the same sample after ~ 38 months storage at room temperature. The inset Fourier transforms show for (a) superlattice spots and for (b) both a diffuse and a spotty ring.

When the growth was performed at 480 °C, i.e. the lowest temperature we used in our experiments [21,22], we observed for Stranski-Krastanow grown (In,Ga)Sb quantum dots in GaSb matrix, occasionally local short range order in TEM (as indicated by extra spots and diffuse to spotty rings in the diffraction patterns, images not shown) ~ 14 days after the growth. As the time interval between growth and TEM investigations was short for these OS-QDs, we suggest that the short-range order may have formed during the growth. Since short-range order exists even above T_c [15], we interpret these observations as further indications that structural transformations take place in this quantum dot system and that T_c is in the order of magnitude of the growth temperature.

With the internal energies for 8 more (In,Ga)P phases ranging from 6.2 to 31.1 meV per atom [10], and similar considerations on the entropy, it is obvious, that agglomerates with many different structures, some of which may not have been observed before in semiconductors, may originate from OS-QDs with the sphalerite structure and may co-exist in one sample in close proximity. Some of the agglomerates may have orientation relationships that lead to better lattice fits with the matrix than others and, therefore, grow over time on the expense of domains that have the larger Gibbs free energies. Such a co-existence of novel semiconductor phases has been observed experimentally, as shown in Figure 2. For Stranski-Krastanow grown (Cd,Zn)Se quantum dots in a ZnSe matrix, we observed ~ 38 month after the growth, short-range and long-range order that did not appear to exist (in this amount) at the time of earlier investigations [23] (Figure 3).

Finally, we speculate on the cause for the increase of the lifetime of (In,Ga)As quantum dot lasers (to ~ 9000 hours from ~ 1200 hours) as an effect of embedding the quantum dots sheets in strain reducing (In,Ga)As layers rather than in pure GaAs spacer layers [13]. On the basis of our thermodynamic arguments, we suggest that external lattice mismatch strain driven cooperative atomic ordering (or phase separation) phenomena took place in the more severely strained QDs while the lasers were tested at room temperature. Since atomic ordering (and phase separation) reduces the bandgap [11,12], the wavelength of the emitted infrared light shifted over time to the red and the fixed 800 µm cavity lengths were more and more out of tune with the lasing condition.

CONCLUSIONS

Thermodynamic arguments for the formation of atomic order in heteroepitaxially grown semiconductor quantum dots were advanced. These arguments have implications that were supported by transmission electron microscopy evidence from different III-V and II-VI compound semiconductor quantum dot systems. The key ingredients for the future development of a qualitative model for structural transformations in ordinarily strained quantum dots were identified.

ACKNOWLEDGMENTS

PM and TT were supported by the National Science Foundation project no. DMR-9733895. Prof. William A. Jesser from the Department of Materials Science and Engineering of the University of Virginia is thanked for bringing the seminal Bragg-Williams paper to our attention. Prof. Jacek K. Furdyna and Dr. Malgorzata Dobrowolska from the Physics Department of the University to Notre Dame are thanked for the supply of samples.

REFERENCES

1. T.P. Pearsall (editor), *"Quantum Semiconductor Devices and Technologies"*, (Kluwer Academic Publishers, 2000).
2. H.J. Kim and Y.H. Xie, *Appl. Phys. Lett.* **79**, 263 (2001).
3. P.B. Joyce, P.B. Joyce and T.J. Krzyzewski, G.R. Bell, B.A. Joyce, and T.S. Jones, *Phys. Rev.* **B 58**, R15981 (1998).
4. M. Strassburg, V. Kutzer, U. W. Pohl, A. Hoffmann, I. Broser, N. N. Ledentsov, D. Bimberg, A. Rosenauer, U. Fischer, D. Gerthsen, I. L. Krestnikov, M. V. Maximov, P. S. Kop'ev, and Zh.I. Alferov, *Appl. Phys. Lett.* **72**, 942 (1998).
5. T. Walter A.G. Cullis, D.J. Norris, and M. Hopkinson, *Phys. Rev. Lett.* **86**, 2381 (2001).
6. P. Möck, T. Topuria, N.D. Browning, M. Dobrowolska, S. Lee, J.K. Furdyna, G.R. Booker, N.J. Mason, and R.J. Nicholas, *Appl. Phys. Lett.* **79**, 946 (2001).
7. P. Möck, T. Topuria, N.D. Browning, L. Titova, M. Dobrowolska, S. Lee and J.K. Furdyna, *J. Electron. Mater.* **30**, 748 (2001).
8. P. Möck, T. Topuria, N.D. Browning, G.R. Booker, N.J. Mason, R.J. Nicholas, L.V. Titova, M. Dobrowolska, S. Lee and J.K. Furdyna, *Mater. Res. Soc. Symp.* **640**, P6.3.1 (2000).
9. T. Topuria, P. Möck, N.D. Browning, L.V. Titova, M. Dobrowolska, S. Lee and J.K. Furdyna, *Mater. Res. Soc. Symp.* **640**, P8.3.1 (2000).
10. N. Marzari, S. de Gironcoli, and S. Baroni, *Phys. Rev. Lett.* **72**, 4001 (1994).
11. D.B. Lanks, S-H. Wei, and A. Zunger, *Phys. Rev. Lett.* **69**, 3766 (1992).
12. A. Zunger and S. Mahajan, *"Atomic ordering and phase separation in epitaxial III-V alloys"*, in *Handbook on Semiconductors* (Elsevier Science B.V., 1994), Ed. T.S. Moss, Vol. 3, Volume Ed. S. Mahajan, pp. 1447-1514.
13. H-Y. Liu, B. Xu, Y-Q Wei, D. Ding, J-J. Qian, Q. Han, J-B Liang, and Z-G. Wang, *Appl. Phys. Lett.* **79**, 2869 (2001).
14. W.L. Bragg and E.J. Williams, *Proc. Roy. Soc. (London)* A **145**, 699 (1934).
15. H.A. Bethe, *Proc. Roy. Soc. (London)* **A150**, 552 (1935).
16. L.A. Girifalco and D.O. Welch, *"Point Defects and Diffusion in Strained metals"*, (Gordon and Breach, 1967).
17. O. Madelung, ed. *"Semiconductors Group IV Elements and III-V Compounds, Data in Science and Technology"* (Springer, 1991)
18. *Internet based semiconductor data base* at http://www.ioffe.rssi.ru/SVA/NSM/Semicond/InP/thermal.html. At the date this paper was written, this URL was deemed to be useful as source of data. Neither the author nor the Materials Research Society warrants or assures liability for the content or availably of this URL.
19. A.H. Cottrell, *"Theoretical Structural Metallurgy"*, (Edward Arnold Publ.1965).
20. W. Seifert, chapter 14 in ref. [1], pp. 139-181.
21. P. Möck, G.R. Booker, N.J. Mason, E. Alphandéry, and R.J. Nicholas, *IEE Proc.-Optoelectron.* **147**, 209 (2000), *and unpublished material.*
22. E. Alphandéry, R.J. Nicholas, N.J. Mason, P. Möck, and G.R. Booker, *Appl. Phys. Lett.* **74**, 2041 (1999).
23. C.S. Kim, M. Kim, S. Lee, J.K. Furdyna, M. Dobrowolska, H. Rho, L.M. Smith, H.E. Jackson, E.M. James, Y. Xin and N.D. Browning, *Phys. Rev. Lett.* **85**, 1124 (2000).

Metallic Nanoparticles

Mat. Res. Soc. Symp. Proc. Vol. 707 © 2002 Materials Research Society   A5.3

Eluding Metal Contamination in CMOS Front-End Fabrication by Nanocrystal Formation Process

Zengtao Liu, Chungho Lee, Gen Pei, Venkat Narayanan and Edwin C. Kan
School of Electrical and Computer Engineering, Cornell University
Ithaca, NY 14850

ABSTRACT

A technique to form metal nanocrystals on silicon or thin SiO_2 film by Rapid Thermal Annealing (RTA) of thin metal film is developed and integrated into standard CMOS processing to make EEPROM devices and improve metal-semiconductor contact resistance. I-V and C-V measurements are carried out on MOSFETs and MOS capacitors containing Au, Ag, Pt, and Si nanocrystals as floating gate for universal mobility and minority carrier lifetime extraction. Mobility around 300 cm^2/V-sec and minority carrier lifetime within $0.02 \sim 0.1$ μsec are observed for all cases including the control samples that do not go through the metal nanocrystal formation process, which suggests that the substrate is virtually free from metal contamination. Using this technique, thicker metal film can potentially be achieved as well by stitching thin metal layers on top of the nanocrystals.

INTRODUCTION

Use of metals in nano-scale CMOS structures offers many attractive device features. Conventional CMOS uses channel doping as a means for threshold voltage adjustment and the control of short channel effect. However, as the devices scale into nanometer regime, the random position/number distribution of the channel dopants introduces large fluctuations in the threshold voltage [1]. Metal gate can alleviate this problem by providing threshold voltage adjustment through the design choice of work function, with extra benefits of eliminating poly depletion and dopant penetration [2]. Metal gate with mid-gap work function is also a necessity for fully depleted ultra-thin SOI devices because of their threshold insensitivity to channel doping [3]. Metal nanocrystal floating gate EEPROM cells are much less subject to interface states and offer lower voltage/longer retention operations than Si/Ge nanocrystal ones [4]. Metal source/drain contacts by thin films [5] or nanocrystals [6] can reduce the sheet/contact resistance in Schottky-barrier MOSFET [7] and potentially eliminate the need of doping entirely in CMOS.

However, Metals are conventionally forbidden in the front-end fabrication before the relatively thick CVD oxide (>100nm) is deposited over poly gate layers, especially for short-channel and shallow-junction devices. Metals can contaminate oxide and Si by junction spiking, grain boundary worming, mismatched expansion and deep diffusion during later thermal cycle, which can cause serious hazard to channel mobility, minority lifetime, interface states and oxide quality [8]. In the present study, we have found that if metals are introduced as ultra-thin films (<5nm), a Rapid Thermal Annealing (RTA) can relax the film stress and transfer it into thermally and chemically more stable nanocrystals, which will prevent metal contamination of oxide or Si substrate from occurring in the subsequent processing. The nanocrystal formation is studied using Scanning Electron Microscopy (SEM) and the contamination is monitored by I-V and C-V measurement of MOSFETs and MOS capacitors containing Au, Ag, and Pt nanocrystals as floating gate.

TECHNIQUE FOR METAL NANOCRYSTAL FORMATION

Metal nanocrystals can be formed by RTA of thin metal film on silicon or SiO_2 layer. On top of a cleanly prepared substrate, a thin layer of metal (1~5nm) is deposited on gate oxide by e-beam evaporation. Then the sample is annealed at temperatures close to its eutectic temperature with the substrate in an inert ambient to transfer the film into nanocrystals. This process is achieved through the relaxation of film stress and limited by the surface mobility. Some long-range forces such as the dispersion force and the electrical double layers will also affect the nanocrystal size and location distributions [9,10].

Figure 1 shows some SEM pictures of Au films of various thicknesses on 8nm thermal oxide before and after RTA. It could be seen that the as-deposited film comes naturally with some thickness perturbation. When it is RTA treated to give the atoms enough surface mobility, the film will be relaxed into a lower-total-energy state. To reduce the elastic energy carried by the stress built into the film during the deposition process, the film tends to break into islands along the initial perturbation. However, minimization of the surface energy and the dispersion force between the top and bottom interfaces can help stabilize the film. So the final geometry will depend on the balance of these driving forces. Once the nanocrystals have formed, the work function difference between the metal and the extrinsic substrate generates localized depletion or accumulation region in the substrate. The repulsion force between those regions helps stabilize the nanocrystals and keep a uniform spacing.

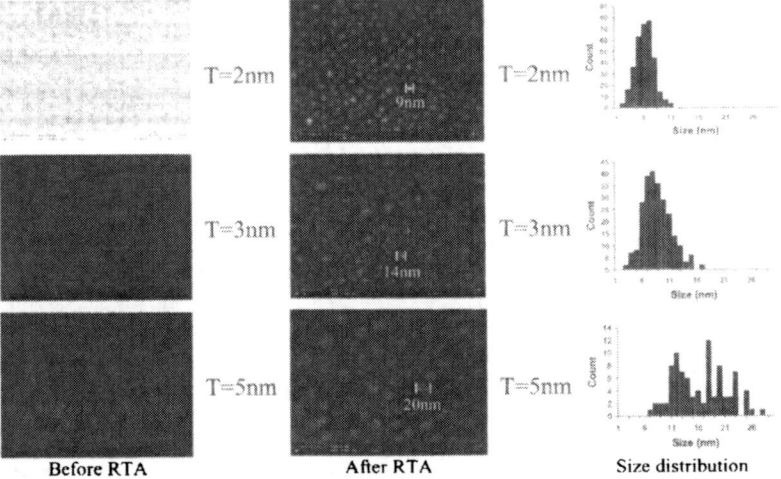

Figure 1. SEM pictures of Au films of various thicknesses on 8nm thermal SiO_2 before and after RTA. RTA is performed at 550°C for 5 minutes. Also shown is the size distribution of the nanocrystals.

From Figure 1 we can see that the nanocrystal density and size distribution can be controlled by the initial film thickness. Because this self-assembly is a thermodynamic process in nature, the annealing temperature and profile will also affect the nanocrystal geometry. Besides Au, we have also demonstrated Ag, Pt and W nanocrystal formation on both silicon and SiO_2 substrate. Nanocrystal density and size distribution can be controlled to some extent through the initial film thickness and annealing profile in a similar fashion, with typical nanocrystal density around mid $10^{11}cm^{-2}$ and nanocrystal size in the 3~20nm range. This process can be easily integrated into standard CMOS processing to make nanocrystal memories [4] or to improve the metal-semiconductor contact resistance [6]. Based on the nanometer size and the thermodynamic nature of their formation process, we assume the metal clusters be in single crystalline forms with significant distortion in surface atomic sites. However, this point needs further experimental corroboration, e.g. high resolution TEM.

DEVICE CHARACTERIZATION AND DISCUSSION

NMOSFETs and MOS capacitors containing Au, Ag and Pt nanocrystals as floating gate are fabricated, together with control devices with Si nanocrystals or without nanocrystals at all. Some sample process parameters are listed in Table I. Both I-V and C-V characterizations are performed for all devices to extract channel universal mobility and minority carrier lifetime.

Table I. Major process parameters for fabricated MOSFETs and MOS capacitors.

Nanocrystals	Tunnel oxide thickness	Control oxide thickness	RTA profile for nanocrystal formation	Control gate material	S/D dopant activation profile
Au	8nm	~ 30nm	550°C, 30sec	WSi_2	800°C, 60sec
Ag	8nm	~ 30nm	500°C, 30sec	WSi_2	800°C, 60sec
Pt	8nm	~ 30nm	900°C, 30sec	WSi_2	800°C, 60sec
Si	8nm	~ 30nm	1000°C, 1hr	WSi_2	800°C, 60sec
None	8nm	~ 30nm	N/A	WSi_2	800°C, 60sec

I-V characterization

If metal atoms penetrate the gate oxide and contaminate the silicon substrate, it can be reflected in the MOSFET I-V characteristics in two ways. First, this penetration process will deteriorate the oxide integrity and induce excessive gate leakage current; secondly, metal atoms in the inversion layer can serve as deep traps, which, after occupied by electrons or holes, will introduce extra Coulomb scattering and reduce the channel universal mobility. Then by measuring the MOSFET I-V characteristics, we can infer the level of metal contamination.

Figure 2 shows the measured MOSFET I-V characteristics for various devices and the extracted channel universal mobility. The universal mobility is extracted from the device drain conductance and carrier concentration following Takagi, et al [11]. The drain conductance is measured under 50mV V_{ds}, while the carrier concentration is determined through gate capacitance measurement. From Figure 2a we can see that the gate leakage current remains low for all the devices, which suggest decent gate oxide quality. All the data shown in Figure 2a are taken from fresh devices before injecting charges into the nanocrystals. The higher threshold voltage of the devices with Si nanocrystals in comparison with those with metal nanocrystals is mainly affected by their thicker effective oxide thickness, because Si nanocrystals act as an extra layer of dielectric. The threshold voltage distribution extracted from I-V and C-V methods in Fig. 2(a) and Fig. 3, on the other hand, has to be explained by the process variations. These include variations in the control oxide thickness across different wafers/dies, the bird's beaks caused by the LOCOS isolation, and short-channel and narrow-width effects. The oxide charges and interface states are unlikely to cause this kind of threshold voltage variation, because the C-V results in Fig. 3 shows rather consistent flat-band voltage for all samples. Figure 2b shows that the universal mobility for different devices almost falls on the same curve, which is an indication that the inversion channel is free from metal contamination.

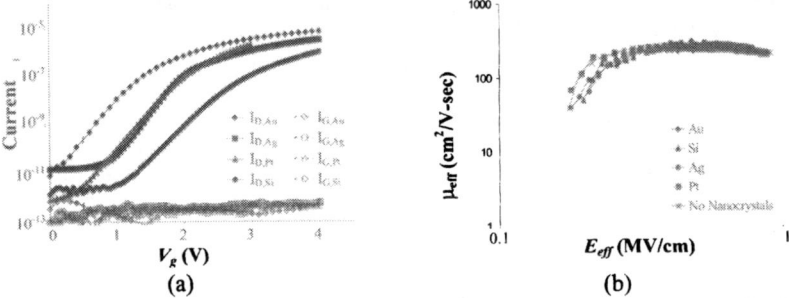

Figure 2. MOSFET I-V characteristics and extracted channel universal mobility for devices with different nanocrystals. The dimensions for Au, Ag, Pt, and Si devices are 3μm/9μm, 3μm/12μm, 3μm/12μm, and 3μm/4μm, respectively.

C-V characterization

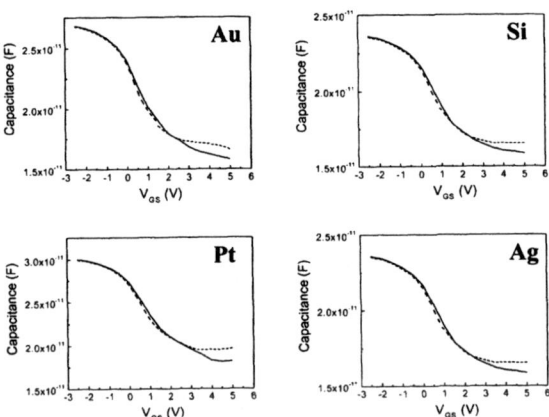

Figure 3. High frequency C-V measurement of MOS capacitors 200μm in diameter containing different nanocrystals. Deep depletion effect can be readily observed by sweep rate variation. Measurement is performed at 1MHz with sweep rate ----- 0.1V/sec, and ——— 1V/sec.

C-V measurement on MOS capacitors is also carried out to estimate the minority carrier lifetime. Because metal atoms in the silicon substrate serve as R-G (recombination-generation) centers, the minority carrier lifetime will be dramatically reduced if excessive contamination occurs. The minority carrier lifetime can be measured by MOS capacitor high-frequency C-V (HFCV) characteristic in steady state and deep-depletion sweeps. If excessive R-G centers present in the depletion region, it will be more difficult to drive the HFCV into deep depletion with a slow DC sweep of the gate bias because charges can be provides by the increased R-G rate. Figure 3 shows the high frequency C-V curves obtained on MOS capacitors with various nanocrystals at different sweep rate. Deep depletion is readily observed on all samples with 1V/sec sweep rate.

The minority carrier lifetime can be estimated from the difference between the deep depletion and quasi-static C-V characteristics [12]. The extracted lifetimes are listed in Table II. It can be seen that minority carrier lifetime for all samples are within $0.02{\sim}0.1\mu$sec range and no apparent degradation can be observed in samples containing metal nanocrystals in comparison with the control devices.

Table II. Minority carrier lifetime extracted from C-V characteristics.

Samples	Au	Ag	Pt	Si	No Nanocrystals
τ (μsec)	0.02~0.07	0.04~0.06	0.06~0.12	0.03~0.05	0.04~0.1

Both the I-V and C-V measurements suggest that no serious contamination is introduced by metal nanocrystals even with thin gate oxide and high temperature source/drain dopant activation. Our speculation is that the local energy minimization in the self-assembled nanocrystal formation process pushes the system into an energetically stable state. Therefore, metal atoms prefer staying in the nanocrystals to penetrating the oxide and contaminating the substrate because any such perturbation involves overcoming a higher dissolution energy barrier, thus is energetically unfavorable.

CONCLUSIONS

In this paper we describe a self-assembled metal nanocrystal formation process that can be easily integrated into CMOS front-end processing. Using this technique, nMOSFETs and MOS capacitors containing metal nanocrystals as floating gate are fabricated. Through I-V and C-V characterization we demonstrated that metal contamination is virtually eluded by this technique. This finding sheds light on the potential of including metals into CMOS front-end processing to alleviate many of its scaling limitations. With this technique, thicker metal film may also be achieved by stitching thin layer on top of the nanocrystals.

ACKNOWLEDGEMENTS

The research project is supported by the National Science Foundation (NSF). The device fabrication is carried out at Cornell Nanofabrication Facilities (CNF). The authors would like to express their great appreciations to all the CNF staff for their kind help and support.

REFERENCES

[1] H.-S. P. Wong, D. J. Frank, P. M. Solomon, C. H. J. Wann, and J. J. Welser, *Proc. IEEE*, **87**, 537 (1999).

[2] Q. Lu, Y. C. Yeo, P. Ranade, H. Takeuchi, T.-J. King, C. Hu, S. C. Song, H. F. Luan, and D. Kwong, *Proc. Symp. VLSI Technology*, 72 (2000).

[3] H. Shimada, Y. Hirano, T. Ushiki, K. Ino, and T. Ohmi, *IEEE Trans. Electron Devices*, **44**, 1903 (1997).

[4] Z. Liu, V. Narayanan, M. Kim, G. Pei, and E. C. Kan, *Tech. Dig. Device Research Conference*, 79 (2001).

[5] J. Kedzierski, P. Yuan, E. H. Anderson, J. Bokor, T.-J. King, and C. Hu, *IEDM Tech. Dig.*, 57 (2000).

[6] V. Narayanan, Z. Liu, Y. N. Shen, M. Kim, and E. C. Kan, *IEDM Tech. Dig.*, 365 (2000).

[7] J. R. Tucker, Proc. Advanced Workshop on Frontiers in Elec., 97 (1997).

[8] S. A. Campbell, *The Science and Engineering of Microelectronic Fabrication*, Oxford University Press (1996).

[9] Z. Suo and Z. Zhang, *Phys. Rev. B*, **58**, 5116 (1998).

[10] D. A. Bonnell, Y. Liang, M. Wagner, D. Carroll, and M. Bühle, *Acta Mater.*, **46**, 2263 (1998).

[11] S. Takagi, A. Toriumi, M. Iwase, and H. Tango, *IEEE Trans. Electron Devices*, **41**, 2357 (1994).

[12] R. F. Pierret, *IEEE Trans. Electron Devices*, **ED-19**, 869 (1972).

Mat. Res. Soc. Symp. Proc. Vol. 707 © 2002 Materials Research Society   V2.1

Self-Assembly of Metal Nanoclusters in Block Co-Polymers

Erica H. Tadd[1], John Bradley[1], Eugene P. Goldberg[2] and Rina Tannenbaum[1]*
[1]School of Materials Science and Engineering, Georgia Institute of Technology,
Atlanta, GA 30332. [2]Department of Materials Science and Engineering, University of
Florida, Gainesville, FL 32611.

ABSTRACT

This paper describes the formation of cobalt and iron metal nanoclusters in
various polymeric domains. The size of the particles, their size distribution and their
geometry is controlled by the extent of the interfacial interactions between the polymeric
phase and the growing metal fragments. Iron oxide particles are shown to exhibit various
geometries as a function of the polymer medium and the temperature at which they are
formed. The selective phase separation and particle confinement of cobalt clusters in the
presence of PS_{25300}-b-$PMMA_{25900}$ block co-polymer was achieved due to the different
reactivities of the functional groups in the blocks towards the metal fragments.
Transmission electron micrographs showed that cobalt clusters aggregated primarily in
the poly(methyl methacrylate) block, while no cobalt nanoclusters were observed in the
polystyrene block, thus creating a patterned distribution that coincided with the
morphology of the block copolymer.

INTRODUCTION

Hierarchical materials are ordered on the molecular (10-100Å), nano (10nm-
100nm) and meso (1μm-10μm) scales. This unique level of organization leads to
specialized material properties that significantly differ from those of less ordered phases.
However, optimal performance requires a degree of control over domain size and
distribution, on all length-scales, which is not easily obtainable with current synthetic
methods.

The development of a synthesis methodology to control the structure and
properties of metallic nanocluster-polymeric composites requires detailed understanding
of the interactions between the metal clusters and their polymeric environment, as well as
the relationship between polymeric parameters and cluster formation. Although various
studies have shown that polymers control cluster formation and properties [1-4], little is
known regarding the details of this relationship. The large number of parameters (e.g.
polymer molecular weight, chemistry and concentration) requires a systematic
investigation. This paper describes the development of a *versatile and controllable*
synthetic process for the formation of *three-dimensional, self-assembled nanoparticle
arrays* in, and aided by, a *polymeric medium*. The development of appropriate synthetic
and processing methods, which can produce finite, ordered domains with a given
geometry, is a central theme in the design of hierarchical ordered structures [5-8].

EXPERIMENTAL METHOD

The solution decomposition of $Co_2(CO)_8$ and $Fe(CO)_5$ to metal nanoclusters was
performed in a sealed three-neck round-bottomed reaction vessel that was first evacuated
and then flushed with N_2. A side neck was equipped with a thermometer, the middle neck
with a reflux condenser, and the other neck with a gas inlet-outlet glass fitting. The
thermometer outlet was also used for sampling. In order to drive the reactions away from

equilibrium and toward completion, the carbon monoxide gas formed during the reaction had to be continuously flushed away by the N_2 stream, while avoiding bubble formation. The jacketed reaction flask was equipped with tubing to circulate heated ethylene glycol around the reaction flask, thus keeping the contents of the flask at a constant temperature. The flask was connected to a vacuum line via the reflux condenser, which allowed strict control of the environment inside the flask at all times. The initial concentration of the metal carbonyl in the solution was 5×10^{-3} M. The solution was allowed to react under a continuous nitrogen stream with constant stirring.

The decomposition of the metal carbonyl precursors in both polystyrene and poly(methyl methacrylate) homopolymers, and in the PS_{25300}-b-$PMMA_{25900}$ block co-polymer was monitored via Fourier transform infrared spectroscopy (FTIR) at low temperatures. The decomposition in poly(styrene), $\overline{M}_n$ =250,000, was carried out in toluene at 90 °C and the decomposition in poly(methyl methacrylate), $\overline{M}_n$ =300,000, was carried out in methylene chloride at 40 °C. The decomposition in the PS_{25300}-b-$PMMA_{25900}$ block co-polymer solution was carried out in toluene at 90 °C as well. The reaction at the low temperature spanned more than two weeks, but data was collected for only the first 192 hours. Aliquots from the reaction solution were placed into a demountable infrared liquid cell with NaCl windows and a 0.2 mm optical path. The flask was placed under nitrogen during sample removal to prevent oxidation of the metal clusters. The cell was not demounted during the experiment, and the sample removal and cell washing was performed by suction. FTIR spectra were collected on a Nicolet Nexus 870 spectrophotometer, with a resolution of 2 cm^{-1} and 50 scans/spectrum.

After the decomposition reactions of $Co_2(CO)_8$ in 0.38 wt. % PS-b-PMMA co-polymer was complete, TEM samples were made by placing an aliquot of the solution onto a carbon coated TEM grid. These films constituted "bulk" block copolymer films. A high quality, free-standing film was also prepared to be microtomed by combining 1 mL of the block copolymer solution with 2 mL of a 30 wt. % solution of PS in toluene, $\overline{M}_n$ = 300,000. The final composition of these films was then 33% PS_{25300}-b-$PMMA_{25900}$ containing cobalt nanoparticles, and 67% PS homopolymer. The solution was added dropwise onto a glass slide until a film covered the entire surface. After allowing the solvent to evaporate, the film was removed from the slide by peeling off one corner of the film and applying water between the film and the slide. The hydrophobic interactions between the film and water caused the film to peel off easily. These films were microtomed and constituted "thinly sectioned" block copolymer films. The TEM imaging was conducted on both Hitachi HF-2000 field emission gun (FEG) and JEOL 2010, and a high resolution JEOL 4000EX. The operating voltage was 200 keV for all three microscopes.

RESULTS AND DISCUSSION

The chemical reactivity of metal clusters is largely determined by the size of the particles, and therefore the ability to control particle size, particle size distribution and dispersion would imply an ability to determine the reactivity and interfacial behavior of the clusters in the polymer-metal cluster interfacial systems. The synthesis of metal clusters, which is designed to accommodate these size manipulations, consists of the decomposition of organometallic complexes under controlled conditions (the energy source could be thermal, UV radiation or E-beam), to form uniform metal dispersions of very small particle size [9,10]. The process is best described in the following three stages (Figure 1): (a) The preparation of homogeneous solutions of metal complexes in a carefully selected solvent; (b) The mixing of the organometallic complex solution with a polymer solution in which the polymer of choice has been dissolved in the common solvent; and (c) The energy-induced decomposition of the organometallic complexes to form uniform metal dispersions of very small particle size in the polymer solution. The overall reaction may be described as follows:

$$nM_x(L)_y \xrightarrow{\text{\textit{Energy, Controlled Atmosphere, Polymer, Solvent}}} M_k^{(0)} \text{ and / or } M_j^{(i)}O_{ij/2} + nyL$$

(where M=metal; L=organic ligand; and k=nx).

This approach may afford unique opportunities for investigating fundamental aspects of nucleation and growth of metal clusters, as well as the properties of such nanoparticles as functions of cluster size, concentration and environment. Metallic fragments created by the energy-induced decomposition of metal complexes are highly reactive [11,12], which constitutes the driving force for the nucleation and growth mechanism to form nanocrystals. The size of the clusters formed is significantly influenced by the following parameters: (a) Reducing or oxidative atmospheres, (b) The diffusion of small cluster fragments through the medium and (c) The viscosity of the medium in which the diffusion takes place. Chemical reactivity, for instance, is found to be strongly related to the size of these metal clusters, since the size of the clusters largely determines their crystal structure.

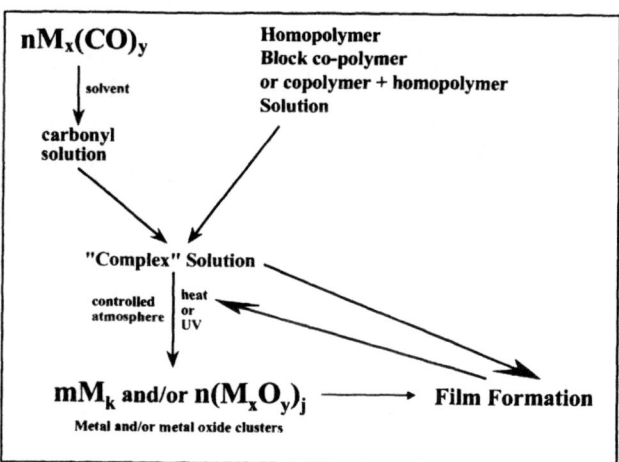

Figure 1: Schematic description of the synthesis methodology for the self-assembly of metal nanoparticles.

The thermal decomposition of cobalt carbonyls to metallic Co in polystyrene (PS) solutions has been used as a *model system* to test the chemical method under investigation. This process consists of a stepwise colloidal reaction mechanism [13-15], which is highly facilitated by the presence of macromolecules that provide the necessary solid state support and microenvironment which constitutes the driving force for cluster aggregation. As particle size grows, the mobility of the reactive metal fragments and their ability to diffuse through the solution and collide with each other decreases, and equilibrium is reached when the diffusion of the particles is too slow for observation in real time.

Experimental results of this model system (Figure 2), show that there is an inverse correlation between cobalt cluster size (measured by *Transmission Electron Microscopy*, TEM) and the PS concentration in the reaction solutions [16], accompanied by a considerable particle size distribution narrowing. When the excess polymer is washed off and the solvent is removed, the particles do not undergo additional growth, which indicates that they have been sufficiently coated by the "capping" polymer to prevent further aggregation.

When all the solvent has been removed from the polymer-containing solutions prior to decomposition (to form a polymer film), the smallest cobalt nanoparticles are formed. The ultimate result is a phase separated material in which the metal cluster and polymer phases are held together via the irreversible interfacial bonds created during the decomposition process. Moreover, the metallic nanoparticles are homogeneously dispersed throughout the polymer film, and given their small size, they exhibit high surface reactivity, and ensure good mechanical coupling of the particles to the polymer matrix.

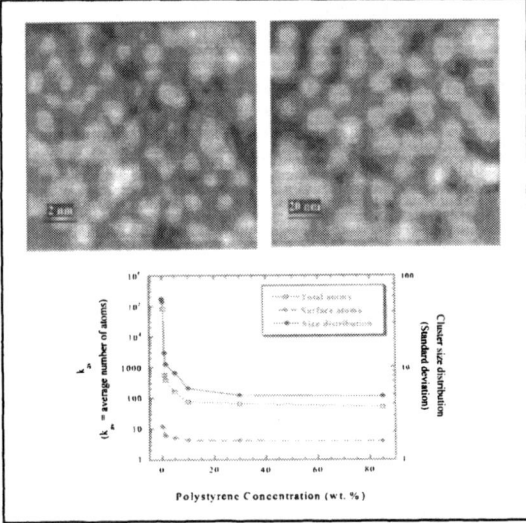

<u>Figure 2</u>: Co particles formed in polystyrene film (top left) and in a hydrocarbon solution (top right). The dependence of Co particle size, size distribution and abundance of surface atoms as a function of initial polymer concentration.

The synthetic method that we have developed for the controlled synthesis of metal nanoclusters can be extended to include also the manipulation of not only cluster size, size distribution and composition, but also geometrical shape. Biological systems have a unique ability to control crystal structure, phase, orientation and nanostructural regularity of inorganic materials [see, for example 17,18]. In biological hybrid systems it has been shown [17,18] that selected peptides can specifically bind to zinc-blende III-V semiconductor surfaces by discriminating between various crystallographic faces. These peptides are being used to grow nanoparticles and nanowires of specific crystallographic structure and orientation. Using these molecular interactions and specific nanoparticles, organic/inorganic hybrid materials may be organized into supramolecular architectures [for example, 19-21].

The preferred adsorption of the polymer chains to a particular crystalographic surface of the growing metal nano-clusters is clearly demonstrated in several Fe-polymer systems, as shown in Figure 3. The preferential adsorption of synthetic polymers to distinct crystallographic faces of the growing metallic fragment causes a distortion of the cluster shape, since the growth directions become differently hindered by the polymer. Hence, the result is a high degree of anisotropy in the cluster shape. In these systems, the protocol by which the decomposition reaction of $Fe(CO)_5$ in the presence of polymers is carried out is essential in determining the metal nanoparticle shape. Systems in which the same polymer has been used but the conditions of the thermal decomposition reaction were different, exhibited different particulate shape. For example, in the pesence of poly(vinylidene difluoride) (PVF2), the thermal treatment of the metal carbonyl-polymer complex solution that is concurrent with solvent removal and polymer film formation ("hot" method), gives rise to nanopyramids. On the other hand, the thermal treatment of the metal carbonyl-polymer complex solution that is performed after solvent evaporation at room temperature ("cold" method), gives rise to nanospheres [9]. In the former case, the solvent is present in the initial stages of the decomposition process, allowing a higher degree of mobility of the growing nanoclusters and the polymer chains, while in the latter case the metal carbonyl precursors are immobilized in the polymer film. Additional

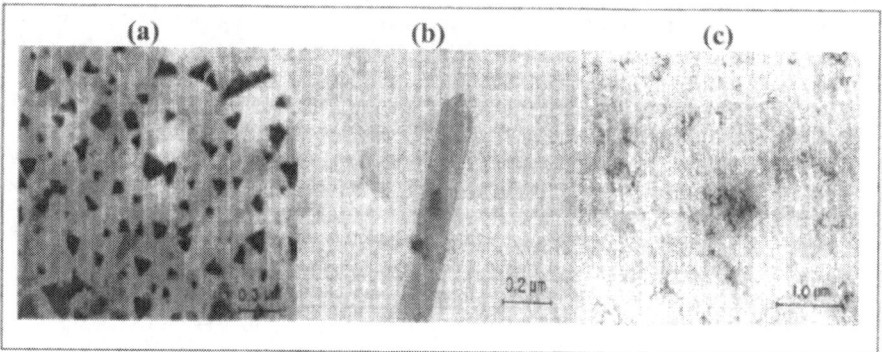

Figure 3: Shape control in the synthesis of iron nanoparticles via the decomposition of $Fe(CO)_5$ in the presence of (a) PVF2-nanopyramids; (b) PMMA-nanorods; and (c) Poly(carbonate)-nanostrings [9].

examples include the formation of Fe nanorods via the decomposition of $Fe(CO)_5$ in the presence of poly(methyl methacrylate) (PMMA), and the formation of nanostrings in the presence of poly(carbonate) (PC). This latter nanostructure is accompanied by extensive scission of the polymer chains, which may be partly responsible for the formation of the small Fe particles organized as nanostrings [9]. These examples are comprised of preliminary results, and a comprehensive study of these phenomena are currently underway. It is our intention to concentrate on the Fe and Co systems in the presence of PVF2, PMMA and PC, in order to establish the protocols for specific nanocluster polymer-induced geometry control as a function of polymer concentration and polymer molecular weight.

A direct extension of the methodology developed here is the creation of *polymer-induced, self-assembled, multi-functional nanoparticulate materials,* i.e. materials in which the metal clusters are phase separated within the polymer according to a predetermined spatial architecture. This spatial architecture can be achieved by the utilization of the phase separation and microdomain formation in multi-component polymers with self-organizing properties, i.e. *block copolymers,* for the anisotropic synthesis of nanoparticulate composites [22-24]. Hence, these block copolymers will be used as a structural and chemical template to produce a controllable, predetermined, self-developing spatial arrangement of clusters [24]. This implies that if the metal clusters are to aggregate in a particular polymeric domain, there should exist a driving force that would direct the phase separation such that certain areas in the polymer will be more reactive than others toward the metal clusters. This can be achieved by using *di-block copolymers* as a model system, in which one block will have reactive groups and the other block will be "inert" toward the metal clusters. One of the main features of block copolymers is the incompatibility between different parts of the same polymeric chain and the formation of microdomains [25]. The microdomain structure may consist of spheres, cylinders or lamellae, depending on the molecular weight of the blocks, their relative concentrations, the solvents used and the interfacial properties of the blocks. By carrying out the decomposition of the metal carbonyls in a di-block copolymer which was designed to phase separate in one of the main microdomain structures, we expect to obtain a high concentration of metal clusters in the domains occupied by the reactive block, and hence the spatial distribution of the clusters will conform to the spatial pattern created by the block copolymers. In this manner, it will be possible to concentrate the metal clusters in an ordered array of complex and highly ordered structures, as shown in Figure 4a. The homogeneous dispersion of the metal nanoclusters within the reactive block will provide certain physical and chemical properties inherent to metallic systems (e.g. mechanical strength, electrical conductivity, etc.), and thus create distinct functional regions within the polymeric material.

The selective incorporation of metal clusters into block copolymers may be achieved by three main pathways: (a) The exposure of block copolymer films to metal vapors, resulting in the selective adsorption of the metal atoms onto the more reactive block; (b) The impregnation of block copolymer films with a metal salt, with a subsequent reduction of the salt to form metallic clusters in the reactive block domains; and (c) The homogeneous mixing of the block copolymers with the organometallic precursors followed by the in-situ phase separation and self-assembly of both metal clusters and copolymer microdomains [22-24,26,27]. In the first two approaches, the block copolymers are first allowed to self assemble into their distinct microdomains, and are subsequently exposed to the metallic moiety. Under these conditions, there are significant differences in the size and dispersion of the metallic clusters between the bulk copolymer films and the thin (microtomed) films, due to the rates of diffusion and penetration depth of the metallic precursors. Moreover, if the polymers are below their T_g, the interfacial adhesion between polymer chains and the growing clusters is severely hindered, and therefore, the limitation on metal cluster size is due to nucleation and growth kinetics and not to a polymer "capping" effect. The third method, and the one that we have developed, is designed to circumvent these issues by performing the nucleation and growth of the metal clusters and the phase separation of the block copolymers *in-situ*, in a homogeneous solution of both components. Under these conditions, nucleation and growth of the clusters will be limited not only by reaction kinetics, but also by the direct interaction with the available reactive sites on the polymer. Hence, the difference in the reactivity of the two blocks toward the metal will be fully exhibited.

Preliminary results for the cobalt-poly(styrene-*b*-methyl methacrylate) block copolymer system (Figure 4c), show a distinct segregation of the cobalt clusters in defined regions, similar to the "worm-like" distribution achieved by Cohen et al. in their microtomed Ag-block copolymer samples [24]. The actual microstructure exhibited by the metal-containing block copolymers is strongly influenced and complicated by the phase behavior of block copolymer in solution prior to metal incorporation, irrespective to the method used. In dilute solutions, the blocks form micelles [27] where one of the blocks forms the core of the micelle (the block less favored by the solvent used), and the other forms the shell. After the incorporation of the metal nanoparticles, the solvent is evaporated, and it is expected that a phase transition will take place resulting in a microstructure that corresponds to the block copolymer composition. However, our own preliminary results and the results in the literature [22-24,26,27] suggest that this phase transformation is incomplete, either due to some degree of crosslinking caused by the metal nanoparticles or by the change in phase behavior of the block copolymers as a result of the presence of the nanoclusters (Figure 4b,c).

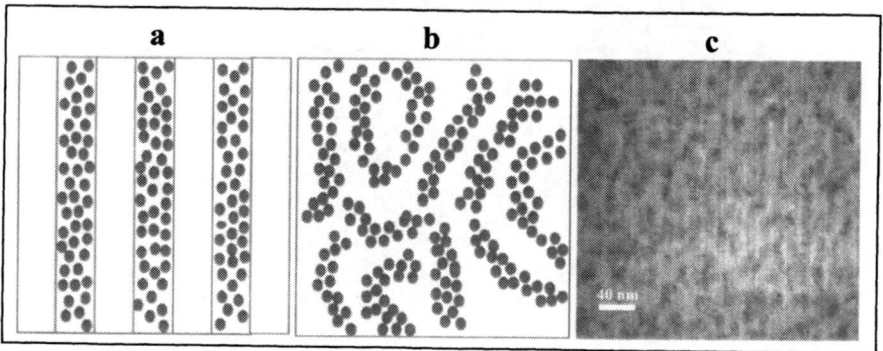

Figure 4: Cobalt nanoparticle size distribution in the PS-PMMA block co-polymer domains. (a) Idealized lamellar phases; (b) Arrested phase transformation; (c) TEM micrograph of particle distribution.

The average particle size is ~ 50 Å, and it does not change upon film formation. Moreover, particle size and microdomain distribution is independent of film thickness, and is a function of the initial polymer solution concentration in the decomposition reaction.

Additional insight into the strength of the interaction between the cobalt nanoparticles and the PMMA block was obtained by probing the diffusion of the cobalt particles from the PMMA domain into a large PS domain. For this purpose, the $PS_{25300}\text{-}b\text{-}PMMA_{25900}$ block copolymer solutions used in this work were doped with high molecular weight polystyrene homopolymer. The addition of a homopolymer to a diblock copolymer has the effect of increasing the domain size of the common block for a particular morphology [27-30], if the total concentration of the homopolymer component is within the concentration regime commensurate with that morphology. In this particular case, since the molecular weight of the polystyrene homopolymer far exceeded (six fold) the combined molecular weights of both blocks, the block copolymer and the homopolymer behaved as a macrophase and microphase separated blend [31-33]. Figure 5 shows a HRTEM image of a "thinly sectioned" microtomed film of a high molecular weight polystyrene ($\overline{M}_n = 300,000$) blended with the $PS_{25300}\text{-}b\text{-}PMMA_{25900}$ block copolymer, that contained cobalt precursors which were then decomposed *in situ* according to the cluster synthesis used in these experiments. The decomposition of the cobalt precursors in the polymer medium occurred either before or after the addition of the PS homopolymer. The distribution of the cobalt particles within the polymeric matrix was similar in both cases. The particles were concentrated in the block copolymer, specifically in the PMMA microdomain, without any traces of cobalt clusters in the polystyrene homopolymer phase. No cobalt particle diffusion from the poly(methyl methacrylate) domain to the polystyrene domain was observed, as evidenced by the irreversible confinement of the cobalt particles in the more reactive block.

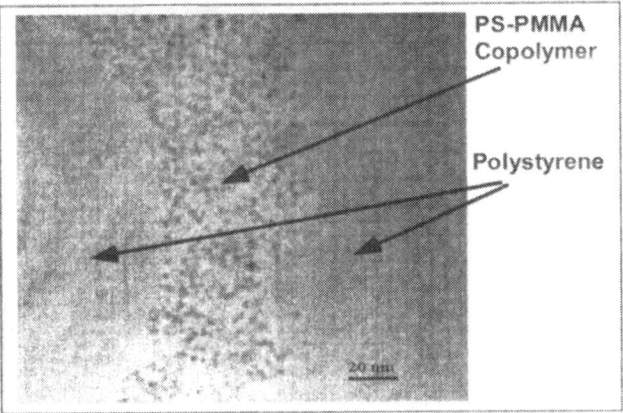

Figure 5: HRTEM image of a microtomed film of the homopolymer/block co-polymer blend showing cobalt particle confinement in the block phase.

SUMMARY

The results presented in this paper demonstrate that the incorporation of metal nanoclusters, synthesized via an *in situ* decomposition of metal carbonyl precursors into a block copolymer, generates an uneven dispersion of the particles, with a preferential aggregation in the poly(methyl methacrylate) block. This can be explained by the reactive adsorption of the PMMA block onto the metal cluster surface, through the interaction of the carbonyl group of the polymer with the metal. This is supported by experimental results obtained with Fourier Transform Infrared (FTIR) spectroscopy that are described elsewhere [34]. The morphology of the system did not conform to the expected lamellar structure, due to the fact that the PS_{25300}-b-$PMMA_{25900}$ block copolymer that was used forms micelles in the solvent of choice (in this case toluene), with the poly(styrene) block constituting the shell and the poly(methyl methacrylate) block constituting the core of the micelles. The preferential incorporation of the cobalt nanoclusters into the poly(methyl methacrylate) block has the effect of immobilizing the micellar structure, thus preventing the expected phase transformation to the expected lamellar morphology upon removal of the solvent. Moreover, when this block copolymer was mixed with a high molecular weight poly(styrene) homopolymer, the common poly(styrene) moieties did not enhance the mutual solubility of the homopolymer and the block copolymer, and the material behaved as a macrophase and microphase separated blend.

ACKNOWLEDGEMENTS

The authors thank NSF-ERC (through the Packaging Research Center), the Georgia Tech Foundation, and the College of Engineering at Georgia Tech for financial support.

REFERENCES

1. Addadi, L.; Weiner, S. *Angew. Chem. Int. Ed. Engl.* **1992**, *31*, 169.
2. Belcher, A. M.; Hansma, P. K.; Stucky, G. D.; Morse, D. E. *Acta Mater.* **1998**, *46*, 733-736.
3. McGrath, K. M. *Adv. Mater.* **2001**, *13*, 989.
4. Pileni, M.-P.; Ninham, B. W.; Gulik-Krzywicki, T.; Tanori, J.; Lisiecki, I.; Filankembo, A. *Adv. Mater.* **1999**, *11*, 1358.
5. Alivisatos, A. P.; Barbara, P. F.; Castleman, A. W.; Chang, J.; Dixon, D. A.; Klein, M. L.; McLendon, G. L.; Miller, J. S.; Ratner, M. A.; Rossky, P. J.; Stupp, S. I.; Thompson, M. E. *Adv. Mater.* **1998**, *10 (16)*, 1297-1336.
6. Stupp, S. I.; Keser, M.; Tew, G. N. *Polymer* **1998**, *39(19)*, 4505-4508.
7. Stupp, S. I.; Pralle, M. U.; Tew, G. N.; Li, L.; Sayar, M.; Zubarev, E. R. *MRS Bull.* **2000**, *25(4)*, 42-48.
8. Grier, D. G. *MRS Bull.* **1998**, *23(10)*, 21-23.
9. Tannenbaum, R.; Goldberg, E. P.; Flenniken, C. L. "Metal-containing polymeric systems", Eds. Carraher C., Pittman C. U. and Sheats J., Plenum Press, New York **1985**, p. 303-340.
10. Tannenbaum, R.; Flenniken, C. L.; Goldberg, E. P *J. Polym. Sci. Phys. Ed.* **1990**, *28*, 2421.
11. Klaubunde, K. J.; Tanaka, Y. *J. Molec. Catal.* **1983**, *21*, 57.

12. Kanai, H.; Tan, B. J.; Klaubunde, K. J. *Langmuir* **1986**, *2(6)*, 760.
13. Kernizan, C. F.; Klabunde, K. J.; Sorensen, C. M.; Hadjipanayis, G. C. *J. Appl. Phys.* **1990**, *67(9)*, 5897.
14. Tannenbaum, R. *Langmuir* **1997**, *13(19)*, 5056, and pertinent references therein.
15. Rotstein, H. G.; Novick-Cohen, A.; Tannenbaum, R. *J. Stat. Phys.* **1998**, *90(1/2)*, 119.
16. Rotstein, H. G.; Novick-Cohen, A.; Tannenbaum R. *J. Phys. Chem. B* **2001**, to appear.
17. Smith, B. l.; Schäffer, T. E; Viani, M.; Thompson, J. B.; Frederick, N. A.; Kindt, J.; Belcher, A.; Stucky, G. D.; Morse, D. E.; Hansma, P. K. *Nature* **1999**, *399*, 761-763.
18. Whaley, S. R.; English, D. S.; Hu, E. L.; Barbara, P. F.; Belcher, A. M., *Nature* **2000**, *405*, 665-668.
19. Ahmadi, T. S.; Wang, Z. L.; Green, T. C.; Henglein, A.; El-Sayed, M. A. *Science* **1996**, *272*, 1924-1925.
20. Petroski, J. M.; Wang, Z. L.; Green, T. C.; El-Sayed, M. A. *J. Phys. Chem. B* **1998**, *102*, 3316-3320.
21. Bradley, J. S.; Tesche, B.; Busser, W.; Maase, M.; Reetz, M. T. *J. Am. Chem. Soc.* **2000**, *122*, 4631-4636.
22. Morkved, T. L.; Wiltzius, P.; Jaeger, H. M.; Grier, D. G.; Witten, T. A. *Appl. Phys. Lett.* **1994**, *64(4)*, 422-424.
23. Caruso, F.; Möhwald, H. *Langmuir* **1999**, *15(23)*, 8276-8281.
24. Clay, R. T.; Cohen, R. E. *Supramolecular Science* **1995**, *2*, 183-191.
25. Bates, F. S.; Fredrickson, G. H. *Ann. Rev. Phys. Chem.* **1990**, *41*, 525-557.
26. Hashimoto, T. ; Harada, M. ; Sakamoto, N. *Macromolecules* **1999**, *32*, 6867-6870.
27. Spatz, J. P.; Mössmer, S.; Hartmann, C.; Möller, M.; Herzog, T.; Krieger, M.; Boyen, H. –G.; Ziemann, P.; Kabius, B. *Langmuir* **2000**, *16*, 407-415.
25. Li, R.R., Dapkus, P.D., Thompson, M.E., Jwong, W.G., Harrison, C., Chaikin, P.M., Register, R.A., Adamson, D.H. *Applied Physics Letters* **2000**, *76 (13)*, 1689-1691.
26. Bronstein, L.M., Valetsky, P.M., Solodovnikov, S.P., Seregina, M.V., Register, R.A. *Macromol. Symp.* **1996**, *106*, 73-86.
27. Winey, K. I.; Thomas, E. L.; Fetters, L. J. *Macromolecules* **1992**, *25(1)*, 422-428.
28. Winey, K. I.; Thomas, E. L.; Fetters, L. J. *Macromolecules* **1991**, *24(23)*, 6182-6188.
29. Winey, K. I.; Thomas, E. L.; Fetters, L. J. *Macromolecules* **1992**, *25(10)*, 2645-2650.
30. Whitmore, M. D.; Noolandi, J. *Polym. Eng. Sci.* **1985**, *25 (17)*, 1120-1121.
31. Pepin, M. P.; Whitmore, M. D. *Macromolecules* **2000**, *33(23)*, 8644-8653.
32. Jeon, K. –J.; Roe, R. –J. *Macromolecules* **1994**, *27(9)*, 2439-2447.
33. Roe, R. –J. *Polym. Eng. Sci.* **1985**, *25(17)*, 1103-1109.
34. Tadd, E. H.; Bradley, J.; Tannenbaum, R. *J. Phys. Chem. B* **2001**, *105*, 0000-0000 (to appear).

Mat. Res. Soc. Symp. Proc. Vol. 707 © 2002 Materials Research Society

Characterization of Self-Assembled SnO$_2$ Nanoparticles for Fabrication of a High Sensitivity and High Selectivity Micro-Gas Sensor

R.C.Ghan, Y. Lvov, and R.S.Besser
Louisiana Tech University
Institute for Micromanufacturing
911 Hergot Avenue,
P.O.Box 10137,
Ruston, LA, 71270.
Fax: (240) 255-4028
Email: rbesser@coes.latech.edu

ABSTRACT

In order to refine further the material technology for tin-oxide based gas sensing we are exploring the use of precision nanoparticle deposition for the sensing layer. Layers of SnO$_2$ nanoparticles were grown on Quartz Crystal Microbalance (QCM) resonators using the layer-by-layer self-assembly technique. Scanning Electron Microscopy (SEM), Transmission Electron Microscopy (TEM), and Electron Diffraction Pattern (EDP) analyses were performed on the self-assembled layers of SnO$_2$ nanoparticles. The results showed that SnO$_2$ nanoparticle films are deposited uniformly across the substrate. The size of the nanoparticles is estimated to be about 3-5 nm. Electrical characterization was done using standard current-voltage measurement technique, which revealed that SnO$_2$ nanoparticle films exhibit ohmic behavior. Calcination experiments have also been carried out by baking the substrate (with self-assembled nanoparticles) in air at 350°C. Results show that 50%-70% of the polymer layers (which are deposited as precursor layers and also alternately in-between SnO$_2$ nanoparticle monolayers) are eliminated during the process.

INTRODUCTION

Solid-state gas sensors find applications in automobiles, toxic and domestic environments, the chemical industry, and elsewhere. The gas-sensor market is fast burgeoning and was estimated to be about $0.9 billion at the end of the last decade [1]. Ceramic SnO$_2$ has been used extensively as a sensor element in semiconductor gas-sensors for detecting a range of gases such as carbon monoxide, oxides of nitrogen, hydrogen sulfide, freon and many others [2]. SnO$_2$ is the prime choice for semiconductor sensors because of its bulk-material stability and resistivity characteristics. Sensors with SnO$_2$ as sensing element function on the principle of surface chemical reaction between an analyte gas and the sensing element, which causes a change in the resistance of the element. Thus the sensing characteristics depend on the surface properties of the element [3]. While performing sufficiently for commercial deployment, these sensors displayed a variety of material issues including the degree of crystallinity of SnO$_2$, crystallite size, density of lattice defects, surface area, and surface structure. These issues translated to low or varying selectivity and sensitivity of the SnO$_2$ sensors. Some of the techniques that have been implemented for improving the selectivity include cyclic manipulation of the sensor temperature, doping of the SnO$_2$ with various additives like Pt and noble metals [3], surface modification of the base metal oxide with hydrophobic groups, calcium oxide, zinc oxide and sulfur [4]. The method of fabrication is also a variant in the process to improve the performance of SnO_2 sensors. The different routes for fabrication adopted for metal-oxide thin film sensors are sol-gel method of deposition of SnO_2 thin films, reactive sputtering [5], and sintering and annealing [6]. These fabrication methodologies attempted to deposit thin films of SnO_2 for manifesting improved selectivity and sensitivity. Decreasing the particle size of SnO_2 is another factor which could contribute to improving performance. In the present study, we attempt to circumvent many of the limitations of the traditional preparation methods by depositing SnO_2 nanoparticles using the Layer-By-Layer (LBL) self-assembly technique which is based on alternate adsorption of oppositely charged components [7].

Nanocrystalline structure is known to impart high sensitivity and selectivity to gas-sensors [8]. Nanoparticles offer a high degree of structural control due to their well-defined size and shape, and the LBL assembly method allows them to be uniformly and precisely ordered on the substrate with the precision of a few nanometers [9]. We therefore anticipate the resulting films to possess high reproducibility of thickness, of crystallite size, and of the geometry of intergranular contact, which are known to be key factors in the gas sensing mechanism [8]. In addition, the LBL process permits the architecture of the layer to be engineered by insertion of monolayers of alternate materials (e.g. SiO_2 to enhance activity), doped nanoparticles, or nanoparticles of varying radius to modulate the film density [4]. We report here on the characterization of the SnO_2 layers formed by the LBL method. SEM, TEM, and EDP analysis were done on the nanolayers of SnO_2, in order to determine the size of the nanoparticles and thickness of the nanofilm. Electrical characterization was done to observe the current-voltage characteristics of the nanofilms. Calcination studies were also performed to investigate the elimination of the polymer layer, which is a by-product of the LBL process.

EXPERIMENTAL DETAILS

SnO_2 nanoparticles were obtained from Nyacol® NanoTechnologies (Nyacol Colloidal Tin Oxide) as two separate samples, SN-15 (negatively charged, pH 8.0-9.0, specific gravity 1.15, viscosity 5-10 cps and counter–ion concentration of 0.5% potassium) and SN-15CG (negatively charged, pH 9.0-10.5, specific gravity 1.15, viscosity 5-10 cps and counter ion concentration of 0.23% NH_3). The polyions used were Polydialyldimethyldiammonium chloride (PDDA), Sodium-polystyrene sulfonate (PSS), and Polyethyleneimine (PEI) of 2 mg/mL in aqueous solution (all Sigma-Aldrich. [9]. The initial characterization of the SnO_2 was carried out on a quartz crystal microbalance (QCM, USI-System, Japan) resonator in order to monitor the film growth during each step in the process. The technique initially involves layer-by-layer electrostatic assembly of several pairs of oppositely charged polyion layers as the precursor layers to promote the structural stability of the subsequently assembled nanoparticles. The outermost precursor polyion layer is oppositely charged (positive) with respect to the anionic nanoparticles. Then the nanoparticles and the complimentary polyion layer are alternately self-assembled as shown in Figure 2.

Self-assembly was tried first with PEI/PSS/(PEI/SnO_2)$_7$ and then with PDDA/PSS/(PDDA/SnO_2)$_{7-16}$. The latter sequence was found to result in more reproducible growth of SnO_2 nanoparticle multilayers. Thereafter all self-assembly experiments were done with the same sequence, i.e., PDDA/PSS/(PDDA/SnO_2)$_7$. In order to assess whether steady-state conditions had been reached during the adsorption process, the time of retention of the QCM resonator in the SnO_2 colloidal solution was varied and the results were obtained as change in frequency plotted against the number of monolayers assembled. The self-assembled samples were sent to RIKEN Frontier Research System, Topochemical Design Laboratory in Japan for SEM analysis. For TEM analysis for nanoparticle size measurement, EDP, and High Resolution TEM of nanoparticle grain structure, the samples were sent to Material Characterization Lab at Louisiana State University.

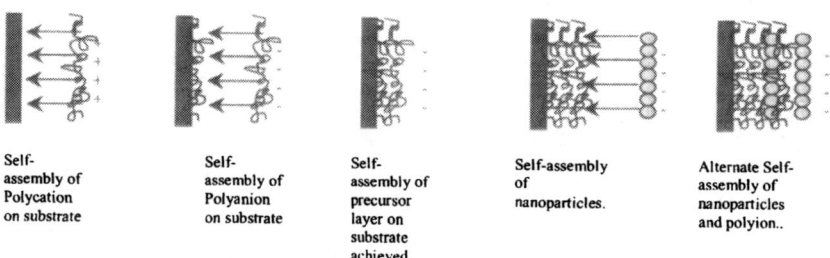

| Self-assembly of Polycation on substrate | Self-assembly of Polyanion on substrate | Self-assembly of precursor layer on substrate achieved. | Self-assembly of nanoparticles. | Alternate Self-assembly of nanoparticles and polyion.. |

Figure 1: Schematic of the Layer-By-Layer process used for self-assembly of SnO_2 nanoparticles.

I-V characteristics were studied using a standard electrical characterization set-up. The samples were prepared as 2.75cm x 3.5cm coupons of Pyrex glass. SnO_2 nanoparticles were self-assembled on the glass substrate. Electrical contact patterns were designed using LEDIT® software. The pattern was transferred onto Mylar® sheets, and cut out to form electrical contact patterns. Patterns for both two-point and four-point tests were created. Using these patterns, Pt contacts were sputter deposited onto the self-assembled glass samples. The samples were then tested for their I-V characteristics.

Calcination studies were carried out by baking a blank silver electrode resonator (resonator with no self-assembled layers), and another with 20 self-assembled layers of SnO_2 in a Thermolyne® Furnace at 350°C for 4 hours in one-hour intervals. The frequency readings for both the resonators were taken after each hour interval.

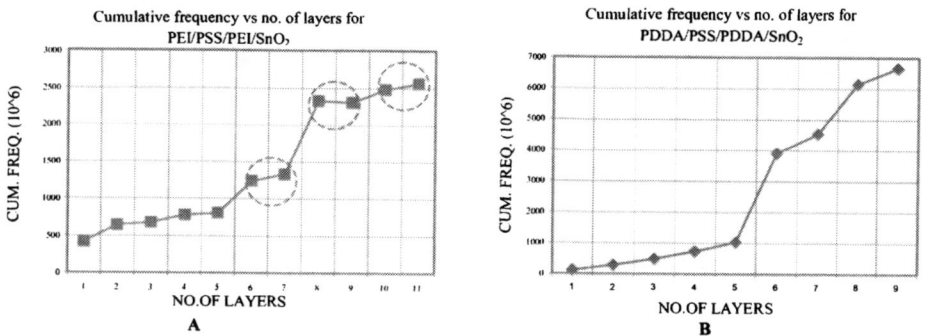

Figure 2: Graphs for cumulative delta F (Hz.) for (A) [PEI/PSS/PEI/SnO_2] and (B) [PDDA/PSS/PDDA/SnO_2].

RESULTS AND DISCUSSION

Self-Assembly of SnO_2 Nanoparticle Multilayers and Characterization

SnO_2 nanoparticles were self-assembled on a QCM resonator using two alternative sequences: 1) PDDA/PSS/ (PDDA/SnO$_2$)$_n$ and 2) PEI/PSS/(PEI/SnO$_2$)$_n$, where n = 5-16. After every adsorption cycle, a sample was dried, and the QCM frequency was measured. The frequency shifts (ΔF) characterized the mass adsorbed on resonator, and an increase of the film thickness [3].

As seen in the Figure 2 (A), the three-circled areas represent poor or irreproducible uptake of the PEI layer. This irreproducibility led to non-uniform deposition of subsequently self-assembled SnO_2 nanoparticle layers. No such phenomenon is observed in the sequence PDDA/PSS/PDDA/SnO$_2$. Therefore it was concluded that the second sequence was better suited for self-assembly of SnO_2 nanoparticles as it exhibited sound linearity in growth characteristic. Further experiments were carried out with the sequence PDDA/PSS/PDDA/SnO$_2$ and the time of immersion in the SnO_2 colloidal nanoparticle solution was varied from 20 minutes, 10 minutes, and 15 minutes down to 5 minutes. Figure 3 shows that the self-assembly of SnO_2 follows an approximately time-independent behavior i.e. addition of nanoparticles is essentially complete in less than 5 minutes of immersion time. This behavior yields a high precision of thickness control in the growth process. The anomalous behavior in the 18th growth layer of the 20-minute immersion film can be attributed to manual errors in handling the resonator while transferring it from one solution to another and/or handling while drying.

High-resolution imaging reveals that the particle size is approximately in the range of 3-5 nm. The thickness of the film deposited was calculated to be 490 nm using the QCM frequency data and the relation $\Delta t = -0.016*\Delta F$ [9] and the results obtained from the SEM analysis show that the thickness of the self-assembled films of eight SnO_2 monolayers is approximately 300-400 nm.

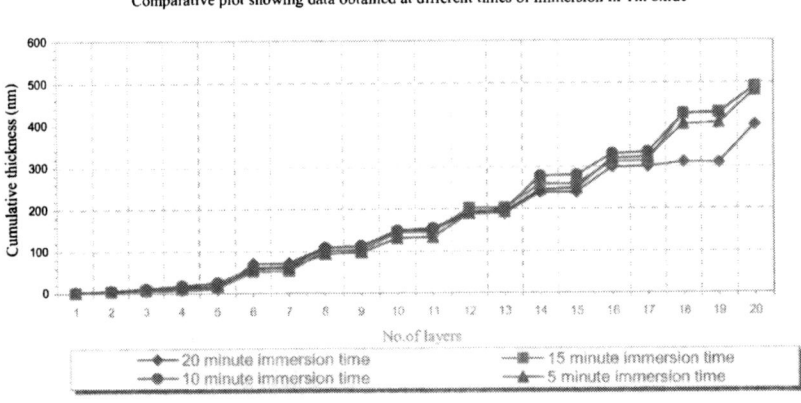

Self Assembly of SnO_2 nanoparticles on QCM resonator (PDDA/SnO$_2$/PDDA/SnO$_2$)
Comparative plot showing data obtained at different times of immersion in Tin oxide

Figure 3: Comparative plot showing data obtained at different times of immersion in tin-oxide colloidal nanoparticle solution.

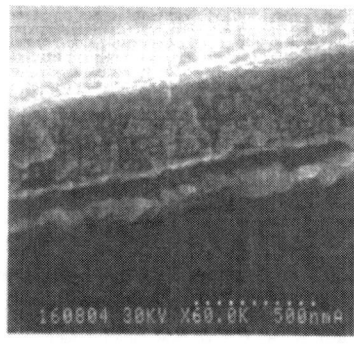

Figure 4: SEM image of cross-section of QCM resonator with self-assembled SnO_2 nanoparticles.

Electrical Characterization

Coupons of Pyrex glass of the same dimensions as the electrical patterns were prepared and Pt contacts 2000Å were sputter coated onto the coupons. One coupon was calcined for 1 hour to eliminate interlayer polyions. Two-point I-V results showed linear (ohmic) behavior of the deposited SnO_2 layers (Figure 5) for the calcined sample. The I-V characteristics for the uncalcined sample showed several spurious readings which are linked to the presence of the polyion layers in between the SnO_2 nanoparticle layers. We believe polyion layers impose regions of parallel conductance that undergo electrical breakdown to a high-impedance state thereby resulting in what appears to be a I-V characteristic superimposed with noise. Measured resistivities were 0.18 ohm-cm for the uncalcined sample, and 0.63 ohm-cm for the calcined sample.

Calcination

Calcination effectively eliminates the spurious points and also results in a higher net resistivity corresponding to the substantial reduction in polyion conductivity contribution .In the first calcination experiment, a self-assembled QCM resonator with 8 layers of SnO_2 (and total of 20 layers including the polyion layers), and a blank resonator control were used for

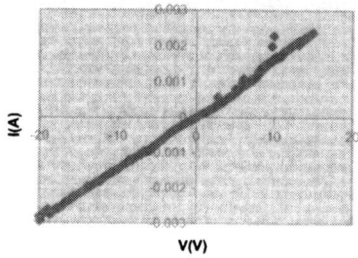 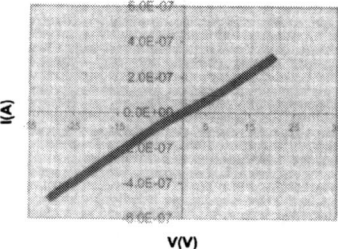

Figure 5: I-V characteristics for uncalcined (left) and calcined (right) samples.

calcination studies. Both the resonators were calcined for 4 hours and frequency measurements were recorded after every 1-hour interval. The initial frequency was measured for both the resonators before calcination. After every 1 hour, frequency deviation was recorded for both the resonators. Mass change was calculated using the relation [3]:

$$\Delta m(ng) = -0.87 \cdot \Delta F(Hz) \tag{1}$$

The self-assembled resonator lost approximately 2000 ng in the first hour, and less than 300 ng thereafter. The initial loss corresponded to 50% loss of total polyion deposited by self-assembly.

These results are in agreement with earlier study of temperature decomposition of the linear polycation / polyanion multilayers [10] which have shown that calcinations at 350° C for three hours is necessary to eliminate these films.

CONCLUSION

We have demonstrated self-assembly of SnO_2 nanoparticles on QCM resonator and glass substrates. SEM, TEM, and HRTEM analysis have been performed which give an estimate of the SnO_2 nanoparticle size, which is approximately 3-5 nm. The SnO_2 nanoparticle films deposited by LBL process are uniformly deposited across the substrate and the thickness of these films can be precisely controlled. Electrical characterization of self-assembled SnO_2 substrates has been performed and results show that self-assembled samples, which are calcined, show smoother ohmic behavior as compared to self-assembled samples, which are not calcined. Preliminary calcination studies show that approximately 55% to 70% of polyion mass is eliminated in the process.

ACKNOWLEDGMENTS

We are grateful to Dr. Izumi Ichinose, of RIKEN-Frontier Research System, Japan for making the SEM micrograph of the film cross-sections. This work was partially funded by NSF-EPSCoR research infrastructure grant "Chemical and Biochemical Micro- and Nanosytems."

REFERENCES

1) M. Madou, "Fundamentals of Microfabrication," (CRC press, 1997), pp. 495-500.
2) B. Hoffheins, "Resistive gas sensors," *Chemical and Biological Sensors*, ed. R.F Taylor, J.S Schultz, (IOP publishing Ltd., 1996) pp. 371-377.
3) S. Morrison, "Chemical Sensors," *Semiconductor Sensors,* ed. S. M. Sze, (John Wiley & Sons, Inc., 1994), pp. 383-412.
4) K. Ihokura, J. Watson. "Stannic Oxide Gas Sensor," (CRC press Inc., 1994) pp. 1-5.
5) S. Davis, A. Wilson, J. Wright, *IEE Proc.- Circuits Devices Syst.*, **145** (5), pp.379, (1998).
6) M. Ippommatsu, H. Ohnishi, H. Sasaki, T. Matsumoto, *J. App.Phys.* , **69**, pp.8368, (1991).
7) G. Decher, *Science*, **277**, 1232, (1997).
8) M. S. Hettenbach, "SnO_2 (110) and Nano-SnO_2: Characterization by Surface Analytical Techniques," pp. 18-20, 2000.
9) Y. Lvov, K. Ariga, I. Ichinose, T. Kunitake, *Langmuir,* **13**, 6195 (1997).
10) T. Farhat, G. Yassin, S. Dubas, J. Schlenoff, *Langmuir*, **15**, 6621 (1999).
11) R.C.Hughes, A.J.Ricco, M.A.Butler, S.J. Martin, *Science*, **6**, 74, (1991).

Mat. Res. Soc. Symp. Proc. Vol. 707 © 2002 Materials Research Society   AA8.6/V10.6

Investigating Catalytic Properties of Composite Nanoparticle Assemblies

M.M. MAYE, J. LUO, Y. LOU, N. K. LY, W.-B. CHAN, E. PHILLIP,
M. HEPEL[a], C.J. ZHONG*
Department of Chemistry, State University of New York at Binghamton, Binghamton,
NY 13902.   [a] Department of Chemistry, State University of New York at Potsdam,
Potsdam, NY 13676.   [*] cjzhong@binghmaton.edu

ABSTRACT

We present herein recent findings of an investigation of catalyst assembly and activation using metallic nanoparticles encapsulated with organic monolayers. Gold nanocrystals (2~5 nm) encapsulated with thiolate monolayers assembled on electrode surfaces, were found to be catalytically active towards electrooxidation of CO and MeOH upon activation. The activation involved partial removal of the encapsulating thiolates and the formation of surface oxygenated species. A polymeric film was also used as a substrate for the assembly of the nanoparticle catalysts. When the polymer matrix was doped with small amounts of Pt, a remarkable catalytic activity was observed. These catalysts were characterized utilizing cyclic voltammetry and atomic force microscopy.

INTRODUCTION

The pioneer work of two-phase synthesis of gold nanoparticles with a few nm core size stabilized by alkanethiolate monolayers has led to increasing research and development interest in the field of composite nanomaterials [1]. The possibility of further processing of these particles into highly monodispersed, larger sized, and stable nanoparticles has enabled the ability to probe size-dependent reactivity, as recently demonstrated in our laboratory [2]. These nanoparticles can be effectively linked to form thin films using molecular crosslinking agents. There are several routes reported for crosslinking. One involves a stepwise "layer-by-layer" assembly method [3], and another involves one step "exchange-crosslinking-precipitation" route developed recently in our laboratory [4]. The nanostructured thin films have potential applications in microelectronics, optics, biomimetics, molecular recognition, drug delivery, chemical and environmental sensing, and catalysis [5,6,7].

Gold is traditionally considered as catalytically inert. The recent finding by Haruta and co-workers [8] demonstrated that the catalytic ability for gold increases as the size is reduced to nanometer scales [9]. Gold nanoparticles supported on oxides show high catalytic activity to CO oxidation. Although the idea of using small sized particles as catalysts has been known for a long time, problems faced when using bare nanoparticles include aggregation, short life times, and propensity of poisoning. We recently hypothesized that the core-shell nanoparticles (CSNs) could be used to solve some of these problems. Part of the concept is related to the high stability and the reactivity of CSNs by which they can be assembled in a controllable way. While the use of surface protected nanoparticles as catalysts has the effect of preventing particles from aggregation, catalytic activity may become hindered due to possible inhibiting surface materials. To demonstrate the viability of the CSN based catalysis, we recently explored pathways that take advantage of the CSNs solubilities and functionalities to assemble thin films, and the controllable activation by core-shell surface re-constitution. The formation of surface oxygenated species is found to play an important role in the effective catalytic abilities of such thin film catalysts. In this paper, our latest results of an investigation of issues related to the catalytic activation are described.

EXPERIMENT

Synthesis. The 2-nm gold nanoparticles (Au_{2-nm}), and 2.5 nm Au/Pt nanoparticles (($Au/Pt)_{2.5-nm}$), were encapsulated with alkanethiolate monolayer shells were synthesized by the standard two-phase method [10]. Briefly, $AuCl_4^-$, or $AuCl_4^- + PtCl_6^{2-}$ (5:1 feed ratio), was transferred to organic solvent by phase transfer agent (tetraoctylammonium chloride), and reduced by sodium borohydride in the presence of decanethiols (DT). The reaction was allowed to proceed under stirring at room temperature for 4 hours, producing a dark-brown solution of DT-encapsulated nanoparticles that was then cleaned in ethanol or used in the heating treatment.

Processing. Highly-monodispersed Au particles (5.3 ±0.3 nm) were prepared by thermally activated treatment of the pre-synthesized 2-nm Au nanoparticles [2]. Briefly, the 2 nm particles were pre-concentrated by a factor of ~15, heated to 140 ^{0}C, and annealed at 100 ^{0}C for ~2 hours. The resulting red nanoparticle solution was then cleaned in ethanol.

Thin film Assembly. The nanoparticles were assembled as thin films on electrode surfaces using molecular linkers via one-step exchange-crosslinking-precipitation route [4]. In a typical experiment, 1,9-nonanedithiols (NDT) were mixed in a hexane solution with DT-encapsulated nanoparticles (0.1~10 µM) and NDT (0.5~5.0 mM). The film thickness was controlled by immersion time. The films were thoroughly rinsed with pure solvent before characterizations.

Pt impregnation in Conducting Polymer Matrix. Polyaniline films with Pt loading were prepared by electrochemical method as reported by Lamy and co-workers [11]. Briefly, 0.1 M aniline was dissolved in 0.5 M H_2SO_4 solution. The polyaniline film was deposited by cyclic potential sweeping between −200 and +1000 mV at 50 mV/s. Polymerization was terminated when the oxidation peak current of ~7 mA/cm^{-2} was achieved. Pt was then deposited into the film at a potential of −200 mV (vs. Ag|AgCl|Sat'd KCl) for 5 minutes in a 10^{-4} M K_2PtCl_6 solution. The activation of the film involved thermal activation at 300 ^{0}C (instead of electrochemical polarization), details of which will be reported elsewhere.

Measurements. Electrochemical measurements (EG&G Potentiostat/Galvanostat 273A) were performed in a standard 3-electrode system using Ag|AgCl|Sat'd KCl as reference electrode, Pt as counter electrode and thin film coated glassy carbon (GC) as working electrode (0.5cm^2). Cyclic voltammetry was performed for characterizing the electrooxidation of both methanol (MeOH, Aldrich) and carbon monoxide (CO, Linde Gas) in alkaline electrolyte (0.5M KOH) with scan rate 50 mV/s. Atomic Force Microscopic (AFM) images were acquired using a Nanoscope IIIa (Digital Instruments).

RESULTS AND DISCUSSION

The existence of thiolate encapsulation and NDT-linkage in the nanoparticle films has been characterized in recent publications [1-2,4]. Figure 1 shows a representative tapping-mode AFM image for a NDT-Au_{2-nm} thin film assembled on GC. The particle size and distribution are relatively uniform. The assembled nanoparticles appear to be individually-isolated. The existence of nanoporosity is also evident. The particles appear

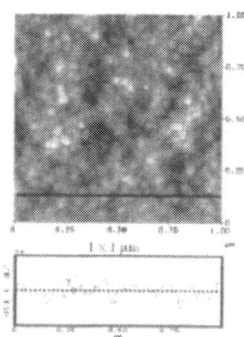

somewhat larger than the core-shell particle size due to tip-sample convolution, but a cross-section view reveals an average height as expected for the particle size. Similar morphology has also been observed for NDT-Au_{5-nm} film.

Figure 2 shows a typical set of cyclic voltammetric curves for NDT-Au_{5-nm} thin film electrode in 0.5 M KOH in the presence of CO (saturated). The catalytic activity is dependent on the activation of the film. The thin film shows a featureless characteristic when it is cycled between -400 and +400 mV (A). In contrast, the film becomes catalytically active to CO oxidation when the electrode is subject to a positive polarization to ~ +700 mV. Following the large oxidation current, a large anodic wave is observed in the negative sweep (B). This wave is attributed to CO electrooxidation to CO_3^{2-} in the alkaline condition. This wave was found to be proportional to both

Figure 1. AFM image of NDT-linked Au_{2-nm} film.

scan rate and CO concentration [12]. The need for a positive potential polarization is believed to be associated with the participation for oxygen species near the catalytic sites at the Au nanocrystal surface. The polarization therefore likely results in the formation of gold oxide species (AuO_x) and a partial removal of the organic shell molecules. These surface species or sites may be operative in catalysis in two ways. First, they reduce the barrier nature of the shell component and increase the conductivity of the thin films. Secondly, the reconstituted shell may preserve the nanocrystal core size.

The electrocatalytic activity of the film towards methanol oxidation was examined. Figure 3 shows a representative set of CV curves for a NDT-$(Au/Pt)_{2.5-nm}$ thin film in the presence of 5 M MeOH in 0.5 M KOH. A similar effect of catalytic activation is observed. In the absence of activation (A), the voltammetric curve displays featureless characteristic. Upon

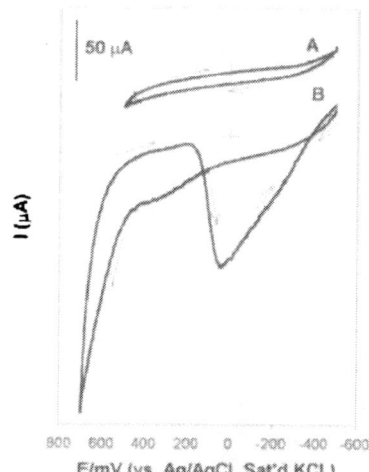

Figure 2: Cyclic Voltammograms of NDT-Au_{5-nm}/GC; (a) unactivated, (b) activated. Electrolyte: 0.5 M KOH, sat'd CO, Electrode area: 0.5 cm^2 (50mv/s).

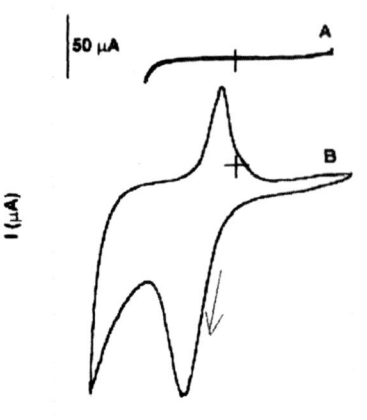

50 μA

A

B

I (μA)

1000 800 600 400 200 0 -200 -400 -600

E/mV (vs. Ag/AgCl, Sat'd KCL)

Figure 3: Cyclic Voltammograms of NDT-$(Au/Pt)_{2.5-nm}$/GC; (a) unactivated, (b) activated. Electrolyte: 0.5 M KOH + 5.0 M MeOH, Electrode area: $0.5cm^2$ (50 mv/s).

electrochemical activation of the film to a positive polarization potential (+800 mV), a large anodic wave is evident at +300 mV. The peak potential closely matches the potential for Au oxide formation, suggestive of the participation of Au oxide in the overall catalytic oxidation mechanism. An integration of the charge from the cathodic wave translates to $\sim9\times10^{-9}$ moles/cm^2 for the amount of reactive Au. An estimate of the catalytic peak current and the quantity of metals yields ~5 mA/mg. Through a systematic study of the concentration and scan rate dependencies, we found two remarkable voltammetric features [13]. First, the anodic peak current increases with increasing methanol concentration, exhibiting a linear relationship. Second, in contrast to the trend for the anodic wave, the peak current for the cathodic wave decreases with increasing methanol concentration, which also exhibits a linear relationship [13]. These two features form an important set of evidence demonstrating that methanol is oxidized at the nanostructured Au catalyst. The opposite trend between the oxidation and the reduction peak currents as a function of methanol concentration is suggestive of a catalytic mediation mechanism involving redox of the surface Au oxide species. Contribution from Pt oxide may be minimum because its redox potential is more positive than Au oxide and the alloyed Pt is a very small fraction (~5%). The shell encapsulation may become partially open as a result of either the surface oxide formation or a change in shell packing due to possible thiolate desorption or reorganization. The participation of surface oxide species in the above reactions is supported both by the occurrence of the oxidation wave at the potential of gold oxidation and the suppression of the gold oxide reduction wave [12,13a], and by our recent electrochemical quartz-crystal microbalance detection of mass increase in the oxidation process [13b].

It has been demonstrated that the oxidation of MeOH often involves CO as an intermediate species. The CO intermediates are often the cause of poisoning of the Pt-group catalyst. It appears that for our catalyst films the catalytic activity is relatively stable over repetitive cycling up to 50 cycles in the presence of methanol or CO. This finding is consistent with the high catalytic activity of CO oxidation observed on bare gold nanoparticles supported on oxides [8]. While a detailed investigation of the reconstituted surface species in the activation and oxidation processes is in progress, we believe that the assembled gold nanoparticles are effective catalysts for both CO and MeOH oxidation in alkaline solution. This assessment may prove extremely important as we develop high performance fuel cell catalysts that have a long lifetime. A further assessment of possible changes of the catalyst morphology due to the activation and the formation of oxygenated species is under way with the aid of in-situ AFM technique [14].

In view of the high catalytic activity of Pt in methanol oxidation [14], we examined a different approach to incorporate Pt component in the catalytic film. In this approach, we first prepared a polyaniline thin film that was loaded with Pt. The Pt loading was accomplished via electrochemical deposition from $PtCl_6^{-2}$ anions in solution into the conductive polymer by reducing Pt^{IV} into Pt^0 particles. The NDT-Au_{2-nm} film was then assembled on the surface of the polymer thin film. Figure 4 shows a preliminary set of CV data for this "layered" nanoparticle thin film in 2.5 M MeOH + 0.5 M KOH electrolyte. In the absence of activation, the voltammetric characteristic is basically silent, similar to the observation for NDT-Au_{2-nm} film. Upon thermal activation, the film shows a large oxidation wave at a potential of -180 mV, much more negative than those observed earlier. We attribute the shift of the

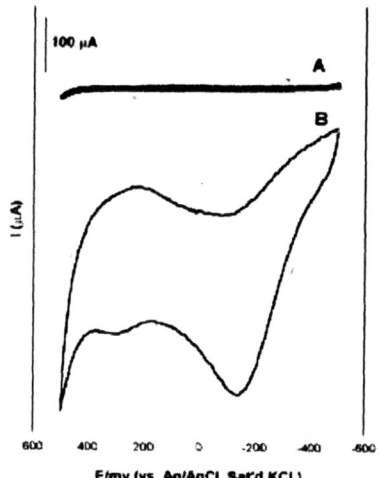

Figure 4: Cyclic Voltammograms of NDT-Au_{2-nm} + PANI (Pt) thin film at a GC electrode; (a) unactivated, (b) thermally activated. Electrolyte: 2.5 M KOH + 2.5 M MeOH; Electrode area: 0.5 cm^2 (50 mv/s).

oxidation wave to the catalytic oxidation of methanol on Pt particles. While this peak potential corresponds closely to that of bulk platinum in alkaline solutions, the oxidation wave traditionally observed for the Pt catalyst on the return negative sweeping is largely absent, even after ~50 cycles. This may suggest that the traditional poisoning effect may be suppressed by the presence of gold nanoparticles. It is also possible that the intermediate CO species is oxidized by gold nanoparticles. In fact a small anodic wave is identifiable at ~+300 mV, corresponding to methanol oxidation at gold nanoparticle sites.

An estimate of the catalytic current vs. the quantity of metals yields ~4 mA/mg, (based on quantities of Pt and Au deposited in the film). This is qualitatively consistent with observations reported for Pt-Ru catalysts loaded in polyaniline film [11], where a ~3 mA/mg current density was found under the condition of controlled potential electrolysis. The two-component system using different approaches is another viable pathway towards the development of a poison-free catalyst.

CONCLUSION

In conclusion, the catalytic activity of the nanostructured catalyst materials was found to be tailorable by three types of interfacial chemistries. First, the utilization of functional shell of the CSNs system is important for the assembly of nanostructures that protect particles form aggregation. The second finding for these systems reinforces the belief that the CSNs catalytic activity can be activated by activation strategies that involve reconstitution of the core-shell structure and composition. Thirdly, the Au-Pt two-component system can effectively increase the catalytic activity. The role of the organic encapsulation is important in two aspects. First, it allows controllable fine-tuning of the core size and composition via synthesis and processing. Secondly, it allows thin film assembly at any substrates. How does the shell and network encapsulation evolve and reconstitute during the catalytic activation and oxidation is a subject of our on-going investigations.

ACKNOWLEDGMENTS

Financial support of this work is gratefully acknowledged from the ACS Petroleum Research Foundation and 3M Corporation.

REFERENCES

1. M. Brust; M. Walker; D. Bethell; D. J. Schiffrin, R. Whyman, *Chem. Commun.*, 801, (1994).
2. (a) M. M. Maye; W. X. Zheng; F. L. Leibowitz; N. K. Ly; C. J. Zhong, *Langmuir*, , **16**, 490, (2000) (b) M. M. Maye; C. J. Zhong, *J. Mater. Chem.*, **10**, 1895. (2000).
3. (a) D. Bethell, M. Brust, D.J. Schiffrin, C. Kiely, *J. Electroanal. Chem.* **409**, 137, (1996). (b) J.K.N. Mbindyo; B.D. Reiss; B.R. Martin; C.D. Keating; M.J. Natan; T.E. Mallouk, *Adv. Mater.*, **13**, 249. (2001).
4. F. L. Leibowitz; W. X. Zheng; M. M. Maye; C. J. Zhong, *Anal. Chem.*, **71**, 5076, (1999).
5. A. C. Templeton; W. P. Wuelfing, R. W. Murray; *Acc. Chem. Res.* **33**, 27; (2000) and references therein.
6. F. Caruso, *Adv. Mater*, **13**, 11; (2001), and references therein.
7. (a) J. J. Storhoff; C. Mirkin, *Chem. Rev.* , **99** , 1849, (1999). (b) S. Mann; W. Shenton; M. Li; S. Connolly; D. Fitzmaurice, *Adv. Mater.*,**12**, 147 (2000).
8. (a) M. Haruta, *Catalysis Today*, , **36**, 153. (1997) (b) P. C. Biswas; Y. Nodasaka; M. Enyo; M. Haruta, *J. Electroanal. Chem.*, **381**, 167. (1995).
9. (a) G. C. Bond and D. T. Thompson, *Catal. Rev.*,**41,** 319 (1999). (b) G. C. Bond, D. T. Thompson, *Gold Bulletin*, **33**, 41 (2000).
10. (a) M. J. Hostetler; J. E. Wingate; C. J. Zhong; J. E. Harris; R. W. Vachet; M. R. Clark; J. D. Londono; S. J. Green; J. J. Stokes; G. D. Wignall; G. L. Glish; M. D. Porter; N.D. Evans; R. W. Murray, *Langmuir, 14*, 17. (1998).
11. A. Lima, C. Coutanceau, J.-M. Leger, C. Lamy; *J. Appl. Electrochem.*, **31**, 379, (2001)
12. M. M. Maye; Y. Lou; C. J. Zhong, *Langmuir*, **16**, 7520. (2000).
13. (a) Y. Lou; M. M. Maye; L. Han; J. Luo; C. J. Zhong, *Chem. Comm.*, 473, (2001). (b) J. Luo; Y. Lou; M. M. Maye; C. J. Zhong; M. Hepel, *Electrochemistry Communications,* , **3**, 172. (2001).
14. M. M. Maye, J. Luo, L. Han, C.J. Zhong; *Nano Letts.*, **1**,10, 575, (2001).

Mat. Res. Soc. Symp. Proc. Vol. 707 © 2002 Materials Research Society　　　　　　

Fabrications and Electron Transport Properties of One Dimensional Arrays of Gold and Sulfur Containing Fullerene Nanoparticles

Sheng-Ming Shih[1,2], Wei-Fang Su[1], Yuh-Jiuan Lin[2], Cen-Shawn Wu[3], Chii-Dong Chen[3]
[1]Institute of Materials Science and Engineering, National Taiwan University,
Taipei, Taiwan, R.O.C.
[2]Biomedical Engineering Center, Industrial Technology Research Institute,
Hsinchu, Taiwan, R.O.C.
[3]Institute of Physics, Academia Sinica,
Taipei, Taiwan, R.O.C.

ABSTRACT

Novel arrays of gold nanoparticles with sulfur containing fullerene nanoparticles were self-assembled through the formation of Au-S covalent bonds. Disulfide functional groups were introduced into C_{60} molecule by reacting propyl 2-aminoethyl disulfide with C_{60}. The two dimensional(2D) arrays were formed at the interface of aqueous phase of gold particles and organic phase of fullerene particles as a blue transparent film. TEM images showed that the fullerene spacing between adjacent Au($\sim$10 nm) particles was about 2.1±0.4 nm, which was consistent with the result of 2.18 nm by molecular molding calculations(MM^+). The arrays were deposited on the top of pairs of gold electrodes to form 2D colloidal single electron devices. The electrode pairs were made by electron beam lithography techniques, and the separation between tips of the two electrodes in a pair was less then 100 nm. Transport measurements at low temperatures exhibited Coulomb-Blockade type current-voltage characteristics, the lower the temperature the more pronounced the Coulomb gap. Also, step-by-step method was used to assemble one-dimensional(1D) array of gold nanoparticles with fullerene derivative between two electrodes spaced with 15 nm. The Coulomb blockade behavior of 1D arrays was clearer than that of 2D arrays.

INTRODUCTION

The single electron transistor is a potential candidate for the next generation of electronic devices. It possesses great advantages in low power consumption and high packing density. Since the operating temperature of a single electron transistor is determined solely by the geometrical size of the island(s) between the electrodes. The island(s) should be as small as only a few nanometers, it presents a challenge to the modern nanofabricaton technology. Even with present day state-of-the-art electron beam lithographic technology, this is still a very demanding requirement. Thus, self-assembly of nano-structured materials provides a potential process to fabricate high temperature operating nano-devices.

Various nanomaterials have been synthesized. However, finding a method to fabricate such materials in an organized fashion and on an appropriate location of circuitry, effectively and efficiently is still a major challenge to scientists. The self-assembly methodology has recently emerged as a useful technique. There are numerous reports regarding the assembly of two-dimensional arrays of quantum dots by lithography and epitaxy depositions.[1] Many metal, insulator and semiconductor nanoparticles have been assembled together with organic molecules such as alkyl dithiol,[2-4] surfactants,[5-7] organic polymers,[8] conjugated DNA[9,10] or biomimic conjugated systems.[11] Two or three dimensional nanoparticles arrays have been constructed by the formation of covalent bonds, hydrogen bonds or van der Waals forces.

In this report, novel nanoparticles arrays were assembled from modified fullerene(C_{60}) derivatives and gold nanoparticles. By combining advanced electron beam lithography and nanophased-material assembling techniques, single electron devices have been fabricated and characterized. The results are presented and discussed in the following sections.

EXPERIMENTAL

Synthesis of fullerene derivative C_{60}-S

Ground C_{60} (0.10 g) (Aldrich Chemical) was added to 3.4 g of propyl 2-aminoethyl disulfide. The propyl 2-aminoethyl disulfide was prepared from dipropyl disulfide and 2-mercaptoethylamine according to the procedure reported by K. Kitsuta.[12] The mixture was stirred for three days at room temperature until all C_{60} had reacted. Methanol (20 ml) was added into the mixture and brown precipitates were formed. The precipitates were collected by filtration and washed with methanol several times. The product was dried with a yield of 0.096 g. Anal. C 71.34; H 4.12; N 4.22.

Self-assembly of gold nanoparticles

An aqueous solution of gold nanoparticles was synthesized by reducing tetrachloroauric acid (Across Chemical) with trisodium citrate (Across Chemical) in distilled water. Fullerene derivatives (C_{60}-S, 1 mg) were dissolved in 10 ml of chloroform. When a gold solution (10 ml) was poured gently into 10 ml fullerene solution, a film was formed at the interface of two solutions after two to seven days. The film was transferred on copper grid for transmission electronic microscopy (TEM, JOEL 100 CX II) study (Figure 1).

Fabrication of nano-device with 2D nanoparticles array

The chip that possesses gold electrodes, was dipped into the interface of gold and C_{60}-S solutions. Thus the film of two-dimensional array of C_{60}-S with gold nanoparticles was

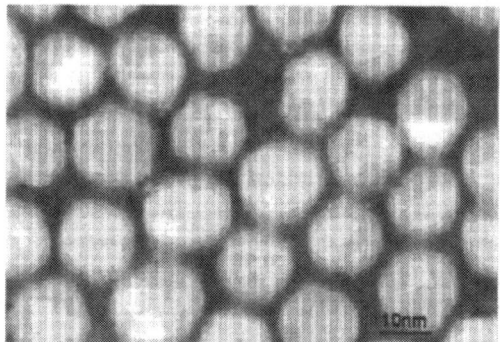

Figure 1 Micrographs of transmission electronic microscopy of self-assembly of gold nanoparticles and C_{60}-S. The average size of gold particles is 14.5±2.3 nm and average space between gold particles is 2.1±0.4 nm. The length of scale bar is 10 nm.

transferred on the chip to linked two electrodes. After evaporating off the solvent, the chip was washed with distilled water for three times and dried by nitrogen. The electrodes are made by standard electron beam lithography and lift-off techniques. The gold electrodes were about 50 nm in width, 25 nm in thickness and the gap between them was about 100 nm. The substrate was a standard Si wafer covered by a thermally grown 300 nm thick SiO_2 layer. To reduce the capacitance of the gold island further, the SiO_2 layer surrounding the junctions was etched away prior to place assembled array. This eliminated the capacitance of the island to the ground and helped to increase the operating temperature.

Fabrication of nano-device with 1D nanoparticles array

To assemble nanoparticles between electrodes on the chip, we deposited 1 і 1 solution of C_{60}-S solution on the chip using micropipet. This solution of fullerene derivatives was prepared by dissolving C_{60}-S (1 mg) in 10 ml of chloroform. After waited a few minute to allow the solvent to evaporate, we then carefully washed the chip with chloroform three times and blew it dry with nitrogen. The chip was then floated on the surface of the aqueous solution of gold nanoparticles. After one day immersion, the chip was removed from the gold solution, washed with distilled water three times and dried with nitrogen.

By the same technique that is described above, the chip with 15 nm gap was fabricated. The Au leads were intentionally made wider so as to sustain themselves. They acted respectively as source and drain of the device.

RESULTS AND DISCUSSION

In order for gold nanoparticles to be assembled by fullerene nanoparticles, we synthesized fullerene derivative with disulfide functional groups. The molecule of C_{60} is a buckyball of carbon, covered on its surface with unsaturated conjugated double bonds. This molecule can be attacked by more than one nucleophile. We have used propyl 2-aminoethyl disulfide as nucleophiles to attack fullerene molecules. The disulfide group of this compound did not react with fullerene and it was reactive toward gold. The reaction of propyl 2-aminoethyl disulfide with C_{60} proceeded as expected. The amino group attached C_{60} to forms sulfur containing C_{60} (C_{60}-S) and the resulting disulfide group of C_{60}-S reacted with gold. The structure of C_{60}-S is a ball with tentacles. The exact positions of amino groups on the surface of C_{60} are not identified. Four tentacles were found on the molecule of C_{60}-S from elemental analysis.

The aqueous solution of gold nanoparticles was synthesized by reducing tetrachloroauric acid with trisodium citrate in water.[13] Uniform size and shape of spherical gold nanoparticles that ranging from 5 to 15 nm were obtained.

When gold nanoparticles solution and C_{60}-S nanoparticles solution were mixed, the disulfide group of fullerene derivatives reacted with gold nanoparticle, the S-S bond broke and two covalent S-Au bonds formed immediately.[14] A transparent film was observed at the interface of the aqueous phase of gold particles and the organic phase of fullerene particles. The molecular modeling calculations (MM^+) showed that the spacing occupied by C_{60}-S between two Au nanoparticles was 2.18 nm. This distance is the sum of the diameter of one molecule of C_{60} and the length of two molecules of 2-mercaptoethylamine. Transmission electronic microscopy studies of the transparent film made from C_{60}-S and gold nanoparticles show that 14.5±2.3 nm gold particles were spaced evenly by 2.1±0.4 nm C_{60}-S(Figure 1). This result indicated that the disulfide group of C_{60}-S reacted with gold particles, as expected, by breaking the disulfide bond into two-sulfur groups that reacted with gold particles individually.

Figure 2a shows the SEM photo of two-dimensional array of gold and C_{60}-S nanoparticle that was transferred to the gap between two gold electrodes by dipping method. Figure 2b shows the current-voltage characteristics of the two-dimensional array. This nonlinear behavior is more pronounced at lower temperature, which is a sign of Coulomb blockade of electron tunneling. The device exhibited a Coulomb gap of about 0.8 V at 4 K. This device is very stable because the shape of the current-voltage characteristics was reproducible for repetitive traces of measurement at a fixed temperature.

For fabricating simpler single electron devices, one-dimensional arrays of gold and C_{60}-S nanoparticles were assembled. They were prepared by depositing gold nanoparticles and C_{60}-S nanoparticles alternatively on a chip circuitry containing pairs of gold electrodes with tip spacing 15 nm. The circuitry was prepared using advanced electron beam lithography. Figure 3a shows a single electron device, which was fabricated using gold nanoparticles connected by C_{60}-S nanoparticles. Figure 3b shows the current-voltage (I-V_b) characteristics taken at three temperatures. The curves are nonlinear with current smaller than the linear ones. At room temperature, a relatively linear current-voltage characteristic was observed. The nonlinear current-voltage characteristics were found when the temperature was lowed to 77K. The nonlinear behavior was more pronounced at lower temperatures, which is a sign of Coulomb

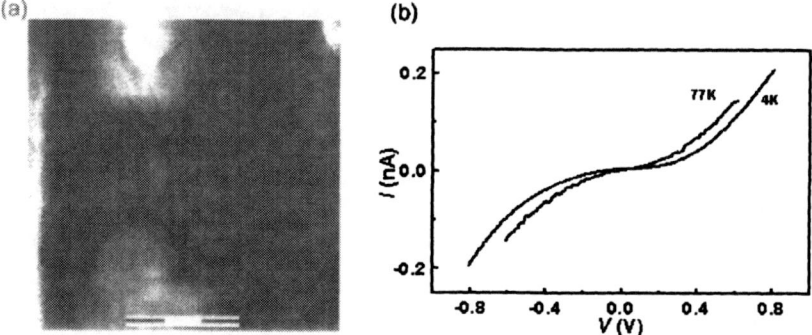

Figure 2 (a) Micrograph of scanning electronic microscopy of two-dimensional array on the chip. The length of the scale bar is 100 nm. (b) Current-voltage characteristics of the two-dimensional array of gold and C_{60}-S nanoparticles in the gap between two gold electrodes.

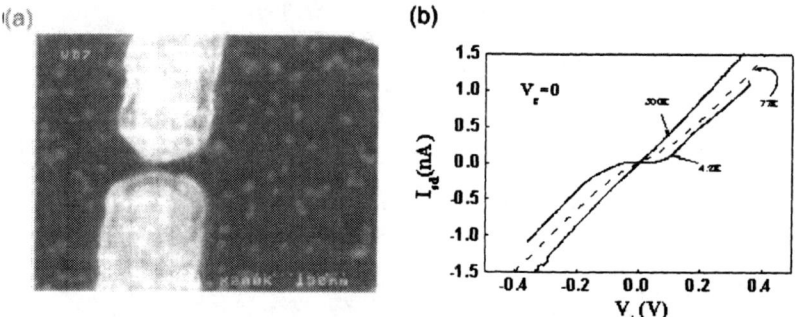

Figure 3 (a) SEM image of a measured sample. The separation between the two Au leads is about 15 nm, and the gate electrode is not shown. (b) $I-V_b$ characteristics measured at 300K, 77K and 4.2K. Suppression of current at low bias voltages is more pronounced at lower temperature, signifying Coulomb blockade of electron tunneling.

blockade of electron tunneling. A clear Coulomb blockade was showed at 4.2K because the thermal energy was much lower than the barrier energy at this temperature. In this device configuration, the self-assembled array overcame the size limitations of lithography, thus it would provide a path for fabricating a single electron device operable at elevated temperature.

CONCLUSIONS

We have successfully bridged a pair of source-drain leads with an Au nanoparticle that was linked by fullerene derivatives. The devices made with both of 1D and 2D nanoparticles array exhibited pronounced Coulomb blockade type current-voltage characteristics . The fabrication of fullerene based nanoparticle arrays offer a new class of materials with tunable electronic properties for nanodevice applications.

ACKNOWLEDGEMENT

The financial supports from Industrial Technology Research Institute (Biomedical Engineering Center) and National Science Council of Republic of China (NSC-88-2216-E-022-041, NSC-89-2218-E-002-072, NSC-90-2218-E-002-049) are highly appreciated.

REFERENCES

1. J. H. Fendler and F. C. Meldrum, *Adv. Mater.* 7, 607 (1995).
2. V. L. Colvin, A. N. Goldstein and A. P. Alivisatos, *J. Am. Chem. Soc.* 114, 5221(1992).
3. R. P. Andres, J. D. Bielefeld, J. I. Henderson, D. B. Janes, V. R. Kolagunta, C. P. Kubiak, W. J. Mahoney and R. G. Osifchin, *Science* 273, 1690 (1996).
4. T. Cassagneau, T. E. Mallouk and J. H. Fendler, *J. Am. Chem. Soc.* 120, 7848 (1998).
5. C. Petit, A. Taleb and M. P. Pileni, *Adv. Mater.* 10, 259 (1998).
6. L. Motte, M. P. Pileni, *J. Phys. Chem. B* 102, 4104 (1998).
7. M. Li, H. Schnablegger and S. Mann, *Nature* 402, 393 (1999).
8. T. von Werne and T. E. Patten, *J. Am. Chem. Soc.* 121, 7409 (1999),
9. C. A. Mirkin, R. L. Letsinger, R. C. Mucic and J. J. Storhoff, *Nature* 382, 607 (1996).
10. R. C. Mucic, J. J. Storhoff, C. A. Mirkin and R. L. Letsinger, *J. Am. Chem. Soc.* 120, 12674 (1998).
11. S. Connolly and D. Fitzmaurice, *Adv. Mater.* 11, 1202 (1999).
12. R. Seshadri, A. Govindaraj, R. Nagarajan, T. Pradeep and C. N. R. Rao, *Tetra. Lett.* 33, 2069 (1992).
13. K. C. Grabar, R. G. Freeman, M. B. Hommer and M. J. Natan, *Anal. Chem.* 67, 735 (1995).
14. R. G. Nuzzo, B. R. Zegarski and L. H. Dubois, *J. Am. Chem. Soc.* 109, 733 (1987).

Mat. Res. Soc. Symp. Proc. Vol. 707 © 2002 Materials Research Society   W6.35

Nanostructuring of multilayers by a thermally driven self assembling process

C. Herweg[1], S. Dreyer[1], P. Troche[1], J. Hoffmann[2], S. Sievers[1], C. Lang[1] and H. C. Freyhardt[1,2]

[1] Institut fuer Materialphysik, Universitaet Goettingen, Windausweg 2, D-37073 Goettingen
[2] Zentrum fuer Funktionswerkstoffe ZFW gGmbH, Windausweg 2, D-37073 Goettingen

ABSTRACT

Multilayers consisting of two immiscible components (e.g. Fe/Ag Fe/Au or Ni/NiO) could by transformed by an annealing process into a nanostructured system of non statistically distributed nearly spherical particles in a surrounding matrix of the complementary component. The non statistical arrangement of the particles and the dynamics of the disintegration process strongly depend on the initial interface energy, i.e. the local interface curvature and the local interfacial stress. Detailed microstructure investigations of the different systems are used to interpret the measured transport properties.

INTRODUCTION

Due to their prominent mechanical, optical, electrical and magnetic properties, materials structured on a nanometerscale are of great interest in modern material science and strong candidates for future applications.

In contrast to other methods, the preparation of nanostructered systems by conventional thin film deposition techniques and a subsequent heat treatment is relatively easy, time- and cost-efficent. A disintegration process caused by interfacial curvature has been first observed in Nb/Cu filament superconductors [1]. Analytical solutions have only been published for cylindrically shaped filaments [2], whereas the surface and changing shapes of interfaces in multilayered films is still a matter actual research [3], [4]. Different systems (Nb/Cu, Fe/Ag, Fe/Au, Cu/Co, Co/C) were studied including multilayers consisting of metals and metaloxids as a second component (Ni/Ni0, Co/CoO) in order to get detailed information on which parameters control the disintegration temperature and the final particle distribution. The transition from an anisotropic layered system to an isotropic granular system, strongly influences the transport properties, e.g. the magnetoresistance. Therefore, these properties give an additional insight in the shape instability of multilayer structures.

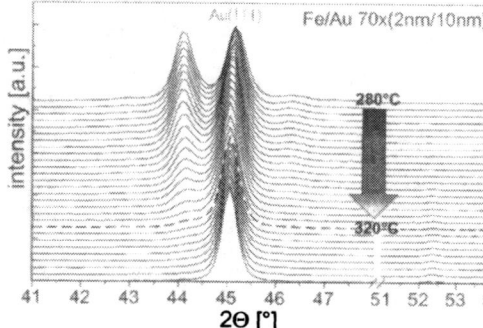

Fig. 1: *XRD scans during annealing. Disintegration of a 70x(2nm Fe/10nm Au) multilayer at 320°C.*

EXPERIMENTS

Metal/metal multilayers were prepared in a UHV-system (background pressure < 10^{-8} mbar) using two indepentent operatable magnetron sputtering sources (working pressure Ar, 10^{-3} mbar) with high purity metal targets. The alternating deposition of the two components was performed by intermittent operation of the two sources, whereby the source not in use was switched off and shutters were employed to ensure a minimum intermixing at the interfaces.

Metal/metaloxid multilayers were prepared by

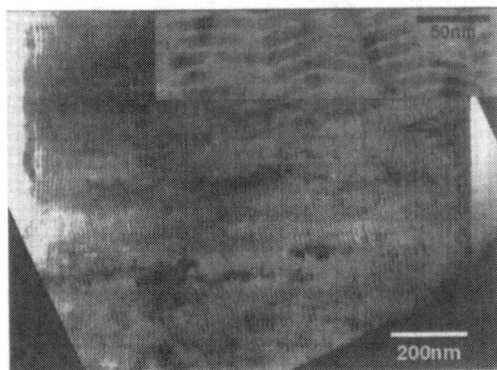

Fig. 2: Z-contrast cross-section TEM of a 100x(2nm Fe/10nm Ag) multilayer as-prepared. Columnar structure, with a column width of 70-90nm.

reactive ion-beam sputtering (Xe, $5 \cdot 10^{-5}$ bar) with Kaufmann ion sources. During reactive sputtering the partial preasure of O_2 in the chamber is $\sim 10^{-4}$ mbar. A metallic target (e.g. Ni) was sputtered continiously during the multilayer deposition. The alternating deposition of the metal and its oxide were realized by switching the oxygen inlet, regulated by a flow controller. Closing the shutter after the deposition of the actual layer, a delay time of ~ 10s is sufficient to add or to remove the oxygen for the following metal resp. metaloxide layer. The depositions on sapphire substrats were performed at room temperature.

In situ resistance measurements during deposition and annealing of the samples were employed using a conventional 4-point system with gold contacts deposited onto the substrates. Heat treatments of the samples were carried out up to 800K in the UHV deposition chamber. Furthermore, XRD measurements were performed in the hot stage of a diffractometer under a vacuum better of 10^{-5} mbar.

The mircrostructure of the samples before and after the annealing was investigated by means of cross-sectional TEM. The magnetoresistance of the multilayers was followed from room temperature down to 4.2K in a helium cryostat. In addition, the angle between substrate and the external field was varied between 0° and 90°.

RESULTS
Microstructure

Fig.1 depicts x-ray diffraction patterns recorded during a heat treatment from room temperature to 350°C of a Fe/Au multilayer. The as prepared intensity distribution reveals a superstructure originating form the periodicity of the multilayer [5]. This superstructure vanishes during the disintegration process and results in the crystaline (111)-peak of the gold matrix. From this the disintegration temperature of the multilayer can be determined to 320°C. The disintegration tempeature increases with increasing thickness of the Fe layer [4].

Fig.2 shows a cross-section of an as prepared Fe/Ag multilayer. Since the hollow-cone-method (z-contrast) [6] has been used the figure depicts a chemical contrast. The 2nm thick iron layers appear dark whereas the 10nm silver corresponds to a bright contrast. The substrate can be seen on the right side of the micrograph. The initial growth of the first layers is comparatively smooth and fine-grained, but after 8 to 10 doublelayers broad growth columns (70-90nm) with convexly

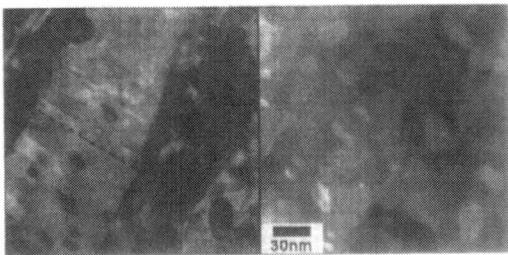

Fig.3:A 100x(2nm Fe/10nm Ag) multilayer after heat treatment for 1h at 500°C (left:bright field picture, right: z-contrast) the line indicates former layer orientation.

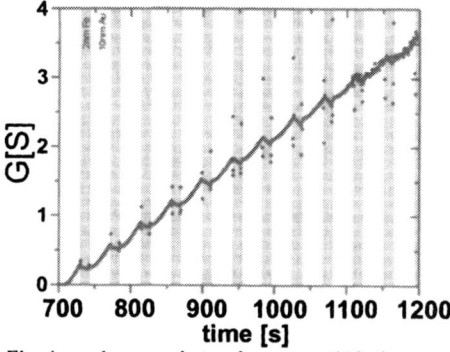

Fig. 4: conductance during deposition of 50x(2nm Fe/10nm Au).

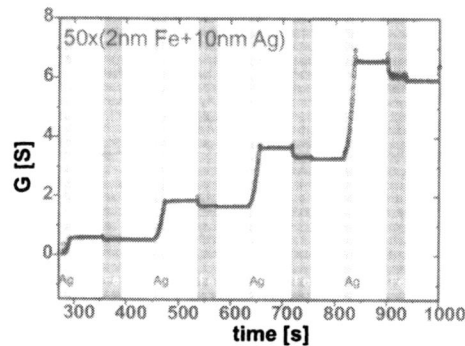

Fig. 5: increasing conductance during Ag deposition, decrease during Fe deposition.

curved interface between the components occur. Therefore, the interfaces reveal a wavy structure with their maximum in the center of the growth columns and their minimum at the confining grain boundaries. Within the columns, Fe and Ag growth, following the Nishiyama-Wassermann relation, i.e. the growth columns are quasi single-crystalline. Similar results were observed for Fe/Au [9] .

After the annealing of a comparative multilayer (Fig. 3) for 1h at 500°C, a columnar structure is still present, but the convex curvature inside the columns vanished. The 2nm iron layers disintegrated into mainly spherical particles (diameter 10-30nm) located in the boundaries of the columns and a few smaller, non-spherical particles inside the columns. The iron-particles in the column borders are arranged in a chain perpendicular to the former layered structure.

TEM investigations of Ni/NiO multilayers reveal a strong interface roughness, but no long length wavy structure. Disintegration of the multilayer is observed in samples annealed up to 550°C . From our investigation on different systems, it is concluded that both, the wavy structure of the interface and the epitaxial strains at the interfaces contributes to the driving force of the disintegration process (for a detailed discussion see [4]). This agrees nicely with the observations of multilayers consisting of Co and amorphous C, which reveal flat interfaces. The layered structure is thermally stable up to temperatures of around 800°C. However, the driving forces of Ni/NiO multilayers is still unclear, the interfaces are rather flat but the disintegration tempeature is moderate.

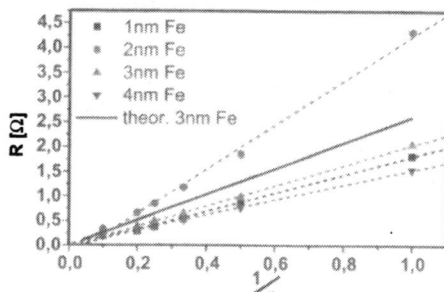

Fig. 6: simple parallel resistor model applied to Fe/Au multilayers.

Diffusion during deposition

During the deposition of Fe/Au and Fe/Ag multilayers, the conductance has been measured continously (Fig. 4). The overall dependence is linear in time an thereby linear in film thickness as the deposition rate is kept constant. Fig. 5 shows a close-up of a comparable multilayer consisting of 50x(2nm Fe/10nm Ag).To rule out the influence of the plasma on the conductance measurement, the source shutters have been closed after the deposition of each metal layers, with a break of 30s for accurate conductance determination. During the

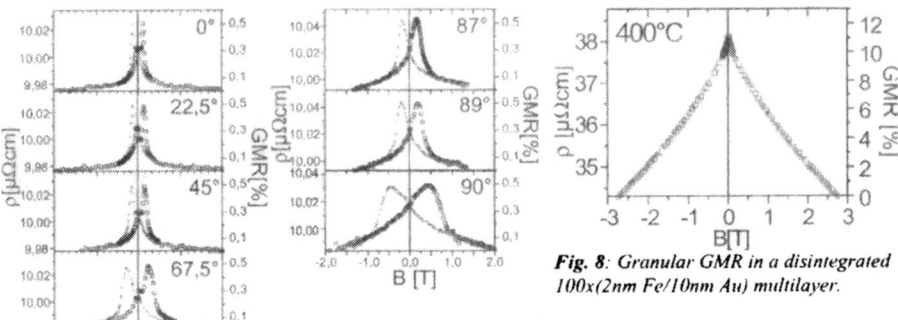

Fig. 7: *Angluar dependence of GMR in a 100x(2nm Fe/ 10nm Au) multilayer as-prepared, T=4,2K.*

Fig. 8: *Granular GMR in a disintegrated 100x(2nm Fe/10nm Au) multilayer.*

deposition of a noble metal layer, the conductance increases monotonously but decreases within the first 10s, if iron is deposited. This is observed for both systems Fe/Au and Fe/Ag.
A simple parallel resistor model leads to the analytical expression:

$$R^{tot} = (\frac{1}{n}) \cdot [\frac{R^{Fe} \cdot R^{Au}}{R^{Fe} + R^{Au}}]$$

n is the number of double layers Fe/Au. This model discribes the obtained data astonishingly well as shown in Fig. 6, i.e. the overall resistivity is simple given by the reduced resistivity of a bilayer. Taken into account that the resistivity of the Fe layer should strongly depend on the thickness ($\rho_{Fe} \propto d_{Fe}^{-x}$, x>1), the slop in Fig. 6 should decrease with increasing Fe layer thickness, which is indeed observed with the exception of the 1nm thick Fe layer. The deviation for very thin layers might be explained by assuming that iron nucleates in the column boundaries, i.e. in the tripple points. Fast diffusion along the tripple junction leads to a wetting of the boundaries in the subjacent gold layer responsible for decrease during the first seconds of iron deposition. After a complete wetting of the boundary with iron the resistance is observed to be constant during further iron deposition. Since the deposition temperature is low, volume diffusion, which takes place at temperatures above 200°C in Fe/Au, can be ruled out.

Magnetoresistance

The magentoresistance of multilayers and granular systems has been of great interest during the last decade. Rotation and reorientation in magnetization of adjacent magnetic layers or particles lead to significant changes in the resistance of these systems [7], [8]. Microstructural

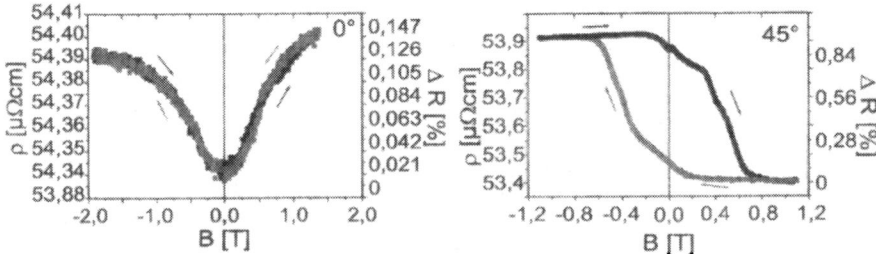

Fig 9: *Positive MR for Θ=0, hysteretic, asymmetric MR for Θ≥22,5° in a 70x(2nm Ni/ 7nm NiO) multilayer.*

changes from a multilayered system (typical GMR system) to a granular system should, therefore, be observable by changes in the GMR.

Fig. 7 visualizes a sequence of GMR curves upon changing the angle, Θ, between the substrate and the magnetic field for a Fe/Ag multilayer. The absolute effect of GMR is rather small due to the large distance between the adjacent magnetic layers of 10nm, preventing the layers from beeing strongly magnetically exchange coupled. Due to the shape anisotropy in the layered sample, the coercetive field, i.e. the field position of the GMR maximum, shifts to higher values with increasing angle. After the disintegration, small but ferromagnetic particles are formed. The surface area between Fe and Ag is effectively reduced, leading to a smaller GMR effect. However, the most important differences occur in the angular dependence [10]. Since the Fe particles are almost spherical, the anisotropy of the coercitive field vanishes.

The disintegration temperature in Fe/Au is substantially lower than in Fe/Ag. Therefore, smaller superparamagnetic Fe particles are formed, a granular GMR is observed (Fig. 8).

Common to all disintegrated systems is that angular dependence of the GMR vanishes. Even Ni/NiO, showing a great variety of characteristics in the magnetoresistance which are far from been fully understood, looses the angular dependence when the multilayer disintegrates. As-prepared Ni/NiO multilayers reveal a reversible positiv magnetoresistance for $\Theta=0°$ but a pronounced asymmetric hysteresis for angles between $22,5°$ and $90°$ (Fig. 9). The latter seems to be related to the antiferromagnetic coupling of interfacial spins. An exchangefield was indeed observed in the magnetization (ZFC, FC).

Low-temperature resistivitiy

The resistance of the metallic multilayers reveals typically the usual behavior: a linear part at high temperatures and a saturation at low temperatures.

However, in some cases a weak lokal minimum at low temperatures was observed. This effect is strongly pronounced in Ni/NiO. Ni/NiO multilayers exhibit always a minimum in the resistance vs. temperature and a logarithmic increase for $T<T_{min}$, (Fig. 10), even if the sample consists only of one bilayer.

The resistance can be modelled by the analytic expression:

$$\rho = \rho_0 + A \cdot T^5 + C \cdot T - B_K \cdot \ln T$$

The logarithmic temperature dependence and the scaling of B_K with T_{Min} is well known for Kondo systems like La(Ce)B$_6$ [11]. We might speculate here, that oxygen solved in nickel changes the electronic struture of nickel [12]. It is possible that oxygen atoms in this enviroment are not fully saturated with 2 electrons, so that the magnetic moments of those oxygen atoms decouple from the nickel. Those decoupled moments might give rise to the effects observed in Kondo systems.

CONCLUSIONS

Thermal activation can be used to transform multilayers prepared by sputtering into nanostructured samples utilizing strong driving force, i.e. interface energy, and the special initial microstructure of the samples. TEM investigations supported by conductance measurements during depositon showed grain boundary resp. tripple junction wetting with iron and subsequent formation of iron particles in the column borders. Vanishing angluar dependence of GMR proofs the

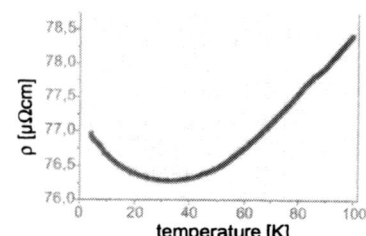

Fig. 10: Resistance rises to low temperatures in a 20x(2nm Ni/7nm NiO multilayer).

evidence of the disintegration of the multilayers into nearly spherical ferromagnetic resp. superparamagnetic particles. The coupling of magnetic moments at the interface Ni/NiO results in rather complex GMR curves.

ACKNOWLEGMENT

The authors like to thank the Deutsche Forschungsgemeinschaft, DFG supporting this work in project No. FR 452/4-2.

REFERENCES
[1] L. Schultz, R. Bormann, J. Appl. Phys., **50** (1979) 418.
[2] J. van Suchtelen, Journal of Crystal Growth 43 (1977) 28-46.
[3] D. Josell, F. Spaepen, MRS Bulletin (1999) 39-43.
[4] P. Troche, Thermisch induzierte Gestaltinstabilitaet von Vielfachschichten, Dissertation, Goettingen, 2000.
[5] J.-P. Locquel, Phys. Rev. B 39 (1989) 13338.
[6] J. C. Ewert, F. Hartung, G. Schmitz, Appl. Phys. Letters. **71**, 10 (1997) 1311.
[7] J. Barnás, A. Fuss, R. E. Camley, P. Gruenberg and W. Zinn, Phys. Rev. B, **42** (1990), 8110.
[8] S. Honda, et al. J. Appl. Phys. **80**, 9 (1996), 175.
[9] C. Borchers, P. Troche, C. Herweg, J. Hoffmann J. Mat. Sci., 37 (2002) 731-736.
[10] C. Herweg, GMR als Sonde zur Untersuchung thermisch induzierter Gestalinstabilitaeten magnetischer Vielfachschichten, diploma thesis (1999).
[11] K. Samwer, K. Winzer, Z. f. Phys. B, **25** (1976).
[12] K. Baberschke, Appl. Phys. A **62** (1996), p. 417-427.

Mat. Res. Soc. Symp. Proc. Vol. 707 © 2002 Materials Research Society

Self-Assembled Structures of Gas-Phase Prepared FePt Nanoparticles

Bernd Rellinghaus, Sonja Stappert, Mehmet Acet, and Eberhard F. Wassermann
Experimentelle Tieftemperaturphysik and Sonderforschungsbereich 445,
Gerhard-Mercator-Universität, D-47048 Duisburg, Germany.

ABSTRACT

We report on a non-lithographic method for the preparation of self-assembled FePt nanoparticles via inert-gas condensation. Prior to deposition the particles can be sintered in flight at temperatures as high as $T_S = 1273$ K. Whereas un-sintered particles have irregular shapes, particles sintered at elevated temperatures $T_S \geq 793$ K show a regular faceting. (High resolution) transmission electron microscopy ((HR)TEM) shows that these regularly faceted particles are of icosahedral structure. When being deposited onto amorphous carbon films, the gas-phase sintered particles are found to have a high mobility. In particular, for the high-temperature sintered FePt nanoparticles, we observe that this mobility leads to the formation of particle arrays with hexagonal close-packed arrangements. Within these ordered patches, the particles are separated from one another. Analytical investigations using energy filtered TEM (EFTEM) show that a carbon layer is formed between the particles. Magnetization analyses give results showing that the gas-phase sintered particles are superparamagnetic at room temperature with a blocking temperature of $T_B = 49$K.

INTRODUCTION

The reduction of size and dimension in solid state materials leads to a significant alteration in their physical properties. Structural, electrical, chemical, optical and magnetic properties may be tailored by simply controlling the size. Owing to the demand for successive integration and miniaturisation in the consumer electronics industry, the magnetism of nanoparticles and nanostructured materials has gained particular importance in recent years. Here, the research on materials, which may serve as information media in high density magnetic data storage, plays a dominant role. Presently, commercial magnetic storage media are nanocrystalline thin films of CoPtTaCr [1]. In this material, the grain size and the (de)coupling between the grains are key parameters for the quality of the material. Whereas in conventional homogeneous media one bit is made up of as many as 100 grains, presently, there are alternative strategies aiming to use single magnetic entities in a regularly arranged array as a bit [2].

Among the methods employed for the preparation of such patterned structures, are optical interference lithography [3], focused ion beam patterning (FIBP) [4], nanoimprint lithography [5], electron beam lithography (EBL) [6], and others. Here, optical lithography provides the advantage of parallel, thus fast structuring of macroscopic samples, while FIBP and EBL suffer from a slow processing due to serial "writing" of individual structures. On the other hand, whereas the smallest feature sizes and periodicities obtained with optical interference methods are in the range 65-125 nm [7], the structures prepared with EBL may be as small as some 10 nm [8]. A promising approach in order to combine both fast processing speed *and* minimum feature size is provided by the self-assembly of nanoparticle arrays [9,10]. The latter method allows for a size reduction of magnetic nanoparticles. Within a regular array, the size can be reduced down to 3 nm $\leq d_P \leq 5$ nm. Depending on the material, these sizes come close to the superparamagnetic limit below which the orientation of the magnetization direction of an individual particle is no longer stable [11].

In order to reduce the superparamagnetic limit to the smallest possible particle sizes (i.e. towards the smallest storage units, thus towards highest storage density) the magnetic materials must have a very high magnetic anisotropy. This can be achieved by an appropriate shape of the magnetic "particles". Self-assembled arrays of cylindrical magnetic nanowires of Fe, Ni, or Co with large aspect ratios meet this requirement [12]. However, the periodicities obtained in such nanowire arrays are as large as 35 – 50 nm. Alternatively, high magnetic anisotropy can be obtained by choosing a material with large magneto-crystalline anisotropy [13]. The magneto-crystalline anisotropy is usually larger in non-cubic crystals as compared to that in cubic crystals. Recently, it has been shown that post-preparation thermal annealing in N_2 of stoichiometric FePt nanoparticles, coated with an organic ligand shell, results in the formation of large-scale periodically arranged and chemically ordered FePt particles ($L1_0$ phase, tetragonal unit cell) [10]. However, as can be seen from results obtained from x-ray diffraction (XRD), this thermal treatment leads not only to the formation of the $L1_0$ order, but also results in a significant growth of the individual crystallites. Furthermore, the influence of the thermal treatment on the organic ligand shell around the magnetic particle cores is not clear.

In the present paper, we report our first results obtained from a gas-phase based method for the preparation of FePt nanoparticles. This work aims at avoiding any a post-deposition annealing processes. The nanoparticles are rather subjected to a thermal annealing ("sintering") as long as they are carried in the gas phase, i.e. prior to being deposited onto any substrate. Our investigations focus on the question as to whether or not this method provides any possibility to form regular particle arrays via self-organization and in how far the formation of such arrays as well as the magnetic properties are affected by the thermal annealing.

EXPERIMENT

FePt nanoparticles are prepared by inert-gas condensation based on a DC-sputtering process in an Ar / He gas mixture. In order to prevent the particles from oxidation, the experimental setup is designed to meet ultra high vacuum (UHV) standards. The system consists of a liquid nitrogen cooled nucleation chamber, a sintering tube, and a deposition chamber. The Ar is fed to the gas inlet of the sputter gun and thus enters the system in the close vicinity of the target. Additional supply of He to the nucleation chamber is used to influence the supersaturation of the metal vapour and thus allows to controll the nucleation and growth of the particles. During particle preparation, the DC power and the pressure are $P_{DC} = 250$ W and $p_{prep} = 0.5$ mbar, respectively. The flow rate of Ar is $r_{Ar} = 4.2$ l/s and that of He is $r_{He} = 2.1$ l/s. After nucleation, the "primary" particles are carried by the Ar / He gas flow through the sintering tube, where they are subjected to sintering at temperatures in the range 293 K $\leq T_s \leq$ 1273 K. The residence time of the particles in the sintering oven is about 1s. In this paper, we refer to the particles which are prepared at $T_S = 293$ K as room temperature (RT) sintered. Finally, the gas-phase sintered nanoparticles are deposited onto substrates which are placed on a liquid nitrogen cooled substrate holder in the deposition chamber. The deposition time for samples used in structural investigations is $t_{sample} = 7$ min, whereas for magnetic measurements it is as high as $t_{sample} = 60$ min in order to obtain a sufficiently large magnetic signal. Further experimental details are published elsewhere [14]. Prior to any subsequent ex-situ characterization, the as-prepared particles are exposed to ambient air after breaking the vacuum of the preparation chamber.

Amorphous carbon films supported by copper grids are used for ex-situ morphological and structural investigations by means of (high resolution) transmission electron microscopy ((HR)TEM). For conventional TEM we use a Philips CM12 Twin (120kV) with LaB_6 cathode. HRTEM investigations are carried out on a Philips Tecnai F20 Supertwin (200kV) with field emission gun. The latter microscope is equipped with a Gatan imaging filter (GIF 2000), which is used for electron energy loss spectroscopy (EELS) and energy filtered imaging (EFTEM). In both microscopes we have the possibility to conduct energy dispersive x-ray analysis (EDX), which is used to determine the chemical composition of the FePt nanoparticles.

Additionally, the gas-phase sintered FePt nanoparticles are deposited onto sapphire single crystals, which serve as substrates for ex-situ magnetization studies by means of a SQUID magnetometer (Quantum Design MPMS 5).

RESULTS

A series of samples of gas-phase sintered FePt nanoparticles is prepared, and the preparation conditions are identical for all the samples except for the sintering temperatures, T_S. T_S is varied in the range $293\ K \leq T_S \leq 1073\ K$. The Fe-to-Pt ratio is measured by EDX both on individual nanoparticles and as average values over large areas. The composition is $Fe_{0.55}Pt_{0.45}$, and no significant deviation is found in individual particles.

In Fig. 1, we show two typical TEM micrographs obtained from a sample of RT-sintered FePt nanoparticles and from a sample of particles which are sintered at $T_S = 873\ K$ for $t_{sample} = 60$ min. The images are recorded at low magnifications in order to allow for representative surveys of the samples. We always observe two different types of areas within the samples. In some parts of the samples, the individual particles are spatially separated from each other, and

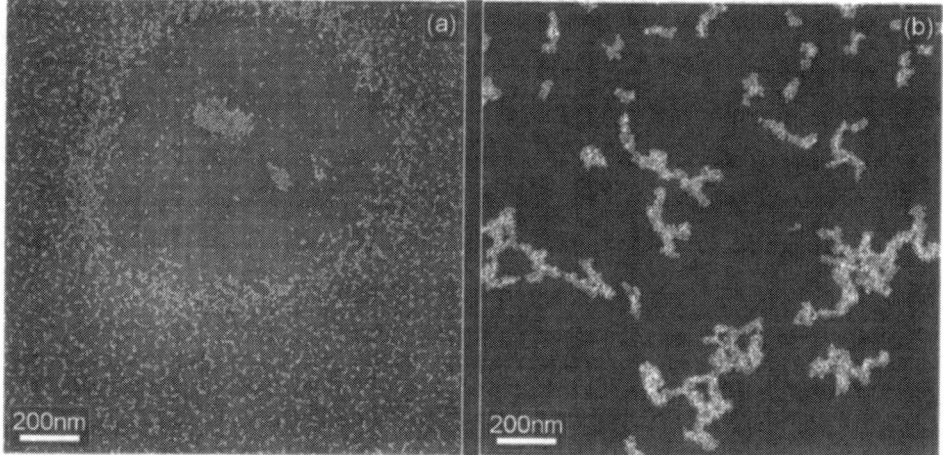

Figure 1: Low-magnification TEM micrographs of gas-phase prepared FePt nanoparticles deposited onto amorphous carbon. The particles are sintered (a) at room temperature for $t_{sample} = 7$ min and (b) at $T_S = 873\ K$ for $t_{sample} = 60$min.

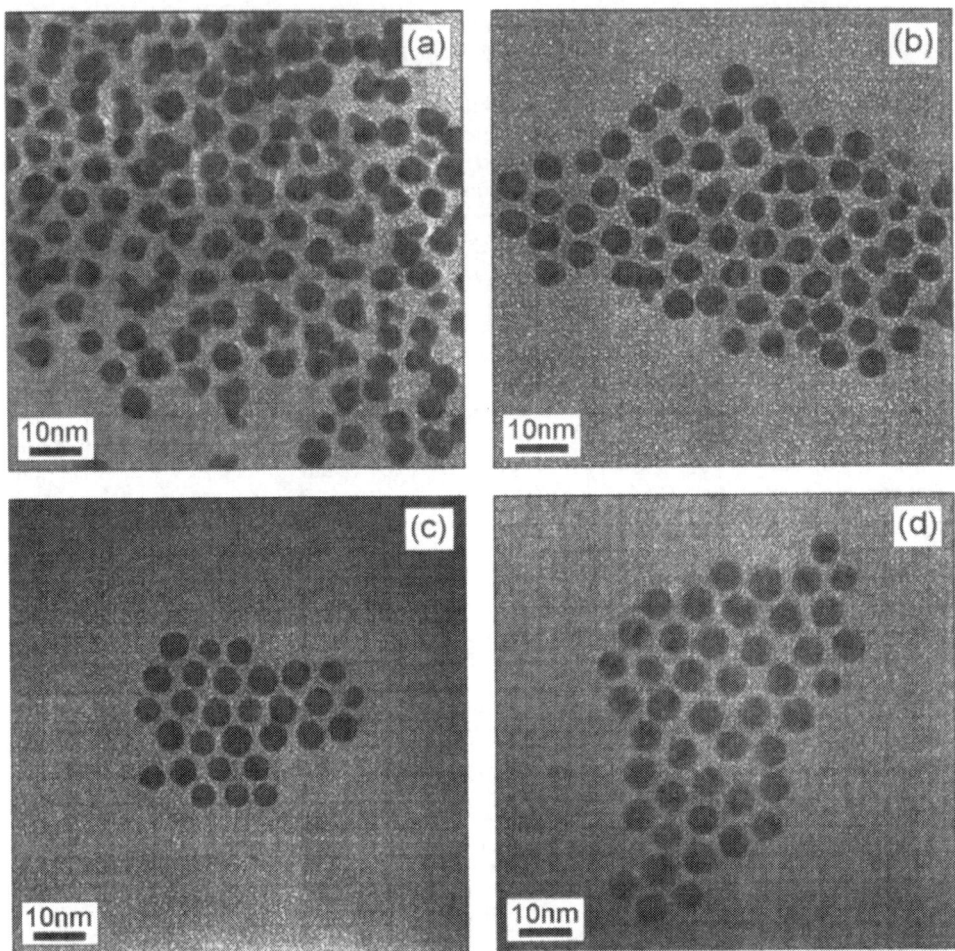

Figure 2: TEM micrographs of patches (aggregates) of FePt nanoparticles sintered in the gas-phase prior to their deposition onto amorphous carbon film. The sintering temperatures are (a) T_S = 293 K, (b) T_S = 673 K, (c) T_S = 873 K, and (d) T_S = 1073 K. Note the concurrent change in the particle morphology and the quality of the order within the self-organized patches.

there are other spots which are predominantly covered with patches of aggregated particles. At low particle coverages, e.g. in samples which are prepared at short deposition times, particle patches is found to be significantly reduced.

In Fig. 2, we present a series TEM micrographs obtained from samples with particles which are sintered at (a) $T_S = 293$ K, (b) $T_S = 673$ K, (c) $T_S = 873$ K, and (d) $T_S = 1073$ K. For all samples, typical particle patches are depicted. The images clearly show that the morphology of the individual particles is significantly changing with increasing sintering temperature. Whereas the RT-sintered particles display irregular shapes, particles sintered at $T_S = 1073$ K are almost spherical. A closer look at the image in Fig. 2d reveals that most of the particles display a 3-fold morphological symmetry. Dark field imaging and HRTEM investigations show that these "spherical" particles are in fact FePt icosahedra; the surface of which is composed of 20 (111) facets [14].

Despite the temperature dependent modification of the particle morphology, the mean particle size, d_P, does not vary with increasing T_S, but remains constant with $d_P = 5.8$ nm. The particle sizes show a log-normal distribution with a logarithmic standard deviation of $\sigma_g = 1.10$.

Concurrently occurring with the morphological changes towards icosahedral, hence almost spherically symmetric shapes with increasing sintering temperature, we find that within the depicted patches, the arrangement of the particles becomes increasingly ordered. Whereas the irregularly shaped particles in Fig. 2a ($T_S = 293$ K) are randomly distributed on the supporting amorphous carbon film, the FePt icosahedra in Fig. 2d ($T_S = 1073$ K) are almost perfectly arranged in a two-dimensional hexagonal close packed pattern. Here, we would like to emphasize that the arrangement of such self-organized patterns is not accidentally induced by the influence of the electron beam while the sample is under investigation. Unless otherwise stated, all (HR)TEM investigations start at low magnification with low intensity and a widely spread beam, and particular attention is paid to any possible modification of the sample during the increase of magnification, thus successive focusing of the beam.

As can be seen from Fig. 2, even within the aggregates, the individual FePt nanoparticles are separated from one another by a distance of some 2 nm. This finding is somewhat striking, since if there is an attractive force between the particles that causes their aggregation, there must be either an additional repulsive force or a mechanical barrier acting to keep them apart. The barrier could be a material wall between the particles, or a shell around their core, that would prevent the particles from contacting. In order to examine if there is any material between the FePt particles, which is not seen in the TEM images, energy filtered TEM (EFTEM) investigations are carried out. We have measured a variety of electron spectroscopic images (ESI) of elements such as O, C, Si, Fe, and others. Fig. 3a shows the carbon mapping of a sample with particles which are sintered at $T_S = 873$ K. The brighter a spot appears in the ESI, the larger is the carbon content at the corresponding position. We observe bright "rings" around the "dark" particles, which indicate that there is an increased amount of carbon between and around the particles. This is further illustrated in Fig. 3b where we show the carbon profile taken along the dash-dotted line in Fig. 3a. A horizontal line indicates the carbon content of the supporting amorphous carbon film. Whereas the carbon content between the FePt nanoparticles (indicated by arrows) is enhanced with respect to the uncovered support, it is reduced at the particle positions themselves. This result is found to be highly reproducible, and similar results are obtained from all samples, independent of T_S. This finding is further evidenced in measurements of carbon profiles by means of EDX in scanning mode TEM (not shown here). Similar material enrichment by other elements between the particles is *not* observed.

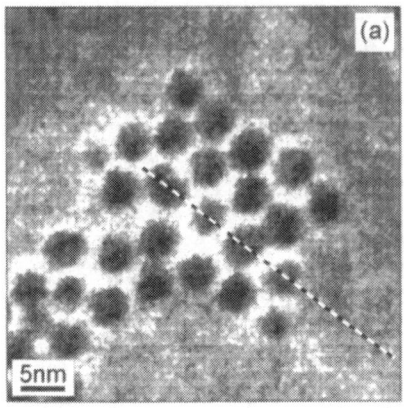
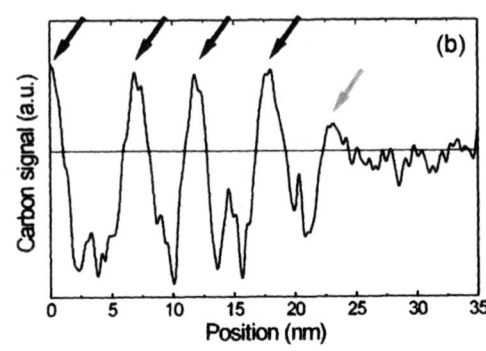

Figure 3: (a) Electron spectroscopic carbon image (K-edge mapping) of gas-phase sintered FePt nanoparticles (T_S = 873 K) deposited onto amorphous carbon film from energy filtered TEM (EFTEM). (b) Profile along the dash-dotted line in (a). The horizontal line indicates the carbon content in the substrate. Arrows denote an increased amount of carbon between the particles.

To study the magnetic properties of the gas-phase sintered FePt nanoparticles, the magnetization M as function of the external magnetic field H is measured at different temperatures in the range 5 K ≤ T ≤ 293 K. Fig. 4 shows a TEM micrograph of a sample that is deposited for t_{sample} = 60 min and sintered at T_S = 873 K. A second sample which is simultaneously prepared on a sapphire substrate is used for magnetization measurements. In Fig. 4b we present the resulting M(H) data obtained after subtracting the diamagnetic background signal of the substrate. We observe a rapid narrowing of the hysteresis loops, thus a decrease of the coercivity with increasing temperature. The hysteresis loops close at roughly T = 60K, whereas the magnetization curves remain non-linear. This behavior is a typical fingerprint for the occurrence of superparamagnetism. In superparamagnetic particles, the (intrinsic) coercivity H_C is expected solely to originate from the magnetic anisotropy, and for a monodisperse fraction of such particles, the temperature dependence of H_C is [15]

$$H_C = H_{C0} [1 - (T/T_B)^{1/2}] . \qquad (1)$$

Here, H_{C0} and T_B denote the intrinsic coercivity at T = 0 and the superparamagnetic blocking temperature, respectively. In Fig. 5 we show the temperature dependence of the coercivity for the sample with T_S = 873 K (solid circles) and for a sample with RT-sintered particles (open squares). In the inset, H_C is plotted as function of $T^{1/2}$. The solid lines represent a fit of equation (1) to the experimental data. The linear decrease of the data in the inset indicates that below T ≈ 50 K, the gas-phase sintered FePt nanoparticles behave as individual and magnetically de-coupled particles. A zero-temperature coercivity of H_{C0} = 619 Oe and a blocking temperature of T_B = 49 K are obtained as fitting parameters. It is remarkable, however, that although the particle morphology changes significantly with increasing sintering temperature, the magnetic response is almost identical for RT-sintered particles and particles which are sintered at T_S = 1073 K.

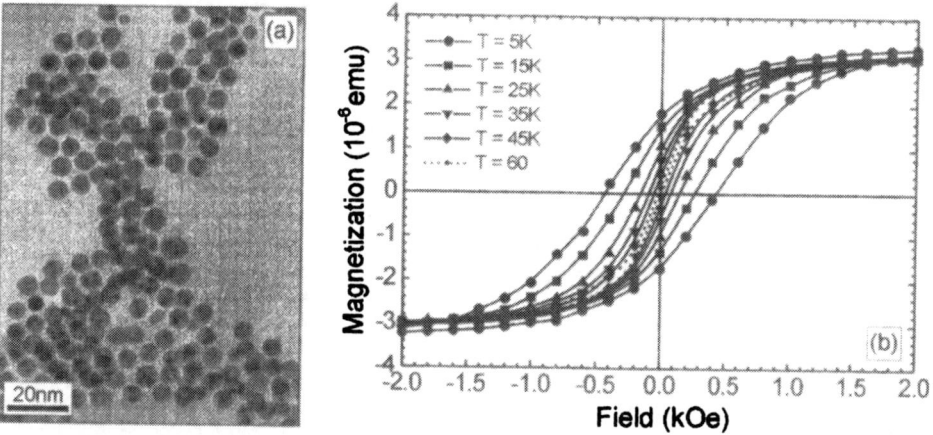

Figure 4: Temperature dependence of the magnetization curves for FePt nanoparticles deposited onto a sapphire substrate. Particles are sintered in the gas-phase at T_S = 873 K. Note the narrowing of the magnetization curves with increasing temperature.

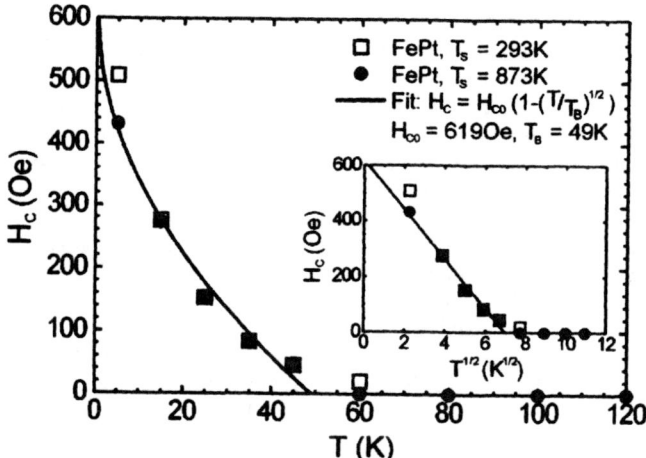

Figure 5: Temperature dependence of the coercivity of gas-phase sintered FePt nanoparticles deposited onto a sapphire substrate. The sintering temperatures are T_S = 293 K (open squares) and T_S = 873 K (solid circles). The solid lines indicate a fit based on a model assuming a monodisperse, non-interacting particle ensemble.

DISCUSSION

An important experimental result, which can be seen in Fig. 2, is the fact that the mean size of the gas-phase prepared FePt nanoparticles remains unaffected by a variation of the sintering temperature. This means that no particle-particle collisions (coagulation) occur along the preparation route, which would result in the formation of complex (partially fractal) particle aggregates [16]. Only intra-particle recrystallization processes (sintering) are to be considered. Additionally, reevaporation from the particle surfaces, which would lead to a significant size reduction of the particles with increasing sintering temperature, can also be disregarded. Thus, our experimental situation is to some extent comparable with molecular beam conditions, although the pressure is as high as $p_{prep} = 0.5$ mbar. These conditions provide the advantage of fast thermalization of the particles, and therefore, allow to impose thermal energy on the particles for sintering in the gas-phase.

On the other hand, the results of the TEM investigations clearly show the occurrence of particle aggregates on the amorphous carbon substrate. Taking into account that there is *no* aggregation in the gas-phase, this leads us to the conclusion that the particles possess a certain mobility on the substrate in order to form more or less regular particle patterns. As a consequence, the question of the origin of this mobility of the nanoparticles arises.

Here, the observation of an enhanced amount of carbon in the immediate vicinity of the FePt particles leads to the assumption of a certain chemical reactivity at the nanoparticle-carbon interface. A confirmation for this assumption can be found from the results of additional experiments in the electron microscope. When the intensity of electron beam is increased, we observe that the amorphous carbon film starts to experience a structural modification which finally results in the complete disintegration of the film *only in the close vicinity of the FePt nanoparticles*. In fact, many transition metals such as Ni, Co, Pd, or even FeNi alloys, are known to be catalytically active for the graphitization of carbon [17]. Pt is a prominent catalyst for the oxidation of CO [18], and, in general, the catalytic activity is found to be significantly increased in nanoparticulate or nanostructured systems [19]. In Pt and Ni, the (111)-facets are the catalytically most active surfaces. Most of the FePt nanoparticles, however, which are sintered at elevated temperatures are icosahedral particles. Such icosahedra are composed of 20 tetrahedra, and their surfaces consist of 20 (111)-facets. The sum of these arguments leads us to the conclusion that the mobility of the FePt nanoparticles on amorphous carbon is likely to be due to an enhanced chemical reactivity at the particle surface.

With increasing sintering temperature, the quality of the self-organized particle patterns is significantly improved. This is attributed to the morphological changes of the particles imposed by the thermal treatment. Particles which are sintered at RT display irregular shapes. This indicates that the coalescence process during growth is incomplete [20], and as a consequence, these particles do not exhibit their equilibrium morphology. The irregular shapes impede a close packed arrangement of the FePt nanoparticles on the substrate. Sintering at elevated temperatures leads to both a completion of the coalescence process and intra-particle recrystallization. This results in the formation of icosahedra, which – though being faceted – are almost ideally spherical. These particles are considered to exhibit their equilibrium morphology.

However, the advantage of the almost spherically symmetric shape of the FePt icosahedra is contrasted by the fact that we do not observe any single crystal $L1_0$ ordered tetragonal particles. The magnetic properties of FePt nanoparticles which are sintered at RT and $T_S = 873$ K are determined by means of SQUID magnetometry. Both samples are superparamagnetic at room temperature with a blocking temperature of T_B = 49 K. The defining equation for the superparamagnetic limit, thus for the blocking temperature, under the constraint that the magnetization remain stable for approximately 10 years is

$$25 \, k_B T_B = K_U V . \tag{2}$$

Here, k_B is the Boltzmann constant, K_U the uniaxial magneto-crystalline anisotropy constant, and V the particle volume. According to equation (2), the magnetocrystalline anisotropy constant for our gas-phase prepared and sintered FePt nanoparticles (d_P = 5.8 nm) is calculated to be K_U = 1.7×10^5 J/m^3. This value is very small compared to the anisotropy constant of L1$_0$ ordered FePt (K_U = 6.6-10$\times 10^6$ J/m^3). However, since no single crystal L1$_0$ FePt particles are observed in our samples, this result is not surprising. Recently, Stahl et al. [21] have measured the blocking temperature of stoichiometric FePt nanoparticles (d_P = 4 nm) which were prepared according to the recipe proposed by Sun et al. [10], but not subjected to a post-deposition thermal treatment. The autors report an anisotropy constant of K_U = 1.1×10^5 J/m^3 for their single crystal, but disordered FePt particles, which is roughly in agreement with our results.

In another paper we discuss the structure model for an icosahedron which is composed of L1$_0$ ordered tetrahedral building blocks [14]. Such a "L1$_0$ icosahedron" should be energetically favored, since the strain energies involved are supposed to be smaller than in regular icosahedra. Nevertheless, at every twin boundary between two neighboring tetrahedral building blocks within an icosahedron, the easy magnetization axes are tilted with respect to one another by approximately 72°. This will necessarily lead to a distribution of easy axes within the particles, thus to a reduction of the effecive anisotropy constant.

CONCLUSIONS AND OUTLOOK

We have employed a gas-phase based method to prepare FePt nanoparticles close to the stoichiometric composition. The method provides the advantage of a thermal treatment (sintering) of the particles in the gas-phase prior to their deposition. We have shown that the gas-phase sintering leads to the formation of monodisperse and almost spherically shaped icosahedral particles. Once being deposited onto amorphous carbon films, the FePt nanoparticles tend to arrange themselves in a close packed two-dimensional pattern via self-organization. The degree of order in these patterns depends on the particle morphology, which can be controlled by choosing an appropriate sintering temperature.

Due to the icosahedral structure of the particles, the formation of single crystal L1$_0$ ordered FePt nano-magnets is inhibited. This results in a relatively low magnetic anisotropy, which does not (yet) meet the requirements for materials in future high density magnetic storage media. However, it is known that preferred faceting of nanoparticles, thus their morphology, can be controlled by a variation of the preparation conditions not only in wet-chemical, but also in gas-phase based methods [22,23]. Therefore, the presented method provides a promising potential for the preparation of tailored nanoparticles.

ACKNOWLEDGEMENTS

Valuable discussions with Hermann Sauer, Fritz-Haber-Insitut of the Max-Planck-Gesellschaft, Berlin, and Dr. Herbert Hofmeister, Max-Planck-Institut für Mikrostrukturphysik, Halle, are gratefully acknowledged. Part of this work is supported by the Deutsche Forschungsgemeinschaft within the Sonderforschungsbereich 445.

REFERENCES

1. E.E. Fullerton, D.T. Margulies, M.E. Schabes, M. Carey, B. Gurney, A. Moser, M. Best, G. Zeltzer, K. Rubin, H. Rosen, M. Doerner, Appl. Phys. Lett. **77**, 3806 (2000)

2. J.F. Smyth, S. Schultz, D.R. Fredkin, D.P. Kern, S.A. Rishton, H. Schmid, M. Cali, and T.R. Koehler; J. Appl, Phys. **69**, 5262 (1991); R. M. H. New, R. F. W. Pease, und R. L. White; J. Vac. Sci. Technol. **B12**, 3196 (1994).

3. H.L. Garvin, E. Garmire, S. Somekh, H. Stoll, and A. Yariv, Appl.Optics, **12**, 455 (1973); D.C. Flanders, H.I. Smith, H.W. Lehmann, R Widmer, and D.C. Shaver, Appl. Phys. Lett. **32**, 112 (1978).

4. J. Lohau, A. Moser, C.T. Rettner, M.E. Best, and B.D. Terris, Appl. Phys. Lett. **78**, 990-992 (2001).

5. S.Y. Chou, P.R. Krauss, P.J. Renstrom, J. Vac. Sci. Technol. B **14**, 4129-4133 (1996).

6. Y. Ochiai, M. Baba, H. Watanabe, and S. Matsui, Jpn. J. Appl. Phys. **30**, 3266-3271 (1991).

7. A. Carl, S. Kirsch, J. Lohau, H. Weinforth, and E.F. Wassermann, IEEE Trans. Mag. **35**, 3106-3111 (1999).

8. S. Zankovych, T. Hoffmann, J. Seekamp, J.-U. Bruch, C.M. Sotomayor Torres, Nanotechnology **12**, 91-95 (2001); A. Tilke, R.H. Blick, and H. Lorentz, J. Appl. Phys. 90, 942-946 (2001).

9. C.B. Murray, C.R. Kagan, and M.G. Bawendi, Science **270**, 1335-1338 (1995).

10. S. Sun, C.B. Murray, D. Weller, L. Folks, and A. Moser, Science **287**, 1989-1992 (2000).

11. C.P. Bean, J.D. Livingston, J. Appl. Phys. **30**, 120S-129S (1959); D. Weller and A. Moser, IEEE Trans. Mag. **35**, 4423-4439 (1999).

12. See e.g.: D.J. Sellmyer, M. Zheng, and R. Skomski, J. Phys.: Condens. Matter **13**, R433-R460 (2001); and references therein.

13. D. Weller, A. Moser, L. Folks, M.E. Best, W. Lee, M.F. Toney, M. Schwickert, J.-U. Thiele, and M.F. Doerner, IEEE Trans. Mag. **36**, 10-15 (2000).

14. S. Stappert, B. Rellinghaus, M. Acet, and E.F. Wassermann, Mat. Res. Soc. Proc. **704**, Nanoparticulate Materials, submitted.

15. B.D. Cullity, "Introduction To Magnetic Materials" (Addison-Wesley, 1972), pp. 410.

16. See e.g.: F.E. Kruis, K. Nielsch, H. Fissan, B. Rellinghaus, and E.F. Wassermann, Appl. Phys. Lett. **73**, 547-549 (1998); and references therein.

17. X. Fan, E.C. Dickey, P. Eklund, K. Williams, L. Grigorian, A. Puretzky, D. Geohegan, R. Buczko, S.T. Pantelides, and S.J. Pennycook, Mat. Res. Soc. Prc. **593**, 129-134 (2000), Warrendale, PA, USA; M. Andersson, P. Alberius-Henning, K. Jansson, and N. Nygren, J. Mater. Res. **15**, 1822-1827 (2000); P.E. Andersson and N.M. Rodrigez, J. Mater. Res. 14, 2912-2921 (1999).

18. S. Wilke, V.Natoli, and M.H. Cohen, J. Chem. Phys. **112**, 9986-9995 (2000); S. Völkening, K. Bedürftig, K. Jacobi, J. Wintterlin, and G. Ertl, Phys. Rev. Lett. **83**, 2672-2675 (1999).
19. D.R. Rolison, in „Nanoparticles: Synthesis, Properties and Applications", ed. by A.S. Edelstein and R.C. Cammarata, (Institute of Physics Publishing, London, UK, 1996), pp. 305; and references therein.
20. R.C. Flagan and M.M. Lunden, Mater. Sci. Eng. **A204**, 113-124 (1995).
21. B. Stahl, N.S. Gajbiye, G. Wilde, D. Kramer, J. Ellrich, M. Ghafari, H. Hahn, H. Gleiter, J. Weißmüller, R. Würschum, and P. Schlossmacher, preprint (2001).
22. T.S. Ahmadi, Z.L. Wang, T.C. Green, A. Henglein, and M.A. El-Sayed, Science **272**, 1924-1926 (1996); J.M. Petroski, Z.L. Wang, T.C. Green, and M.A. El-Sayed, J. Phys. Chem. B **102**, 3316-3329 (1998).
23. D.L. Olynick, J.M. Gibson, and R.S. Averback, Appl. Phys. Lett. **68**, 343-345 (1995).

Magnetic Materials

Self-Assembled Magnetic Dots, Antidots, Dot Chains, and Stripes: Epitaxial Co on

Ru(0001)

Dongqi Li and Chengtao Yu

Materials Science Division, Argonne National Laboratory, Argonne, IL 60439

Lateral magnetic nanostructures have been grown via molecular beam epitaxy in ultrahigh vacuum and characterized *ex-situ* with atomic force and magnetic force microscopy. We observed that epitaxial growth of Co onto Ru(0001) at elevated temperature results in three-dimensional Co islands (dots) or a flat Co film network with deep holes (antidots) in truncated pyramidal shapes. The lateral size of these dots/antidots, in the order of 100 nm, tends to be uniform at each given coverage. We attribute the growth mode mainly to strain relaxation of Co epitaxy on Ru, which has a 8% lattice mismatch. In addition, we have explored the placement of these dots on a grooved Ru(0001) surface. The dots automatically align into linear chains along the asymmetric grooves to form either dot chains or continuous stripes, which would open new opportunities in creating either ordered magnetic arrays or arbitrary arrangements.

1. Introduction

The fabrication of lateral magnetic nanostructures via self-assembly is attracting increasing interests for both basic research and technology development.[1,2,3] The physical size of a magnetic system affects its magnetic properties by altering its dimensionality, structure, surface/interface electronic structure, quantum size effects and transport, and domain/domain wall structure and motion. In addition, patterned media[4,5] with individual magnetic dots has been actively explored as an alternative route for ultrahigh density information storage, since having one bit per dot promises to improve signal-to-noise ratio significantly. Self-assembly not only offers an intrinsically simpler, faster, and less costly alternative route to lithography, but also promises thermodynamically stable structures even beyond the lithographic limits. Self-assembled quantum dots have been realized in semiconductor systems,[6] such as Ge/Si, GeSi/Si, and InAs/GaAs, where a strain-driven Stranski-Krastanov (SK) growth mode results in dots of narrow size distributions. Besides using self-assembled masks for thin film deposition,[3] metal self-assembly during growth has been observed in a few systems with one or several monolayer (ML) thick 2D structures utilizing unique surface phenomena such as step decoration,[7,8,9,10] surface strain[11,12] and reconstructed surfaces as templates.[13,14] In this paper we review our recent work on a new metal-on-metal self-assembly mode, where magnetic lateral structures of 70-600 nm wide and 1-7 nm thick form mainly due to strain relaxation in epitaxy.[15,16] For Co grown on flat Ru(0001) at elevated temperature, we observed both 3D islands (dots) or holes in film networks (antidots in smooth films) with well-defined shape, surprisingly smooth surfaces, and relatively narrow size distribution.[15]

A general challenge in self-assembly is associated with placement and alignment of the structures at predetermined locations, a task that is routine for lithographic fabrication. For example, magnetic patterned media[4,5] consisting of a magnetic dot array would require dot alignment along tracks. It would even be more demanding to produce complex spintronic devices[17] by means of self-assembly. It is, therefore, critical to develop methods to align self-assembled magnetic nanostructures. We have aligned self-assembled magnetic dots along substrate grooves that form due to residual scratches from the mechanical surface polishing used in preparing the substrate.[16] The results offer the promise that magnetic dot arrays or arbitrary arrangements could be fabricated by self-assembly with the assistance of lithographic substrate patterning.

2. Experiment

The experimental details have been reported elsewhere.[15,16] The Co samples with wedge-like thickness gradient of $0 - 420$ nm were grown with molecular beam epitaxy (MBE) on smooth and grooved Ru(0001) crystal substrates at 350°C in a ultrahigh vacuum (UHV) system with base pressure of 6×10^{-11} Torr. The grooves result from the residual polishing lines or unexpected step bunching, which are straight or slightly curving along an arbitrary direction with a period of $\sim 0.5 - 2$ μm and height of $\sim 4 - 16$ nm. The surface was cleaned $in\text{-}situ$ by cycles of O_2-annealing at $1300 - 1500$ K and flashing at $1500 - 1600$ K.[18] The resultant Ru surface, and the subsequent Co surfaces, are free of O and C contamination as determined with Auger electron spectroscopy. Sharp hexagonal low-energy electron diffraction (LEED) patterns were obtained both on the clean substrate and along the clean Co wedge with no apparent broadening. The Co samples were then covered with a thin layer of Au (< 1 nm) at room temperature to reduce the oxidation of Co in air. The morphology and magnetic imaging of the wedges were taken *ex situ* with atomic force microscope (AFM) and magnetic force microscope (MFM).

3. Results and Discussions

Figure 1 shows a variety of self-assembled Co lateral structures, i.e., (a) dots, (b) antidots, (c) dot chains, (d) stripes and (e) dot arrays, grown under different conditions. At elevated temperature, i.e., close to equilibrium growth condition, Co forms quasi-hexagonal three-dimensional (3D) islands (dots). When the nominal thickness is large enough, instead of having the islands to connect into a smooth, continuous film, Co forms a network of flat film with deep quasi-hexagonal holes, i.e., antidots in a film. For both dots and antidots, the edges are along the high-symmetry directions and the surfaces are nearly atomically smooth. The well-defined shape, atomically smooth tops and film network, and the sharp LEED pattern suggest that these islands may be structurally coherent with few defects.

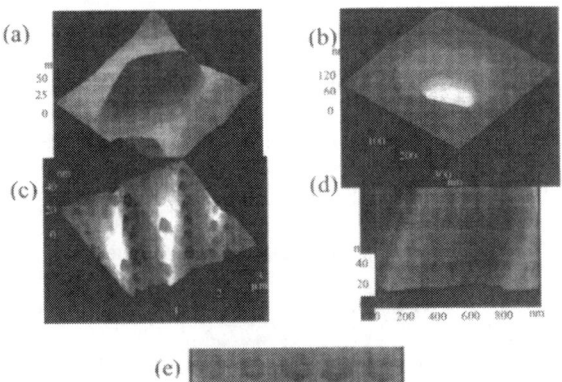

Fig. 1. Typical AFM images of Co (a) dot, (b) antidot, (c) dot chains, (d) stripes, (e) dot arrays.

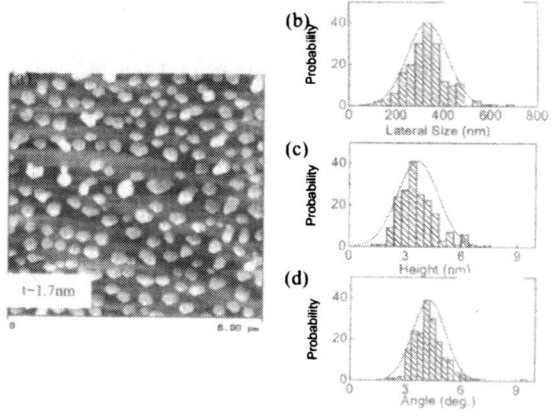

Fig. 2. (a) AFM image of Co dots, and the distribution of (b) lateral size, (c) height, and (d) side-wall angle.

Such a growth is driven by strain relaxation in a highly-ordered epitaxial system. Co has a lower surface energy and a large 8% lattice mismatch with Ru(0001). Instead of forming dislocations in the film, they can form coherent 3D islands to release the strain when growing in S-K mode. As discussed in detail for semiconductor self-assembled quantum dots, such a growth mode prefers uniform dot size since smaller dots grow faster than the large ones, and there exists a minimal size for such dots.[6] And both dots and antidots are equivalent in releasing strain.[19] In Fig. 2, it is shown that the dots have a tendency to be uniform in size and shape, even though the growth conditions are not yet optimized.

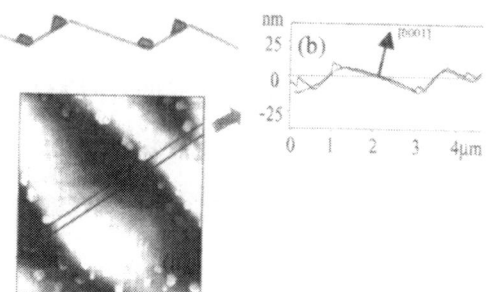

Fig. 3. AFM image of Co dot chains along the grooves on Ru(0001). Line profiles across the grooves, one representing the substrate and the other including Co dots on the grooves, indicate the location of the dots.

While the distribution of the dots on a flat Ru(0001) surface is never completely regular, we have discovered that the placement of these dots can be manipulated with substrate morphology. The dot chains, stripes and dot arrays in Fig. 1 are formed on grooved Ru substrates, where we left some directional polish scratches. After annealing, the surface reveals a grooved structure with asymmetric saw-tooth profile, as seen in Fig. 3b. The dots align along the top and bottom of the grooved structures into linear chains as in Fig. 1(c). Depending on coverage and groove period, stripes and dot arrays are also observed as seen in Fig. 1(d) & (e).

Figure 3 describes the dot chains in more details. Fig. 3a is a typical AFM image of the self-assembled Co dots on the grooved Ru(0001) at a nominal Co thickness of 1.1 nm. The location of the dot chains is most obvious on Fig. 3b, which shows the line profiles along the two parallel lines across the grooves as indicated on Fig. 3a. Along one line that does not cross any dot, an asymmetric saw-tooth profile of the substrate is apparent. The other parallel line runs across several dots, indicating the location of the dots on the saw-tooth profile. It is clear that the dots grow both at the bottom and the top of the grooves, forming two parallel chains of dots along each edge. More specifically, the ones on top are primarily on the short side of the saw-tooth and the ones at the bottom primarily on the long side. Since the dots are driven by strain relaxation, it is therefore not surprising to see a preferential nucleation of these dots near the top and bottom of the grooves, where strain should be the largest. The fact that the dots sit on only one side of the slope suggests that, besides the strain, other factors such as diffusion along stepped surfaces also play an important role for metals. Noted that the individual dots on grooved substrates show asymmetric profiles perpendicular to the grooves with the top of the dots always in parallel with the long edge (Fig. 3b). We postulate that the growth front of the dots tends to be the stable hcp(0001) plane and therefore result in such an asymmetric shape.

These dots and films with antidots are ferromagnetic. All the dots, ~70-600 nm in diameter, exhibit in-plane single domain states, as seen in Fig. 4a. Our micromagnetic simulations indicate that they are mostly metastable against a vortex ground state, but can easily be trapped in single-domain state depending on history.[20] Indeed, the dots not only have single domain virgin states as observed with MFM, but also exhibit the characteristic square loops for single domains.[16] For thicker films with antidots, the Co magnetocrystalline anisotropy, which is along the c-axis perpendicular to substrate surface, overcomes the in-plane shape anisotropy and forms stripe domains with up-and down perpendicular magnetization (Fig. 4b). Fig. 4c shows that along the dot chains, the dots couple ferromagnetically due to inter-dot magnetostatic interactions, which has been discussed in terms of classical 1D Ising chain. It demonstrates that these self-assembled magnetic dots offer model systems in understanding low dimensional physics.

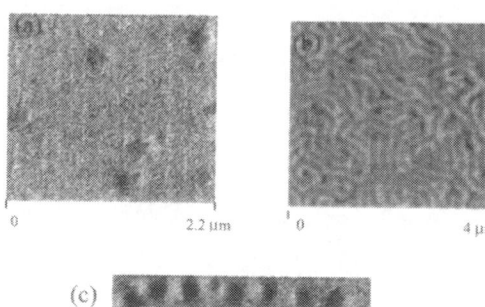

Fig. 4. MFM images for (a) dots, (b) thick film with antidots, and (c) a dot chain.

4. Summary

We have observe a metal-on-metal growth mode in Co/Ru(0001) at elevated temperature, where self-assembled 3D dots and antidots in rather regular truncated pyramidal shapes are mainly attributed to a stress-driven mechanism. While good lattice match has been one of the major criteria to guide epitaxial growth, it is possible that well-chosen lattice mismatched systems could be utilized to fabricate strain-engineered regular magnetic nanostructure arrays with different sizes and periodicity. A linear alignment of self-assembled Co dots and stripes can be created along grooves on a Ru(0001) substrate. Our observations suggest that it may be possible to direct the alignment and positioning in self-assembly of complex patterns by means of substrate templating, which should be of general applicability beyond the Co/Ru system. Magnetically, the dots are ferromagnetic with in-plane single-domain state, while the thick films with antidots exhibit perpendicular stripe domains.

Acknowledgment

The authors thank the coworkers John Pearson, and Sam Bader for their contributions. This work was supported by the US DOE BES-Materials Sciences under contract # W-31-109-ENG-38.

References

1. Shouheng Sun, C.B. Murray, D. Weller, L. Folks, and A. Moser, Science **287**, 1989 (2000); C.T. Black, C.B. Murray, R.L. Sandstrom, and Shouheng Sun, Science **290**, 1131 (2000).

2. P. Gambardella, M. Blanc, H. Brune, K. Kuhnke, and K. Kern, Phys. Rev. B **61**, 2254 (2000); P. Gambardella, M. Blanc, L. Bürgi, K. Kuhnke, and K. Kern, Surf. Sci. **449**, 93 (2000).

3. K. Liu, S. M. Baker, M. Tuominen, T. P Russell and I. K. Schuller, Phys. Rev. B **63**, 060403(2001).

4. Phys. Rev. 63, 060403(2001)

5. Y. Chou, M.S. Wei, P.R. Krauss, and P.B. Fisher, J. Appl. Phys., **76**, 6673 (1994).

6. R.L. White, R.H. New, and R.F.W. Pease, IEEE Transactions on Magnetics, **33**, 990 (1997).

7. P. Politi, G. Grenet, A. Amarty, A. Ponchet, and J. Villain, Phys. Reports **324**, 271(2000), and references therein.

8. F.J. Himpsel, T. Jung, and J.E. Ortega, Surf. Rev. Lett. **4**, 371 (1997).

9. J. Hauschild, H.J. Elmers, and U. Gradmann, Phys. Rev. B **57**, R677 (1998).

10. J. Shen, M. Klaua, P. Ohresser, H. Jenniches, J. Barthel, C.V. Mohan, and J. Kirschner, Phys. Rev. B **56**, 11134 (1997).

11. Dongqi Li, B. Roldan Cuenya, J. Pearson, and S.D. Bader, Phys. Rev. B **64**, 144410 (2001); B. Roldan Cuenya, J. Pearson, Chengtao Yu, Dongqi Li, and S.D. Bader, J. Vac. Sci. Technol. A **19**, 1182 (2001).

12. H. Röder, E. Hahm, H. Brune, J-P Bucher and K. Kern, Nature **366**, 141 (1993); H. Brune, K. Bromann, H. Röder, K. Kern, J. Jacobsen, P. Stoltze, K. Jacobsen, and J. Nørskov, Phys. Rev. B **52**, R14380, (1995).

13. E.D. Tober, R.F.C. Farrow, R.F. Marks, G. Witte, K. Kalki, and D.D. Chambliss, Phys. Rev. Lett. **70**, 3943 (1993).

14. Dongqi Li, V. Diercks, J. Pearson, J. S. Jiang, and S. D. Bader, J. Appl. Phys., **85**, 5285 (1999).

15. D. D. Chambliss, R. J. Wilson, and S. Chiang, Phys. Rev. Lett. **66**, 1721 (1991).

16. Chengtao Yu, Dongqi Li, J. Pearson, and S.D. Bader, Appl. Phys. Lett. **78**, 1228 (2001).

17. Chengtao Yu, Dongqi Li, J. Pearson, and S.D. Bader, Appl. Phys. Lett. **79**, 3848 (2001).

18. For an introduction on spintronics, see K. Hathaway and G. Prinz, Phys. Today, **48**, 24 (1995).

19. R. G. Musket, W. Mclean, C. A. Colmenares, D. M. Makowiecki, and W. J. Siekhaus, Appl. Surf. Sci. **10**, 143 (1982).

20. J. Tersoff and F.K. LeGoues, Phys. Rev. Lett. **72**, 3570 (1994).

21. Chengtao Yu, J. Pearson, and Dongqi Li, J. Appl. Phys., in press.

Mat. Res. Soc. Symp. Proc. Vol. 707 © 2002 Materials Research Society

Fabrication and Properties of Self-Assembled Nanosized Magnetic Particles

G. Salazar-Alvarez, M. Mikhailova, M. Toprak, Y. Zhang, and M. Muhammed.
Materials Chemistry Division, Royal Institute of Technology.
SE-100 44 Stockholm, Sweden.

ABSTRACT

The synthesis and characterisation of gold-coated cobalt nanoparticles, as well as their chemically- and magnetically-induced self-organisation have been studied. Metallic core-shell nanoparticles were prepared using two different experimental techniques: bulk reductive precipitation, with average particles size ~15 nm, and microemulsion confining method, with average particle size of ~6 nm. The self-assembly of prepared nanoparticles on flat substrates was achieved by derivatising the substrate and particle surfaces with bifunctional organic molecules that attaches to both particles and substrates.

Examination of the self-assembled systems was carried out by a number of characterisation techniques including transmission electron microscopy (TEM), UV-visible spectrophotometry (UV-VIS), and atomic force microscopy (AFM).

INTRODUCTION

Over the last decade, control and design of nanoparticles and, in particular, core-shell structured materials have drawn the attention of physicist, chemists and material scientists. The tailoring of the surface properties combined with those shown by the core provides a number of applications that range from catalysis, and coatings to spintronics and magneto-resistants, among others.[1-3] Furthermore, during the last few years, it has been observed that the quantum size effects in low dimensional systems of such materials vary considerably than those in the bulk. Such behaviour has been associated with the degree of organisation, i.e., 0D, 1D, 2D, and 3D self-assembled arrays.[4] The understanding, control, and design of such systems open a true supramolecular engineering of materials, where a defined nanophase grain boundary plays a major role. Among the prospective applications for self-organised systems are: photonic band-gap crystals,[5] ultra high-density data storage,[6] and biomedical applications.[7]

In this work, attempt has been made at studying the controlled preparation of superstructures based on magnetic particles and the chemical parameters that affect them.

EXPERIMENTAL
Reagents

All the salts used were purchased from Aldrich with ACS grade. As solvents c-Hexane (cHex, +99.5%), n-decane (C_{10}, 97%), and n-octane (C_8, 97%) were used as obtained without any further purification. Triton-X 100 (TX100), n-Hexanol (HxOH), and n-Butanol (BuOH) were purchased from Fluka and were used as obtained. 3-aminopropyl trimethoxy silane (APTMS) and 3-mercaptopropyl trimethoxy silane (MPTMS) were purchased from Aldrich. High purity water was used with a specific resistivity of 18 MΩ·cm (Milli-Q, Millipore Inc.).

Colloidal system

Microemulsions were prepared by solubilising aqueous solutions of the salts and aqueous solution of NaBH$_4$ into a TX100-HxOH/cHex system. The control of size was achieved by adjusting the water to surfactant molar ratio (w), while the surfactant to cosurfactant molar ratio (p_0=0.6) was kept constant in all the experiments. N$_2$ was flowed for 1 h before and throughout the experiments to eliminate the possibility of further oxidation. The concentration of the metal salts varied from 0.2 to 0.5 M, and a five-fold molarity of sodium borohydride was employed.

A sequential synthetic procedure, as depicted in Figure 1, was used in order to prepare a core-shell nanostructured material. A metallic core (c) was prepared by contacting two microemulsions with w_1 containing the metal ion (a) and sodium borohydride (b), respectively. The water droplet size was then enlarged to a w_2 by adding an aqueous solution of the shell-forming ion (d) and then reduced (e) by mixing with a second sodium borohydride emulsion at w_2 (b'). The microemulsion was lysed with ethanol (f) and the particles were extracted with oleic acid, and then re-dispersed in toluene for further utilisation (g).

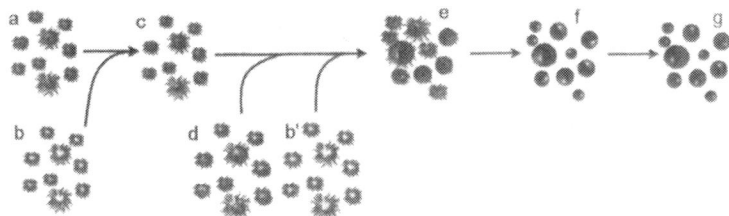

Figure 1. Schematic representation of the sequential synthetic method for the preparation of core-shell nanostructured particles.

Substrate preparation

Silicon wafers p-type were used as substrates for the self-assembly of nanoparticles. The substrates were immersed at 80 C for 20 min in a 1:3 v/v solution of H$_2$O$_2$ and H$_2$SO$_4$, thoroughly rinsed with ethanol and water in sonicated baths and stored in a water-free chamber for further utilisation. Clean substrates were silanised for 2 h under a 0,3 M ethanolic solution of APTMS or MPTMS at 60 C, rinsed with absolute ethanol, 95% ethanol, and finally water until a clean mirror-like surface was obtained. Samples were blown-dried with nitrogen and used immediately for self-assembly experiments.

For the deposition of particles, the substrates were immersed in ethanol to which aliquots containing diluted ethanolic solutions of the particles were added drop-wise with and without an external magnet placed underneath the substrate level.

Characterisation techniques

A JEM 2000EX transmission electron microscope (JEOL) was used to measure the size and morphology of the core-shell nanoparticles. Surface characterisation was carried out using a spectrophotometer Cary 100 (Varian). Topographic images of self-assembled structures were obtained with a Nanoscope III atomic force microscope (Digital Instruments) operated in tapping mode (TMAFM) using a Si$_3$N$_4$ cantilever with a resonance frequency of 286-313 kHz.

RESULTS AND DISCUSSION

   Nanoparticles of gold-coated cobalt were prepared via a sequential nanoemulsion synthesis (typical size 5-7 nm) and bulk synthesis (typical size 12-25 nm).
   Figure 2 shows typical micrographs of particles prepared under both methods, i.e., sequential microemulsion technique and bulk reductive precipitation. From the micrographs it can be seen that the particles prepared by microemulsion exhibit a very narrow size distribution with no agglomeration, while the particles prepared in bulk are nearly the same size but with a broader size distribution and important inter-particle interactions. A colloidal system consisting of H_2O/TX100-HxOH/cHex with a water-to-surfactant ratio of 4.4 was used for the formation of the cores, while the shells were produced using a $w_2=6.6$. Elsewhere it has been described a linear relationship for the dependence of the water pool radius on the water content,[8] thus the shell thickness is expected to be, according to simple calculations, around 2 nm.

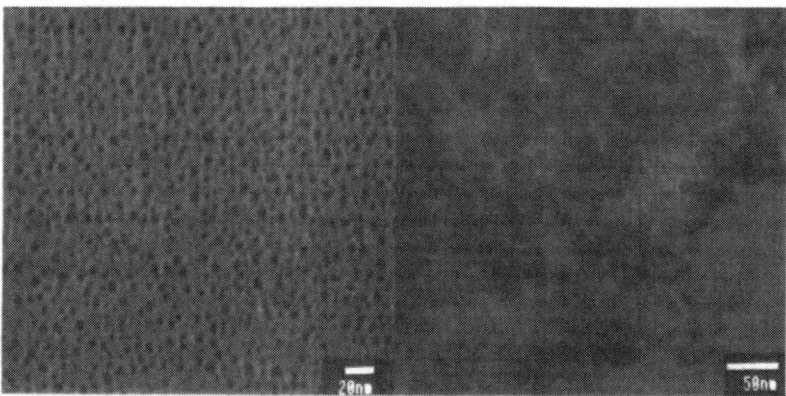

Figure 2. TEM micrographs of Au-coated cobalt prepared by a) microemulsion method, b) bulk reduction.

   A comparison of the absorption spectra for particles synthesised by bulk reduction and nanoemulsion confining is depicted in Figure 3. From the graph, it can be seen that pure gold particles present a typical plasmon resonance peak (PRP) around 520 nm, with some deviations due to their size distribution [9] and the naked cobalt nanoparticles show a monotonous increasing absorption towards UV region, which is characteristic for metallic cobalt. [8] The Au-coated cobalt nanoparticles synthesised by microemulsion confining show the gold PRP while the gold-coated ones show the Au plasmon resonance peak with a shoulder at 350 nm that is likely due to a surface oxidation of the incompletely covered cobalt. [8]

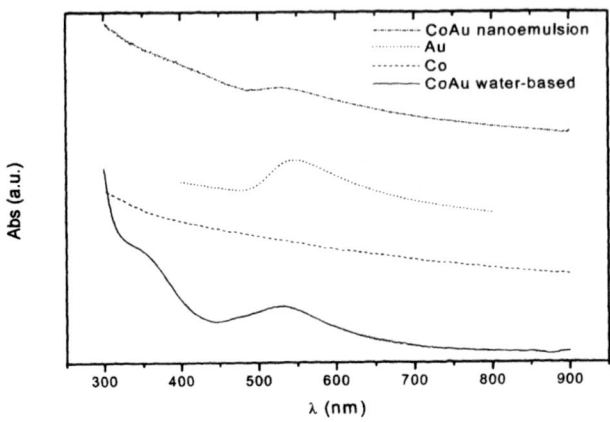

Figure 3. Optical absorption spectra of synthesised nanoparticles

The particles were extracted from toluene solution into an ethanolic solution for faster and more homogeneous complexation with the functional groups at the substrate surface. MPTMS was chosen at first to investigate the pure-chemical interaction of particles and substrate. The results depicted in Figure 4 correspond to the ~7 nm Au-coated Co nanoparticles assembly, where a smooth continuous and complete layer of (Co)Au core-shell structures was obtained. The complete surface coverage, confirms a complete silanisation of the substrate surface. The section analysis of the substrate was in agreement with the particle size results obtained from TEM (Figure 2).

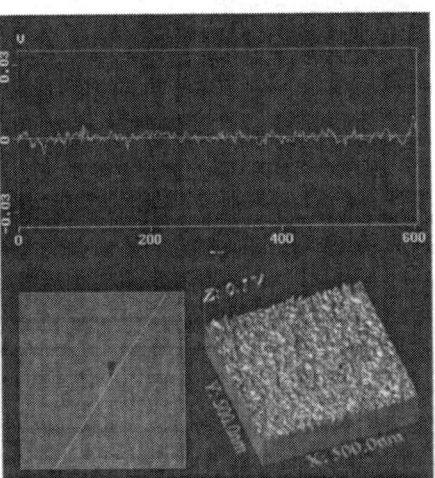

Figure 4. Line section analysis and topographic images of Au-coated cobalt nanoparticles self–assembled on MPTMS-derivatised silicon flat surface. The distance between red arrows indicates 13.672 nm.

The magnetophoretically-induced self-organisation of (Co)Au nanoparticles on a flat APTMS-derivatised silicon substrate was achieved by carrying out the complexation of nanoparticles while applying an external magnetic field *in-situ*. Terminal amine groups were selected for this experiment due to a lower reactivity towards Au, if compared to thiol groups, as stated by the Pearson's HSAB theory. [10] A relatively low concentration of particles was used to contact the substrate surface under slow mechanical stirring and alkaline pH. In Figure 5, it is shown the 1D arrangement of particles into continuous lines along the substrate. The lines consist of single particles oriented along the direction of the magnetic field, as demonstrated by the line profile showed in Figure 5a. This, together with the evidence of self-assembly in absence of magnetic field, yields a proof of the superparamagnetic properties of the synthesised particles, which is in agreement with earlier reports for similar size pure cobalt nanoparticles. [11] Furthermore, in Figure 5c it can be observed that in some regions a 2D assembling consisting of single particles aligned vertically normal to the substrate is obtained.

Such structures are predicted for small dipoles, which due to a minimisation of Zeeman energy tend to align parallel to the stray field of the external magnet, thus reducing the number of dipole charges in the surface. The 2D case, is an extrapolation of the situation observed in 1D: if we consider a 'cone of alignment' induced by the external field, then a parallel arrangement will be the one that will have lower energy, and if we have a high concentration of 'free' particles a newly created shape anisotropy together with the existing dipolar interactions will induce the particles to join the structure where the energy in minimised, i.e., on top of the existing lines.

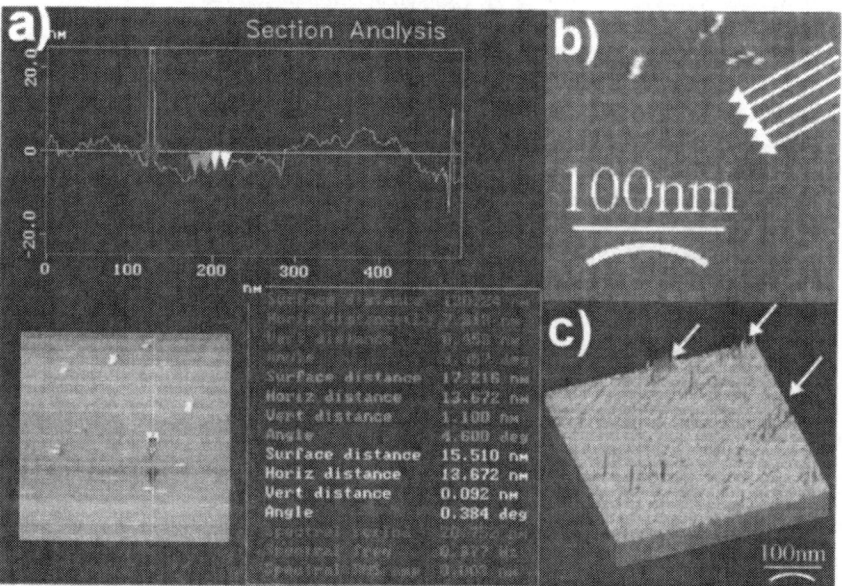

Figure 5. Magnetically induced self-assembled (Co)Au magnetic nanoparticles. a) Line analysis indicating particle diameter, b) higher magnification of the topographic image showing arrangement of particles, and c) perpective image showing localised stacking of particles.

CONCLUSIONS

Colloidal self-assembly is a versatile technique that can be used to organise any type of particles, provided that the surface chemical properties of the particle match those shown by the surface-confined functional groups on the substrate. Particles synthesised by nanoemulsion presented a narrow size distribution when compared to a bulk method. The uniformity of the gold-coating was confirmed by optical absorption, whereas the colloidal system proved to be much more efficient to achieve a homogeneous gold shell. Chemically driven self-assembly was useful to prepare a homogeneous coverage of particles, whereas an external force induced local 2D assemblies and it may used to produce 3D structures. The procedure followed in this work is related to a magnetically-induced self-assembly which, in principle, can be used to generate from 0D to 3D superstructures.

ACKNOWLEDGEMENTS
    The authors would like to thank Doc. A. Zagorodni for his valuable help.

REFERENCES
[1] F. Caruso, *Adv. Mater.* **13**, 11 (2001).
[2] S.D. Bader, *Surf. Sci.*, to be published.
[3] H. Gleiter, *Acta Mater.* **48**, 1 (2000).
[4] M.P. Pileni, *Appl. Surf. Sci.* **171**, 1 (2001).
[5] Y. Saado, M. Golosovsky, D. Davidov, A. Frenkel, *Synth. Mater.* **116**, 427 (2001).
[6] D. N. Lambeth, E. M. T. Velu, G. H. Bellesis, L. L. Lee, and D. E. Laughlin, *J. Appl. Phys.*, **79**, 4496 (1996).
[7] D. K. Kim, et al, *Scripta Mater.* **44**, 1713 (2001).
[8] C. Petit and M.P. Pileni, *J. Magn. Magn. Mater.* **166**, 82 (1997).
[9] C.G. Bohren, D.F. Huffman; *Absorption and Scattering of Light by Small Particles*, (Wiley-Interscience Publication, New York 1998).
[10] R.G. Pearson, *J. Am. Chem. Soc.*, **85**, 3533 (1963).
[11] J. Legrand, C. Petit, D. Bazin and M.P. Pileni, *Appl. Surf. Sci.* **164**, 186 (2000).

Mat. Res. Soc. Symp. Proc. Vol. 707 © 2002 Materials Research Society

Magnetic Interactions in Fe Nanoparticle Arrays

D. Farrell[a], S. Yamamuro[a], Yumi Ijiri[b], and S. A. Majetich[a]
[a]Dept. of Physics, Carnegie Mellon University
Pittsburgh, PA 15213, U.S.A.
[b]Dept. Of Physics, Oberlin College
Oberlin, OH 44074

ABSTRACT

The preparation of monodisperse Fe nanoparticles and self-assembly into hcp and fcc or fcc-like arrays is described. Here dipolar interactions dominate for the interparticle spacings studied (1.4-3.4 nm). Comparison of the low temperature magnetic properties of multilayer arrays with those of dilute suspensions of the same particles show increased coercivity and slower magnetic relaxation in the arrays. Mean field calculations of magnetic interaction fields suggest the type of ordered structures formed.

INTRODUCTION

Here we describe how dipolar interactions between nanoparticles affect the structure and the magnetic properties of self-assembled arrays. Compared with other self-assembling nanoparticle systems, magnetic nanoparticles possess additional magnetostatic forces due to dipolar interactions. The dipolar field B_{dip} surrounding a monodomain particle is anisotropic,

$$B_{dip} = \frac{\mu_0}{4\pi} \left(\frac{3(\mu \bullet r)r}{r^5} - \frac{\mu}{r^3} \right),$$

(1)

and so are the dipolar forces. Here μ_0 is the permeability of free space, μ is the dipole moment, and r is the distance from the center of the dipole. When magnetic nanoparticles have a stable magnetic moment and magnetostatic forces dominate over the dispersion forces, the particles prefer to form chains. This problem can be circumvented if the magnetic interactions are weaker than the other forces driving self-assembly, which occurs when the particles are superparamagnetic. While arrays rather than chains are formed, it is unclear whether the magnetic interactions have more subtle morphological effects due to dipolar interactions during self-assembly.

Dipolar interactions affect the magnetic properties of nanoparticles even under high dilution [1, 2]. It is therefore expected that at high concentration, as in a close-packed array, that there will be significant differences in the behavior. What differentiates these arrays from nanocrystalline magnets is the lack of exchange interactions. Between *atoms* the exchange strength J between atoms is given by

$$J = J_0 \exp[-R/L_{ex}].$$

(2)

Here J_0 is a constant, R is the separation, and L_{ex} is the exchange length. $L_{ex} = (A_{ex}/K)^{1/2}$, where A is the exchange stiffness and K is the magnetocrystalline anisotropy. For bulk Fe, L_{ex} is on the order of 15 nm. Between *particles* the exchange strength depends on the interparticle separation s and a barrier height $\Delta\varepsilon$:

$$J_{interparticle} \sim (1/s^2)exp[-(const.)(\Delta\varepsilon)^{1/2}]. \tag{3}$$

[3]

The estimated exchange field H_{ex} between islands in Reference 17 is 50-100 Oe. The nanoparticles of Figure 2 had a 3.6 nm closest spacing between particles. Here $H_{ex} < 10^{-3}$ Oe, but the dipolar field estimated by including up to fifth nearest neighbors is ~ 800 Oe. For an edge-to-edge interparticle spacing of 3.6 nm, typical of oleic acid-coated nanoparticles in self-assembled arrays, this exchange field $H_{ex} < 10^{-3}$ Oe. While the dipolar interaction field depends strongly on the particle size, it is orders of magnitude larger, typically on the order of hundreds of Oersted.

In the remaining sections, we describe the structures of the self-assembled arrays. Next the magnetic properties of the arrays are discussed, both relative to those of dilute samples of the same particles and as a function of the interparticle spacing. Finally we describe preliminary mean field calculations related to the superferromagnetic ordering within an array, as a function of the array diameter and number of layers.

EXPERIMENTAL

Improved synthetic methods have made it possible to prepare monodisperse magnetic nanoparticles that self-assemble into arrays [4, 5]. Monodisperse Fe nanoparticles were prepared under an argon atmosphere. 4 mg of platinum acetylacetonate and 150 mg of 1,2-hexadecanediol were dissolved in 15 mL of octyl ether. While heating the platinum salt is reduced to form platinum clusters, which later act as nuclei for the iron nanoparticles. To make 8 nm particles, 0.2 mL of $Fe(CO)_5$ was added when the solution reached 100 °C. Heating continued up to 280 °C. The solution was cooled, and an additional 15 mL of octyl ether was added. At 100 °C 1.5 mL of $Fe(CO)_5$ was added, and the heating continued to 260 °C. The amount of iron added determines the final size of the particles. Using this method, particles with average sizes ranging from 4-10 nm were prepared, with typical monodispersities of 10% or less. The fluid was kept under an inert atmosphere, and 0.025 mL each of the surfactants oleylamine and oleic acid were added to the cooled solution. In some cases different surfactant mixtures, such as hexanoic acid and hexyl amine were used. The particles were transferred from octyl ether to a low boiling point solvent such as hexane by adding sufficient ethanol to precipitate the particles, decanting the supernatant, and redispersing in the new solvent. The concentration of surfactant in the solvent was kept constant throughout the washing procedure.

The structural properties of the resulting particles were characterized by transmission electron microscopy (TEM). Here a TEM grid was dipped into the solution and suspended in an argon atmosphere to dry. The grids were then imaged in a 120 keV Philips EM420 electron microscope to reveal the particle size and size distribution, the array structures, and the relative amount of the grid covered by the arrays.

For magnetic measurements we compared two samples prepared from the same batch of particles. In one case the particles were diluted in an organic solvent to 0.01 vol.% Fe, assuming the bulk density, and sealed under argon in a glass ampoule. In the other, arrays were prepared on TEM grids inside the glove box, and then they were stacked and sealed for transfer into the SQUID magnetometer.

The samples were magnetically characterized by regular and remanent hysteresis loops, and by magnetic relaxation measurements. Unless otherwise specified, all measurements were performed at 10 K. For the hysteresis loops, the field was applied and the magnetic moment of the sample was measured; after each at-field measurement, the field was applied and turned off, and the remanent moment M_r was measured. In some cases the magnetization was measured with the applied field perpendicular to the plane of the grids, but unless otherwise specified the field was in-plane. The at-field measurements form the regular hysteresis loop, and the zero field measurements form the remanent hysteresis loop. In magnetic viscosity or relaxation measurements, the sample was first saturated at 50 kOe. The field was then reduced to zero, or a reverse field equal to the coercivity H_c was applied. The moment of the sample was measured at 90 – 600 sec time intervals over a ten hour period at this field. Diamagnetic background was measured for the TEM grids (either Cu or nylon), and for the sealant.

RESULTS AND DISCUSSION

A. Structure

TEM shows that monolayer arrays of Fe nanoparticles always have a hexagonal structure [6], as shown in Figure 1. In thicker nanoparticle lattices both hcp and fcc lattices are observed (Figure 2) [6, 7]. Occasionally defects are seen in which the nanoparticles bond in two-fold bridging sites rather than three-fold sites [8]. As the number of layers increases, the images appear to be more fcc-like because of the possibility of stacking faults during the room temperature self-assembly process. All of these features are seen for self-assembled arrays of other types of spherical nanoparticles coated with surfactants, and are therefore unrelated to the magnetism of the particles.

One feature to our knowledge not observed for nonmagnetic nanoparticle arrays is a preference for an odd number of layers (Figure 2). This is seen in hcp arrays made from larger Fe particles. Black arrowheads represent the boundaries between the different layer thicknesses. For the hexagonal closest packed structure, particles occupy particular

Figure 1. Monolayer array of Fe nanoparticles.

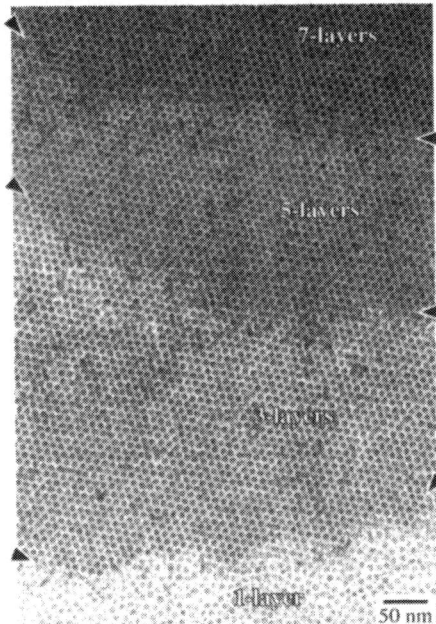

Figure 2. TEM image of an hcp-stacked multilayer of 7.0 nm Fe nanoparticles, showing a preference for odd numbers of layers. The image consists, from the bottom to the top, of a randomly packed monolayer, and ordered regions with 3-, 5-, and 7-layers.

sites *A* or *B* in alternating layers: *ABABAB*.... Particles in odd numbered layers occupy the same type of site, and the images of a monolayer and a trilayer differ mainly in the degree of contrast. An abrupt transition between a monolayer and a trilayer is a common feature of Fe nanoparticle arrays, and in some cases arrays are seen with 1, 3, 5, 7, and 9 layer steps (Figure 2).

B. Magnetic Properties

Over the size of a TEM grid, both fcc (or fcc-like) and hcp arrays are seen, with *average* structural domain sizes on the order of hundreds of nm. Typical thicknesses for arrays used for magnetic measurements were on the order of ten layers. We assume that the samples were identical except for the effect of magnetic dipolar interactions between particles. For the dilute sample, the *average* particle spacing estimated from the mass concentration was greater than 500 nm. Magnetostatic interactions are expected to be much stronger in the arrays than in the dilute samples.

One way to characterize the strength of the magnetic interactions is by comparing the blocking temperature, T_B, for different samples. A sample of nanoparticles is said to be blocked if its magnetization does not relax to the equilibrium value, $M = M_s \tanh(M_s HV/kT)$, within the measurement time, τ. Here M_s is the saturation magnetization, V is the particle volume, and kT is the thermal energy. In the SQUID magnetometer, the typical measurement time per data point is on the order of 20 seconds.

One method of determining the blocking temperature T_B is from the peak in the zero field cooled (ZFC) magnetization curve. In ZFC magnetization experiments, the sample is cooled to low temperature with H=0, and then a small field is applied. The magnetization M is measured as the sample is warmed. Initially M is low, since the particle moments are randomly oriented. As the temperature rises, the ratio of the anisotropy plus magnetostatic energy to the thermal energy drops. The particle moments are more likely to switch directions and become more closely aligned with the applied field, which increases the ZFC magnetization. Eventually thermal fluctuations are large enough to switch the particle moment in any direction, not just toward that of the applied field.

For the arrays T_B was always greater than or equal to that of the corresponding dilute sample. However, it is desirable to have a more quantitative means for comparison of magnetic interaction strengths. T_B depends on size as well as interactions between particles. Ideally $T_B = \ln(\omega_0 \tau^{-1})KV/k$, where ω_0 is the attempt frequency, K is the magnetocrystalline anisotropy, and k is the Boltzmann constant. However, the effective K is likely to differ from the bulk value, and due to surface contributions can be strongly size dependent. We therefore used an alternative method to assess the strength of magnetic interactions in the arrays.

Remanent magnetization hysteresis loops $M_r(H)$ were used to compare the switching field distributions in the dilute and arrays samples. The remanent magnetization removes contributions from rotations and from paramagnetic or superparamagnetic species. These measurements were done at low temperatures (typically 10 K), well below the T_B of the dilute nanoparticle sample. The coercivity of the arrays is always greater than or equal to that of the dilute sample, as shown in Figure 3. In this approach the switching field of particles in the arrays is equal to the vector sum of the intrinsic switching field of the particles plus the effective field of other particles. The average value of the effective field can be calculated from mean field theory. For the sample in Figure 3, there is a broad distribution of switching fields in the arrays sample.

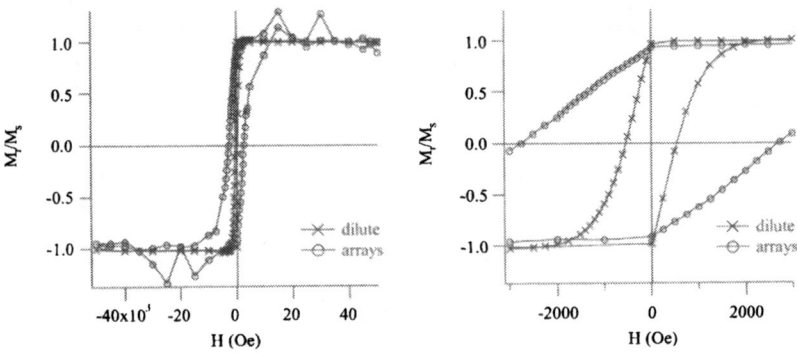

Figure 3. Remanent magnetization loops taken for dilute and arrays samples of 6 nm Fe nanoparticles at 10 K [9]. The magnetizations are normalized to the saturation values. In these measurements, the plane of the grids was perpendicular to the applied field. The larger fields required to saturate the arrays sample reflect a preference for an in-plane magnetization, due to interactions between particles.

To clarify the reason for the distribution, we also measured the time-dependent relaxation of the magnetization, both at zero field and near the coercivity H_c. The results are shown in Figure 4. A distribution of switching fields causes the magnetization to decay logarithmically in time, so that $M = M_0 - S\ln(t/t_0)$ [10,11]. Here, M_0 is the initial magnetization at field, S is the slope of the line, and t_0 is the time of the initial measurement. By different choices of the initial time t_0 the value of the magnetization intercept M_0 is changed, but the value of the slope S is not. The dilute samples show logarithmic decay at all times. Though the particle size distribution is narrow, there is a distribution of energy barriers due to the random crystallographic orientations [12]. The arrays have a sharp initial drop, followed by logarithmic decay. This initial relaxation period typically occurs during the first 2-3 minutes of the measurement.

There appear to be two distinct types of magnetization decay in the arrays sample. In the saturation field, the particle moments are presumably aligned parallel to the applied field. When the field is removed, the rapid initial drop in the magnetization may be due to reorientation in order to form flux closure pathways within the arrays. This is done most easily with an in-plane magnetization configuration.

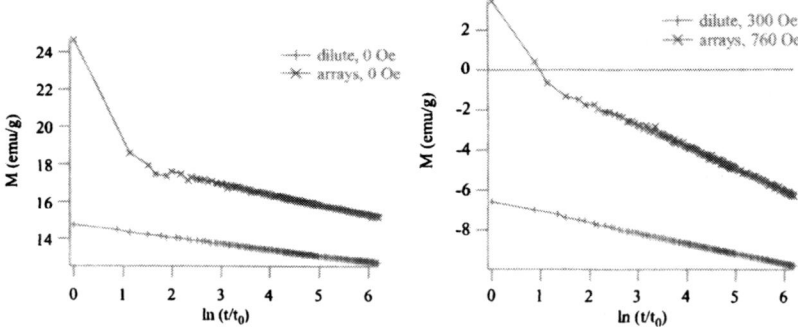

Figure 4. M vs. ln (t/t_0) at 10 K. Left: H = 0; Right: arrays, H = 760 Oe; dilute, H = 300 Oe [9].

To vary the strength of the magnetic coupling between particles, we varied the thickness of the surfactant coating. Here a batch of particles was divided, and different surfactants were added to each solution prior to array formation. For particles coated with a mixture of oleic acid and oleyl amine, the average thickness of the surfactant coating was found to be 1.5 nm. For a coating of hexanoic acid and hexyl amine, the average thickness was 0.7 nm. The remanent hysteresis loops are shown in Figure 5. Surprisingly, we did not observe a significant change in the coercivity. However, the sample with a smaller average separation showed a more gradual approach to saturation. Clearly some particles within the arrays have large switching fields, but again there is a broad distribution. We attribute this to structural inhomogeneity within the arrays. As the length of the surfactant chain is reduced, so is the relative mobility of particles during the self-

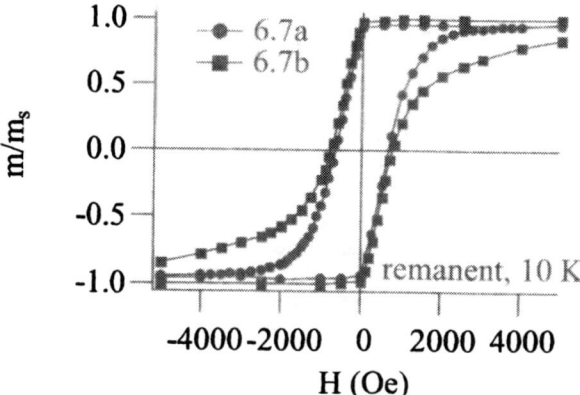

Figure 5. Remanent hysteresis curves at 10 K for arrays of 6.7 nm Fe nanoparticles. Curve 6.7a is for a sample with an oleic acid/oleyl amine coating, and curve 6.7b has a hexanoic acid/hexyl amine coating. Here the applied field was parallel to the plane of the grids.

assembly process, for a given core size. This suggests that alternative methods may be needed to prepare highly uniform arrays with very small spacings.

Slower relaxation in the more closely spaced particles is confirmed by magnetic relaxation measurements, as shown in Figure 6. Here we compare the slow decay components only, by normalizing to in "initial" time t_o of approximately 5 minutes after the saturating field was removed. Stronger interactions between more closely spaced particles increase the magnetic viscosity S.

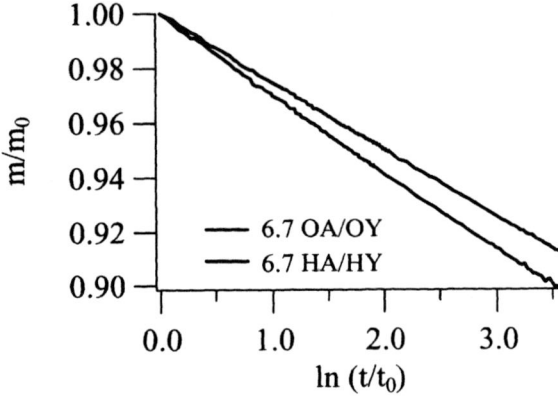

Figure 6. Comparison of long time (5 min-10 hr) zero field relaxation rates for 6.7 nm Fe particles coated with different surfactants, at 10 K. For this sample with OA/OY, a 50/50 molar ratio of oleic acid and oleyl amine, the average edge-to-edge particle spacing is 3.0 nm. For HA/HY, a 50/50 ratio of hexanoic acid and hexyl amine, the average edge-to-edge spacing is 1.4 nm.

C. Modeling

To better understand the magnetic interactions among nanoparticles in the growing arrays, we calculated the local fringe field at a particle on the surface of a growing array, due to its neighbors. For simplicity we assumed that the system was in its ground state, which is clearly not the case at room temperature. We also assumed that the particles were perfectly spherical and the arrays had perfect crystallographic alignment. Similar calculations have been done for a wide variety of *infinite* lattices of magnetic dipoles. Both two-dimensional (2D) square arrays and 3D simple cubic lattices have antiferromagnetic ordering [13-16]. Ferromagnetism is seen for 2D hexagonal arrays, as well as fcc and bcc structures [13, 15-17].

When calculating the ordering within a single finite hexagonal layer, we found that by including contributions from up to fourth or more nearest neighbors was sufficient for a small preference for ferromagnetic ordering. Next we calculated the field experienced by a particle on the surface of a ferromagnetically ordered layer, as shown in Figure 7. Here a finite 2D circular layer of particles is placed in the x-y plane with a radius R_c. Each particle has a saturated magnetic induction B_s and a radius R. Here R refers to the radius of the iron core, while R_{eff} is the radius of the core plus the surfactant layer thickness. The magnetic moments of the spheres are assumed to lie in the positive y-direction. By numerically summing the fringe field from each particle using Eq. 3, the net field at the origin was determined. The x- and z-components of the net field are always zero due to symmetry. This calculation was repeated for multiple layers to find the net field due to the nearest layer, the second nearest layer, etc.

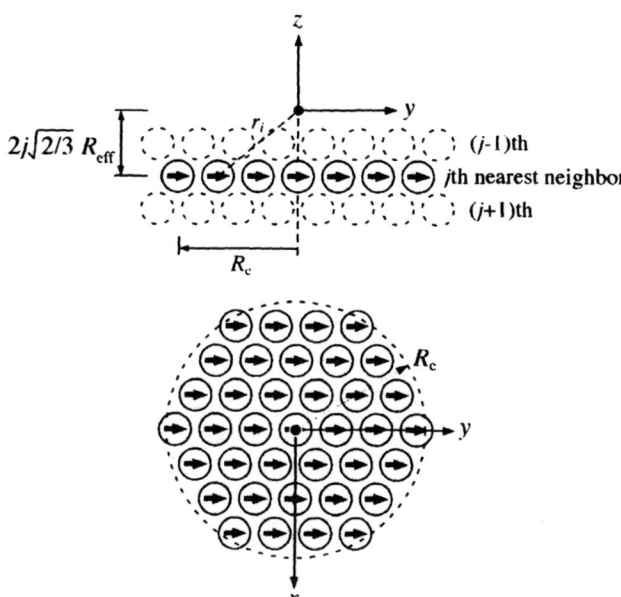

Figure 7. Geometry for calculations of dipolar fringe field caused by particles of j-th nearest-neighbor layers and sensed by a particle on the growing array surface. R_c represents the size equivalent to the layer radius.

Figure 8 shows the dependence of the fringe field on the layer size. A positive value for B_y indicates that the particle in its ground state will align in the positive y-direction. For small layer diameters, all layers have negative values of B_y, and therefore lead to antiferromagnetic alignment of the central particle of the growing layer,

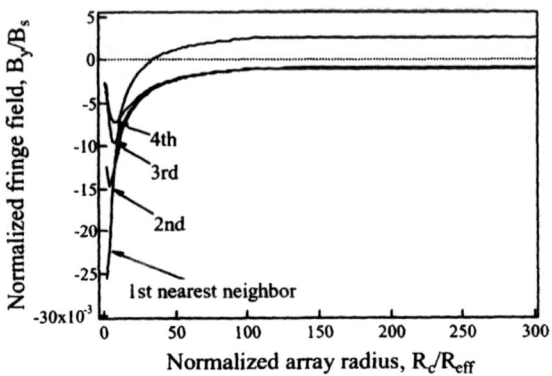

Figure 8. Normalized fringe field B_y/B_s due to aligned, ferromagnetically ordered hexagonal layers vs normalized array radius R_c/R_{eff}. Here the particle-core radius $R = 3.5$ nm and $R_{eff} = 4.7$ nm.

relative to the nth layer below. Only the first layer contribution crosses over from an antiferromagnetic to a ferromagnetic net field as the layer size increases. More distant layers make antiferromagnetic contributions for all layer diameters, and the values converge for large layer sizes.

Once a small layer is formed on the monolayer, the magnetic moment direction of the second layer particles is locked in by the large in-plane Lorentz cavity field, which can be up to ten times the net fringe field. Eventually a bilayer array is formed with both layers ferromagnetically coupled. When a particle deposits at the center of a large bilayer, it experiences a strong ferromagnetic field due to the layer below, plus a weaker antiferromagnetic field from the bottom layer. However, if the second layer is small, there will be a net antiferromagnetic field and the expanding third layer will have an opposite magnetic orientation to those below. In stable trilayer arrays, the second and third layers typically meet this criterion.

When the second and third layers have the same size and are antiferromagnetically coupled, there would be a strong fringe field between their boundaries for flux closure. This could explain the abrupt transition between the monolayer and trilayer in Figure 2. Because the interaction strongly depends on the array diameter and thickness, it becomes rather complicated to calculate the energetically stable magnetic configuration for thicker multilayer arrays. However, if the diameter of the top layer in the array grows large before the next layer nucleates, a similar process will favor the addition of another antiferromagnetic bilayer coupling. This would explain the preference for odd number of layers.

CONCLUSIONS

Surfactant coated monodisperse Fe nanoparticles self-assemble into hcp and fcc or fcc-like arrays. In many cases there is an unusual preference for forming hcp arrays with only odd numbers of layers, a feature not observed in nonmagnetic self-assembled arrays. The blocking temperature and coercivity are increased in the arrays, though the magnitudes of the changes vary. The coercivity in the arrays can be understood qualitatively in terms of the average local field due to dipolar interactions with neighboring particles. Broad distributions of switching fields remain in the arrays, showing the need for improved structural homogeneity. Magnetic relaxation measurements show a rapid decay over a short time period, attributed to the rapid formation of flux closure domains after the saturating field is removed. At longer times logarithmic decay is observed, with a slower decay rate for more closely spaced particles.

Mean field calculations of magnetic interaction fields suggest the type of ordered structures formed for perfect arrays at T = 0 K. Within a layer the average field favors ferromagnetic coupling for all but the smallest diameters. As the array forms multiple layers, the interlayer coupling varies. For very large diameter layers, the second layer is magnetized in the same direction as the first, while additional layers are magnetized in the opposite direction. For medium-sized layers there can be a net cancellation of the fields, a result that could be related to the preference for odd layers in the hcp arrays. More detailed experimental characterization of the magnetic structures within the arrays is needed to verify this hypothesis.

ACKNOWLEDGMENTS

We would like to acknowledge support from the National Science Foundation grant #CTS-9800127 and the American Chemical Society Petroleum Research Fund grant #ACS PRF 33866-AC5.

REFERENCES

1. J. Zhang, C. Boyd, and W. Luo, Phys. Rev. Lett. **77**, 390-393 (1996).
2. R. Chamberlin, K. D. Humfeld, D. Farrell, S. Yamamuro, Y. Ijiri, and S. A. Majetich, J. Appl. Phys. (in press, 2001).
3. V. N. Kondratyev and H. O. Lutz, Phys. Rev. Lett. **81**, 4508 (1998).
4. S. Sun. C. B. Murray, D. Weller, L. Folks, and A. Moser, Science **287**, 1989 (2000).
5. C. Petit, A. Taleb, and M. P. Pileni, J. Phys. Chem. **103**, 1805 (1999).
6. S. Yamamuro, D. Farrell, K. Humfeld, and S. A. Majetich, MRS Symposium Proceedings, **636**, D10.8.1-D10.8.6, (2001).
7. S. Yamamuro, D. F. Farrell, and S. A. Majetich, (unpublished).
8. D. Zanchet, M. S. Moreno, D. Ugarte, Phys. Rev. Lett. **82**, 5277-5280 (1999).
9. D. Farrell, S. Yamamuro, and S. A. Majetich, MRS Symp. Proc. **674**, U4.4.1-U4.4.6 (2001).
10. R. Street and J. C. Woolley, Proc. Phys. Soc. *A* **62**, 562 (1949).
11. R. W. Chantrell, M. Fearon, and E. P. Wohlfarth, Phys. Stat. Sol. (a) **9**, 213 (1986).
12. Keith D. Humfeld, Anit K. Giri, Sara A. Majetich, and Eugene L. Venturini, IEEE Trans. Mag. **37**, 2194 (2001).

13. V. Russier, C. Petit, J. Legrand, and M. P. Pileni, Phys. Rev. B **62**, 3910 (2000).
14. J. F. Fernandez and J. J. Alonso, Phys. Rev. B **62**, 53 (2000).
15. J. M. Luttinger and L. Tisza, Phys. Rev. **70**, 954 (1946).
16. G. Mukhopadhyay, P. Apell, and M. Hanson, J. Magn. Magn. Mater. **203**, 286 (1999).
17. J. E. Martin and R. A. Anderson (private communication).

Nanowires and Chains

Mat. Res. Soc. Symp. Proc. Vol. 707 © 2002 Materials Research Society

SELF - ASSEMBLING OF NANOPARTICLES IN THE FORM OF DOUBLE LINEAR CHAINS AND SUPERLATTICES ON THEIR BASIS

R.T. Malkhasyan*, R.K. Karakhanyan*, M.N. Nazaryan*, C. Sung**.
*Scientific – Production Enterprise "Atom", Tevosyan str. 3/1,
Yerevan 375076, Armenia. rmalkhas@aua.am
**University of Massachusetts, Lowell M.A.

ABSTRACT

The self-assembling systems in nanoscale powders of crystalline MoO_3 are disclosed. For the first time, the double linear chain aggregates of the MoO_3 nanoparticles were revealed in nanoscale powders of MoO_3, treated by vibrationally excited molecules of hydrogen. Double linear chain aggregates form a linear and an orthogonal supperlattices. Both the double linear chain aggregates and the supperlattices formed are being considered as self-assembling systems. The treatment of the powders by the molecules of hydrogen is a method of obtaining long self-assembling chains of inorganic materials and supperlattices on their basis.

INTRODUCTION

Disclosuring the new self-assembling systems is an important problem of materials science [1-4]. The development of the advanced technology for the treatment of the nanosized powders can lead to a formation of self-assembling nanostructures. In [5-8] a new technology is proposed for the preparation of the amorphous single-component metals. This new technology, termed as "quantum-chemical", is based on carrying out nonequilibrium chemical processes where vibrationally excited molecules of hydrogen take part. The purpose of the present work is to obtain the new self-assembling systems in nanoscale powders of MoO_3 treated by quantum-chemical technology, i.e. by vibrationally excited molecules of hydrogen.

EXPERIMENTAL

A fine dispersive powder of molybdenum oxide MoO_3 with grain size less than 50μm had been installed into the inclined evacuated reactor, which was evacuated to pressure below 10^{-4} Torr. Upon gaining the above-mentioned pressure, the vibrationally excited molecules of hydrogen generated in the close proximity from the reactor's active zone were injected and the hydrogen molecules interacted with the MoO_3 nanoparticles. Hydrogen molecules at the third quantum level with the excitation energy 1,5±0.2 eV were used. The investigation of MoO_3 powders was carried out on the transmission electron microscope TESLA BS 500 at the accelerating voltage of 90 kV. To prepare the samples, 1-2 drops of MoO_3 powder suspension in hexane was drifted on the copper grid with carbon coat.

RESULTS AND DISCUSSION

Electron microscopic investigations showed that during an intense electron irradiation, an ejection of the MoO_3 nanoparticles from their agglomerates took place. The MoO_3 nanoparticles, which ejected under the influence of intense electron irradiation on the carbon coat, are distinctly seen in Fig. 1. The minimal size of those nanoparticles is about 5 nm.

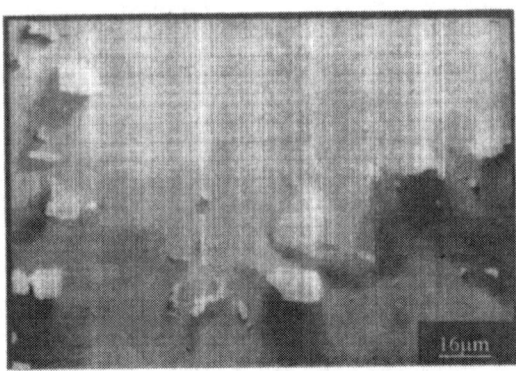

Fig. 1. The MoO₃ aglomerates with ejected nanoparticles.

During the ejection, the microstructure of agglomerates, previously unknown, was revealed. Particularly, the long double linear chain aggregates of MoO₃ nanoparticles were observed. These linear chain aggregates consist of round crystalline MoO₃ nanoparticles with minimal diameter of about 20 nm. The linear double chain aggregates form a linear superlattice with the parameter of about 1 μm from straight chains with the length of about 30 μm that are parallel to each other (Fig. 2). As seen from Fig. 3, which was gained at higher magnification, the linear chain aggregates are double, i.e. they are composed of two separate chains, which are close to each other. Long double chain aggregates of the MoO₃ nanoparticles have transversal branches, which in some places reach the neighboring chains. The branches themselves, in their turn, have smaller branches.

During the ejection of MoO₃ nanoparticles from the treated agglomerates, the orthogonal superlattices, with parameters a=60 nm and b=80 nm, formed by linear chain aggregates of MoO₃ nanoparticles were also revealed (Fig. 4). The microdiffraction patterns of the linear and orthogonal superlattices revealed their polycrystalline character, testifying to the arbitrary orientation of MoO₃ crystalline nanoparticles. The linear and orthogonal superlattices of MoO₃ nanoparticles are very stable under the influence of the electron irradiation, irrespective of its intensity and duration. The superlattices are also stable when the carbon coat is absent. In Fig. 2, the brigther background in the range, which is noted by arrow, corresponds to a case when the carbon coat is absent.

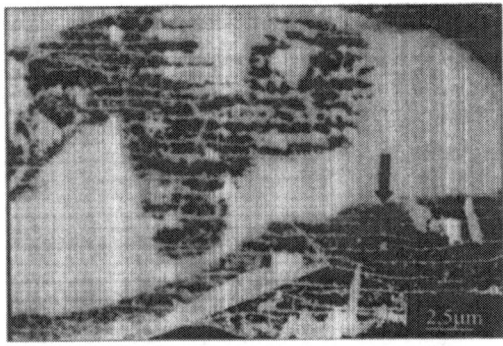

Fig. 2. The MoO₃ nanoparticles linear chain aggregates.

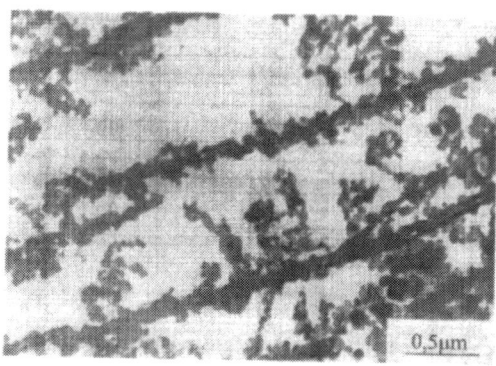

Fig. 3. The double linear chain aggregates of MoO_3 nanoparticles.

Taking into account the fact that linear and orthogonal superlattices were obtained only in the cases of MoO_3 powders which were treated by vibrationally excited molecules of hydrogen, we come to a conclusion that the quantum-chemical treatment of powders is the method of obtaining new inorganic self-assembling systems, i.e. the double linear chains and superlattices on their basis. In [9], the long single chains of Co particles have been observed. Their formation is bound with the ferromagnetic properties of Co. Our results show that the presence of ferromagnetic nanoparticles is not obligatory for the formation of the chain aggregates because MoO_3 is not a ferromagnetic material.

On the basis of the above results, one can conclude that the double linear chain aggregates and the linear as well as the orthogonal superlattices are the new self-assembling systems, with MoO_3 crystalline nanoparticles as building blocks. As known, the self-assembly involves, in particular, hydrogen-bonded building blocks. By taking into account that molecules of hydrogen have treated the MoO_3 nanoparticles, it is evident that the presence of the hydrogen surplus in the reactor's active zone leads to the formation of hydrogen-bonded MoO_3 nanoparticles, to what the electronegative oxygen in MoO_3 promotes. Just hydrogen-bonding leads to the formation of the chain aggregates of MoO_3 nanoparticles, which leads the polymerization process through hydrogen-bonding. It is necessary to note that in [10] the polymer-like behavior of the inorganic nanoparticles chain aggregates has been investigated.

The double linear chain aggregates result from the attraction between the two separate linear chains through hydrogen-bonding interaction. In this case the coupling also takes place between neighboring nanoparticles of the two chain aggregates.

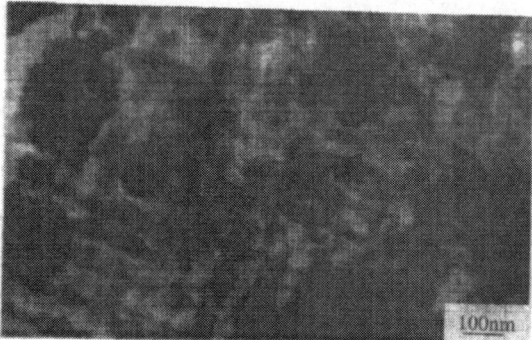

Fig. 4. The orthogonal superlattice of MoO_3 nanoparticles.

The self-assembling of the double linear chain aggregates is similar to the self-assembling of DNA, where the double helix results from the attraction between the two separate molecular strands through the hydrogen-bonding interaction. In our case the two separate chain aggregates interlace, as the detailed analysis of Fig.3 shows, and are closely adjoined to each other at full length with the formation of a straight line. Such interlacing of the two chains with effective averaged uniform cros-section leads to self-assembling along the straight line. The high stability of the double linear chain aggregates and superlattices formed is conditioned by the fact that self-assembling provides a «self-checking» process that ensures the high integrity in the present work self-assembling systems.

The self-assembling systems frequently demonstrate a superlattices structure [1]. In [11], the orthogonal superlattices with the parameter 8 nm of rod shaped iron oxyhydroxide particles embedded in the polysaccaride framework have been obtained. However, the existence of the superlattices on the basis of the double linear chains aggregates has been unknown before as well as themselves double chain aggregates of inorganic materials which, for the first time, have been observed in the present work.

CONCLUSION

Thus one can conclude that the MoO_3 nanoparticles treated by quantum-chemical technology through the molecules of hydrogen serve as the building blocks for spontaneous generation of the double linear chain aggregates and linear as well as orthogonal superlattices. The existence of a long double chain and the superlattices are manifestation of the self-assembling opportunities. One can suppose that the self-assembling of the double linear chains in inorganic materials, as in general in nature, is realized in the form of helix that serves to minimize the free energy of the systems.

ACKNOWLEDGMENT

The authors express their gratitude to Mr. Makaryan V.K., Mrs Grigoryan S.L., and Mrs. Hovhannisyan I.L. for help in preparation of this paper. This work was partially supported by the ISTC Grant A-139.

REFERENCES

1. W.M. Tolles, MRS Bulletin. **25**, 36 (2000).
2. J.-M. Lehn, Angew. Chem. Int. Ed. Engl. **29**, 1304 (1990).
3. G.M. Whitesides, J.P. Mathias and C.T. Seto, Science.**254**, 1312 (1991).
4. I.S. Lindsey, New J. Chem. **15**, 153 (1991).
5. R.T. Malkhasyan and G.H. Movsesyan, Review Scientific Instruments USSR. **4**, 127 (1991).
6. R.T. Malkhasyan and G.L. Movsesyan and V.K. Potapov, High Energy Chemistry. **26**, 3 (1992).
7. R.T. Malkhasyan in Metastable Phases and Microstructures, edited by R.Bormann, G.Mazzone, R.D.Shull, R.S.Averbakh and R.F.Ziolo (Mater. Res. Soc. Proc. **400**, Pittsburgh PA, 1995)pp. 77 – 82.
8. R.T. Malkhasyan, E. Agababyan and R.K. Karakhanyan. Chemical Physic. **30**, 57 (1996).
9. A.R. Tholen, in Nanophase Materials, edited by G.S. Haadjipanayis and R.W. Siegel(Kluwer Academic Publishers, Dordrecht/ Boston/ London, 1994), p.57.
10. S.K. Friedlander, Journal of Nanoparticles Research. **1**, 9 (1999).
11. T.G.St. Pierre, P. Sipos, P.Chan, W.Chua-Anusorn, K.R. Bauchspiess and J. Webb, in Nanophase Materials, edited by G.S. Haadjipanayis and R.W. Siegel (Kluwer Academic Publishers, Dordrecht/ Boston/ London, 1994), p.49.

Mat. Res. Soc. Symp. Proc. Vol. 707 © 2002 Materials Research Society   AA6.2/M5.2

Electrochemical self-assembled of copper/copper oxide nanowires

S. Kenane and L. Piraux
Unité de Physico-Chimie et de Physique des Matériaux, Place Croix du Sud 1, B-1348 Louvain-la-Neuve, Belgium

ABSTRACT

Arrays of Cu/Cu$_2$O nanowires are electrodeposited at room temperature from alkaline solutions of copper lactate in the nanopores of track-etched polymer membranes. Using the appropriate solution, the electrode potential spontaneously oscillates during the application of a constant cathodic current. Both the period of the oscillations and the composition of the nanowires can be controlled by varying the applied current density. A nanocomposite of copper and cuprous oxide is deposited at an applied current over which oscillations occur. In contrast, pure Cu or Cu$_2$O nanowires are obtained at deposition current out of the range of oscillation.

INTRODUCTION

The fabrication of magnetic nanowires and multilayers has attracted considerable interest motivated by applications in the area of ultra-high density magnetic recording and by promising giant magnetoresistance and high-frequency properties [1-2]. Also, ultrathin superconducting wires exhibiting large critical field and anomalous resistive behavior have been investigated recently [3]. In addition, the template method was also employed for the synthesis of electrically conducting polymer tubules with conductivity of one order of magnitude larger than in the bulk form [4].

On the other hand, electrodeposited oxide/metal layered structures would be interesting because tunneling barrier effects could be studied. It was recently demonstrated that layered structures of copper/cuprous oxide can be prepared by electrodeposition [5-6]. The fabrication of multilayers by electrodeposition is usually made by pulsing the applied potential or current during the growth. However, it was shown that, using appropriate solution and pH, the electrode potential may spontaneously oscillate when a fixed current is applied. Evidence for layering was reported with individual thickness layers in the nanometer range [6-8]. The phase composition and layer thicknesses was found to be very sensitive to the applied current deposition and pH of the solution.

In the present study, we report on the fabrication and properties of arrays of self-assembled Copper/Cuprous oxide nanowires made by electrodeposition in nanoporous track-etched polycarbonate membranes. These membranes are produced at the labscale and exhibit improved properties in terms of pore shape, pore size, porosity control and parallel arrangement of the pores [9]. The same fascinating phenomena already reported for films was also observed in the nanowire system, i.e., the electrode potential spontaneously oscillates during the application of a constant cathodic current. Our results provide new insight in the understanding of the oscillating phenomena. Electrical transport measurements were also recently reported in these Cu/Cu$_2$O nanowires.[10]

EXPERIMENTAL DETAILS

Nanoporous membranes (22 microns thick) were used in this study with pore diameter and pore density respectively of 200 nm and 5×10^8 for membrane M_0, 230 nm and 5×10^8 cm^{-2} for membrane M_1 and 550 nm, 9×10^7 cm^{-2} for membrane M_2. The electrolytic bath consisted of a mixture of copper sulfate, lactic acid and sodium hydroxide. To prepare the deposition solution, 126 g of copper(II) sulfate pentahydrate was added to 150 ml of 88% lactic acid under stirring. The solution was first completed with 450 ml of NaOH (5 M) in 50-100 ml increments. The pH of the solution was finally fixed to 9.2 by adding small quantities of sodium hydroxide (5 M). It is noted that the Cu content in the solution was significantly larger than in previous works [5, 6]. Prior to electrodeposition, a gold layer was evaporated on one side of the membrane; this electrode serves as cathode during the electroplating process. The counter electrode was a Ni wire and the reference electrode was a saturated Ag/AgCl electrode. Electrodeposition was carried out at room temperature without stirring. The deposition potential and current were controlled by an EG&G model 263 potentiostat/galvanostat.

Direct observation of the nanowires was made possible after the dissolution of the membrane in dichloromethane and using high resolution scanning electron microscopy (SEM). Figure 1 shows a SEM image of nanowires electrodeposited at a constant current using membrane M $_0$ (200 nm). The length of the nanowires is in the range of 10-20 μm. The diameter of the nanowires produced by this technique depends on the pore diameter of the membranes. We have obtained nanowires with diameter in the range of 200-600 nm. The structural composition of the electrodeposited nanowires was also studied using X-ray diffraction (XRD) with CuK $_\alpha$ radiation.

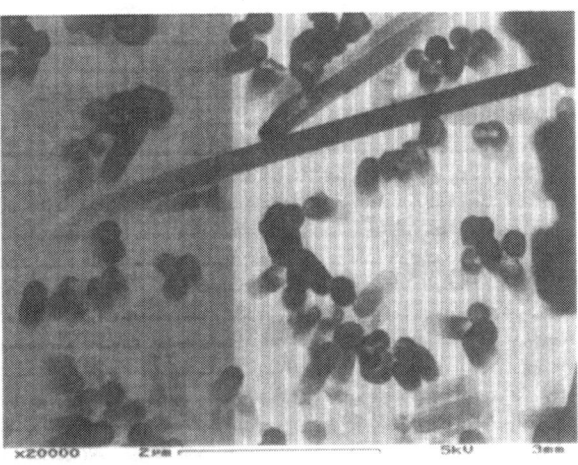

Figure 1. SEM-FEG image of electrodeposited Cu/Cu $_2$O nanowires (M $_0$, ϕ = 200 nm) grown at a constant current of 0.2 mA.

RESULTS AND DISCUSSION

To illustrate the electrochemical properties of samples, two different electroplating deposition procedures were followed. The former consists of making a sweep of current and recording the electrode potential. Increasing the deposition current in nanopores membrane M_0 with a scan rate of 2 μA/s leads to the different features shown in Figure 2 :

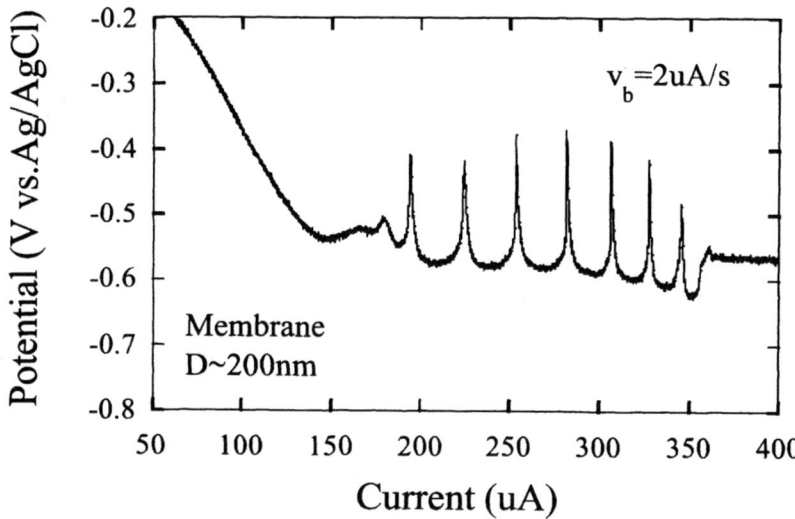

Figure 2. Scan of the applied current showing potential oscillations in some range of current for nanoporous membranes (M_0, $\phi = 200$ nm).

first, a decrease in potential occurs at small currents; then, for intermediate currents, the electrode potential spontaneously oscillates; finally, the potential reaches an almost constant value at large current. Measurements made on several samples have shown that, for a given diameter, the oscillating range can be shifted by ± 10%. For all membranes, the potential oscillations are less pronounced at small current than at larger current. We also found that the peak-to-peak modulation of potential has a maximum value around 180 mV, independently of the sample.

In a second set of experiments, the time evolution of the deposition potential has been recorded for selected values of the applied current. Figure 3 shows the evolution of electrode potential during the application of a constant current of 0.2 mA using membranes M_0. The observed oscillations may persist during several hours. In each experiment, the potential modulations are periodic and the average amplitude is around 170 mV.

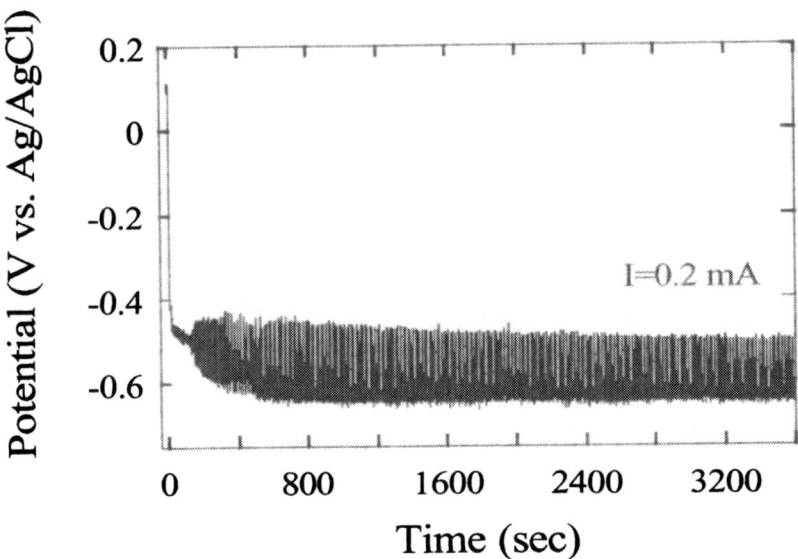

Figure 3. Potential oscillations versus time for applied current of 0.2 mA membrane (M_0, $\phi = 200$ nm).

In their study, Switzer et al [5, 6] have observed that the amplitude and period of the potential oscillations depend on several parameters such as the pH, temperature and applied current density. Several substrates such as gold, silver, and platinum films were used. Oscillation phenomena were reported for a wide range of deposition current density and for a pH range of 8.5-10 [5]. Two different mechanisms have been proposed [5] in order to account for the oscillating phenomena. First, the formation and breakdown of rectifying Cu_2O/solution interface has been invoked as a probable source of potential oscillations [11]. The second mechanism is based on possible oscillations of the surface pH during deposition. Indeed, at low pH, if the complexation of copper (II) by the lactate ion is ignored, Cu is expected to form according to the following reduction process :

$$Cu^{2+} + 2e^- \rightarrow Cu \tag{1}$$

In contrast, at higher pH, the formation of Cu_2O would be favored :

$$Cu^{2+} + 2e^- + 2OH^- \rightarrow Cu_2O + H_2O \tag{2}$$

The reduction of Cu^{2+} on Cu_2O according to Eq 2 would induce the depletion of OH in the diffusion layer, causing the local pH to decrease. The decrease in pH would then favor the formation of copper metal according to Eq 1. However, during the deposition of Cu, the depletion of OH⁻ is progressively eliminated thus increasing the pH. This process may lead to a local oscillating pH at the cathode. Assuming that the reduction mechanism of Cu^{2+} obeys Eqs 1 and 2, copper would be made during the more negative swings of the potential, and cuprous oxide would be made during the sharper positive swings in the potential (see

Figure 3). However, the local pH mechanism was criticized by the authors themself, in part because the oscillation period was found to be independent of the value of the applied current density. This observation seems to contradict the local pH mechanism as the rate of depletion of OH$^-$ at the electrode surface, and thus the oscillation period should depend significantly on the applied current density.

In contrast, we found that the period of oscillation depends on the applied current density (Figure 4). From the experiments performed using membrane M_1, we observed that the period is about twice at low current density ($j \sim 3$ mA/cm^2) what it is at high current density ($j > 4$mA/cm^2). The same feature was observed for membrane M_2. Figure 4 shows the evolution of the oscillation period with current density. It appears that the period is almost constant at high current density and increases rapidly as j is reduced. This is in contrast with the results obtained on planar substrates where the oscillation period was almost independent of the applied current density. This observation gives direct support to the local pH mechanism as the rate of reduction of H$^+$ increases with the applied current so that the oscillation period is expected to be a decreasing function of the applied current density. In contrast, at large current, mass transport of H$^+$ saturates, giving rise to a constant period.

Finally, in comparison with previous works on planar substrates [5, 6], larger copper concentration was required in order to observe the oscillating phenomena in the nanopores. This can be attributed to the fact that the depletion of Cu^{2+} at the interface is likely enhanced in confined media compared to planar surfaces. The range of pH where the oscillations occur is also quite narrow, around 9-9.2.

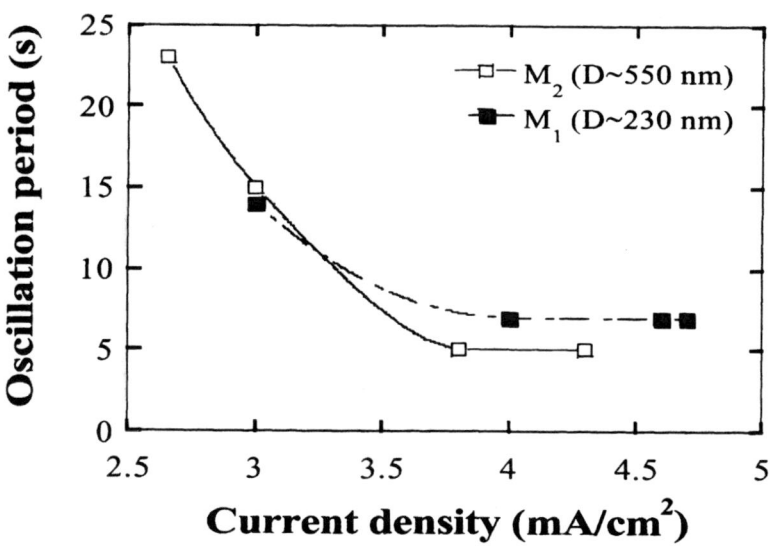

Figure 4. Variation of oscillation period as a function of current density using membranes M_1 and M_2.

CONCLUSION

In conclusion, we have grown arrays of Cu/Cu_2O nanowires using nanoporous membranes and the technique of electrodeposition. Using appropriate solution, the electrode potential spontaneously oscillates during the application of a constant cathodic current. The results of structural composition studies reveal the formation of a nanocomposite of copper and cuprous oxide. The period of these oscillations depends on the applied current density suggesting that a local pH mechanism is responsible for the observed oscillating phenomena. Work is now in progress to investigate in more detail the composition of the nanowires and determine whether the potential oscillations leads to the formation of Cu/Cu_2O multilayered nanowires. Much effort will also be devoted to investigate the transport properties of these nanowires in relation with the structural composition and modulation period.

REFERENCES

1. T. M. Whitney, J. S. Jiang, P. Searson and C. Chien, Science **261**, 1316 (1993).

2. A. Fert and L. Piraux, J. Mag. Mag. Mater. **200**, 338 (1999).

3. S. Dubois, A. Michel, J. Eymery, J.L. Duvail and L. Piraux, J. Mater. Res. **14**, 665 (1999).

4. C.R. Martin, Science **266**, 1961 (1994).

5. Ling-Yuang Huang, Eric W. Bohannan, Chen-Jen Hung, and Jay A. Switzer, Israel Journal of Chemistry **37**, 297-301 (1997).

6. Jay A. Switzer, Chen-Jen Hung, Ling-Yuang Huang, F. Scott Miller, Yanchun Zhou, Eric R. Raub, Mark G. Shumsky, and Eric W. Bohannan, J. Mater. Res. **13** (4), 909-916 (1998).

7. Jay. A. Switzer, Chen-Jen Hung, Ling-Yuang Huang, Eric R. Switzer, Daniel R. Kammler, Teresa D. Golden, and Eric W. Bohannan , J. Am. Chem. Soc. **120**, 3530-3531 (1998).

8. Eric. W. Bohannan, Ling-Yuang Huang, F. Scott Miller, Mark G. Shumsky, and Jay. A. Switzer, Langmuir **15**, 813-818 (1999).

9. E. Ferain and R. Legras, Nucl. Instrum. Methods Phys. Res. **B174**, 116 (2001).

10. S. Kenane and L. Piraux, Journal of materials research (in press).

11. H. D. Dewald, P. Parmananda and R. W. Rollins, J. Electrochem. Soc. **140**, 1969 (1993).

Mat. Res. Soc. Symp. Proc. Vol. 703 © 2002 Materials Research Society   AA8.9/V10.9

Studies of Intersubband Transitions in Arrays of Bi Nanowire Samples Using Optical Transmission

M. R. Black[a)], K. R. Maskaly[b)], O. Rabin[c)], Y. M. Lin[a)], S. B. Cronin[d)], M. Padi[d)], Y. Fink[b)], M. S. Dresselhaus[a,d)]
[a]Department of EECS, Massachusetts Institute of Technology, Cambridge, MA
[b]Department of Material Science and Engineering, Massachusetts Institute of Technology, Cambridge, MA
[c]Department of Chemistry, Massachusetts Institute of Technology, Cambridge, MA
[d]Department of Physics, Massachusetts Institute of Technology, Cambridge, MA

Abstract

This paper reports the fabrication of large diameter pores (> 150 nm) in anodic alumina that can be used to create wire arrays with significant surface effects, but without significant quantum confinement. These wires, therefore, allow us to distinguish between optical absorption spectra features originating from quantum effects and those from surface effects. The paper presents techniques towards fabricating these bismuth wire arrays, and presents optical absorption data from two bismuth nanowire arrays in the semimetal-semiconductor transition diameter regime. The results from previous publications are summarized and future directions are outlined.

Introduction

We are no longer completely limited to the constraints on the properties of 3D bulk materials. By utilizing quantum confinement, we have learned how to engineer the band gap of a material. As a result we can tailor the properties of a material to conform to a desired application. Once the effects of quantum confinement are fully understood, the size of the quantum confined dimension(s) can be selected to achieve the degree of quantum confinement and surface effects desired.

Bismuth nanowires exhibit a transition from a semimetal with a small band overlap ($38\,\mathrm{meV}$ at $0\,\mathrm{K}$) to a semiconductor, as the wire diameter becomes small enough to support significant quantum confinement effects, as shown in Fig. 1. This transition occurs in Bi nanowires at relatively large wire diameters because of its small effective masses and small band overlap. For example, this semimetal-semiconductor transition is predicted to occur at a wire diameter of $16\,\mathrm{nm}$ in the (012) direction (the growth direction of our nanowires) at room temperature, and at $47\,\mathrm{nm}$ at 77K, Fig. 1. The change from a semimetal to a semiconductor has significant effects on the electronic and optical properties of bismuth, which may be desirable for some applications, such as thermoelectricity.[1, 2]

We are working towards measuring the optical properties of bismuth wires in three different size regimes: large wire diameters (~ 200 nm) which are semimetals, small wire diameters (~ 15 nm) which are semiconductors, and an intermediate wire diameter size (~ 40 nm) where the wires are a semimetal at room temperature and a semiconductor at low temperatures. Well ordered arrays of 45 nm pores in alumina have been fabricated and filled with bismuth to create wires in the intermediate diameter regime. These wires are near the semimetal-semiconductor transition at 77K. The 45 nm wires absorb in the far infrared.

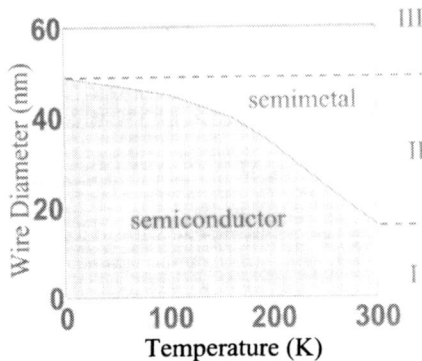

Figure 1: The phase diagram of the semimetal to semiconductor transition in bismuth nanowires is shown as a function of temperature and wire diameter. Depending on the wire diameter, a nanowire is either a semimetal (III), a semiconductor(I), or switches from a semimetal to a semiconductor as the temperature is decreased(II).

This optical absorption is attributed to intersubband transitions.[3] Well ordered arrays of small pore alumina templates (15 nm) have also been fabricated successfully. However, the liquid phase injection technique used to fill the alumina with bismuth is ineffective in filling these small diameter pores. We are therefore developing an electro-chemical method for filling small pores in anodic alumina with bismuth.[4] The process to fabricate the large pore alumina templates (> 55 nm) proved to be the most difficult to develop due to the fact that the process requires a higher voltage and many highly sensitive process parameters needed to be optimized. Optical absorption measurements, like those already reported for 45 nm wires, for both the semiconducting and semimetallic wires will help us better understand the observed absorption spectra. We therefore seek to perform measurements over the whole diameter range to gain a better understanding of the quantum confinement characteristics of the Bi nanowires.

Fabrication and Experimental Details

Since aluminum can be anodized to form an alumina layer with an ordered array of pores[5], all the bismuth nanowires in this work were fabricated by filling porous anodic alumina. Several theories have been presented to explain the formation of the porous alumina, such as Refs. [5] and [6]. In addition, several theories have been recently proposed to explain the ordering of the pores into a hexagonal pattern when very specific conditions for the anodization are used.[7, 8] Many papers present excellent experimental results documenting conditions at which the anodization forms ordered pores and those at which the pores are irregular and not ordered.[9, 10, 11] However, to our knowledge, no complete theory exists to predict the optimal acid type, concentration, and temperature that should be used to obtain well-ordered arrays with a uniform pore diameter for a given anodization voltage and corresponding pore diameter. Therefore, whenever the desired pore size has not yet been well studied, many attempts of trial and error are required to establish the recipe for well-ordered

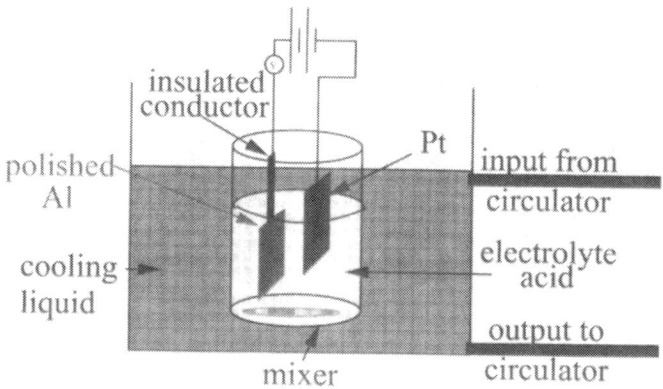

Figure 2: The anodization setup for high voltage anodization. The automatic cooling system, the high cooling liquid level, and the insulated conductor, all prevent breakdown during the anodization process.

pores. To minimize the trial and error required to find the ideal anodization conditions for large diameter pores (> 80 nm), we sought to reproduce the well-ordered arrays with a pore size of 200 nm diameter reported in Ref. [7]. However, when repeating their conditions of 160 V anodization voltage in 10 % phosphoric acid at 3°C, the anodization current became unstable. The current increased during the anodization, increasing the temperature of the acid. The rise in temperature in turn further increased the current, until the reaction became uncontrollable. This rapid rise in acid temperature and anodization current is called "breakdown" in the literature. [8]

In our samples, in addition to the increase in current at breakdown, sizzling occurred at the air/acid interface where the aluminum was exposed to air. Furthermore, the onset of this phenomenon was correlated with the ambient air temperature. This suggests that the sizzling is caused by the higher temperature of the sample at the surface (surface heating). We therefore implemented three techniques to reduce surface heating and the likelihood of breakdown, (see Fig. 2). When all three techniques are used, sizzling under our anodization conditions is eliminated. The first improvement was an automatic cooling system. A refrigerated recirculator is used to keep the cooling liquid at -1.5 ± 0.05 °C. In addition, the acid level is kept below that of the cooling liquid. This helps cool the air around the acid-air interface. Thirdly, the aluminum electrode is completely submerged in the chilled acid. In order to prevent shorting of the contact to the acid, an insulated conductor is used to make contact with the Al sheet inside the acid. The aluminum is anodized using this method for about 3 hours. The alumina is then etched off and the remaining aluminum then undergoes a second anodization for about 4 hours. The pores from a 40 V and a 160 V anodization are shown in Fig. 3 A and B, respectively.

The pores in the alumina are filled with Bi using a pressure injection technique.[12] The alumina template is then etched off using a selective etch, leaving only bismuth wires

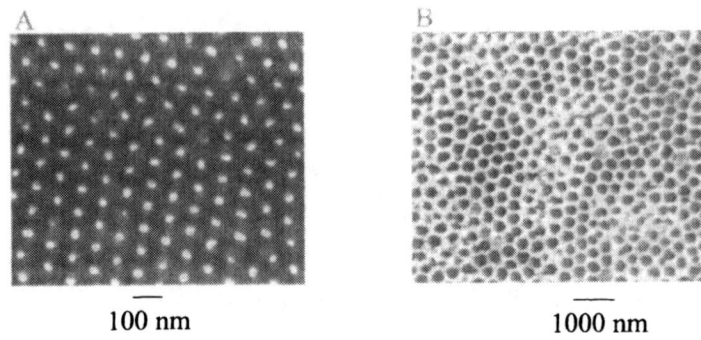

100 nm   1000 nm

Figure 3: Scanning electron micrographs of anodic alumina anodized at (A) 40 V and (B) 160 V with pore diameters around 40 and 200 nm, respectively.

behind. Using the micro-FTIR (Fourier Transform Infrared Technique), the reflection and transmission were measured from free-standing Bi nanowires with a bismuth oxide coating around the nanowires and a thin Bi film holding them together. A schematic diagram of the sample is shown in Fig. 4(C). Bismuth nanowires protrude out of the bismuth film, which is balanced on the edge of a glass slide. Light is transmitted through the sample in the direction of the wires so that the electric field of the incident light is always perpendicular to the wire axis. Since the transmission is proportional to e^{-Kx}, where K is the infrared absorption coefficient and x is the sample thickness, the negative log of the transmission is proportional to the absorption coefficient. This is used to find the absorption spectra of the wires.

Results and Discussion

The absorption coefficients (times the sample thickness) as a function of wavenumber, obtained by taking the negative log of the transmission intensity of $\sim 45\,nm$ and $\sim 30\,nm$ diameter free-standing bismuth nanowires are shown in Fig. 4(A). For comparison, Fig. 4(A) also shows the absorption coefficient of a film of bismuth. The two arrays of wires used for the free-standing nanowire measurements had diameters of 60 nm and 45 nm before the alumina template surrounding the wires was etched away. Since a $\sim 7\,nm$ oxide grows on the free-standing wires after the alumina is selectively etched away [13], the inner bismuth portion of the free-standing wires is expected to have a diameter of around 45 nm and 30 nm. Since the thickness of the thin bismuth film holding the bismuth nanowires and the thickness of the nanowire array are not known, arbitrary units of absorption are used.

The absorption spectra of the nanowire arrays show many more features than that of the bismuth film. The absorption features are predominant for wavenumbers less than $1300\,cm^{-1}$. Simulations show that intersubband absorption tails off for $\omega > 1300\,cm^{-1}$ in bismuth nanowires, because of a fall off in the coupling between the initial and the final states.[14] Figure 4(A and B) also shows that the absorption spectra between the two wire arrays differ significantly. The absorption of each sample was taken several times and found

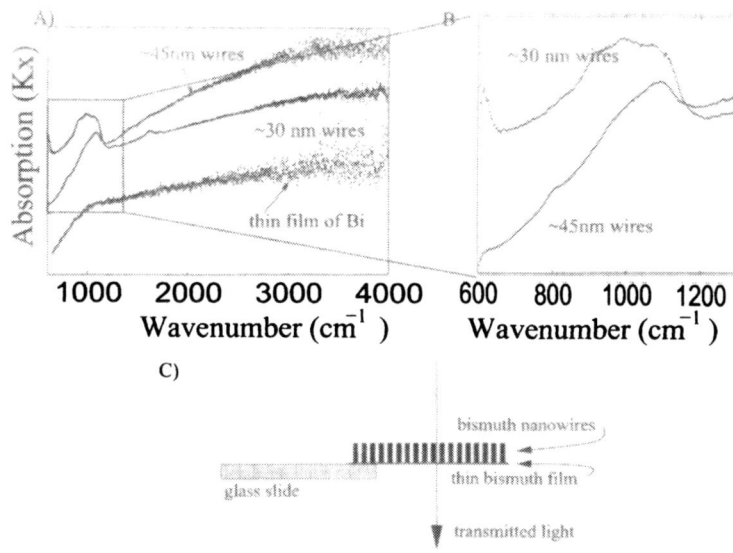

Figure 4: The dependence of the absorption coefficient on wavenumber, found by taking the negative log of the transmission of free-standing bismuth nanowires with diameters around 45 and 30 nm as well as the absorption on a thin piece of bismuth is shown in A) and B). C) shows a schematic of the experimental setup to measure absorption.

to be reproducible. The differences in the absorption spectra for wavenumbers less than $1300\,cm^{-1}$ are attributed to differences in the subband energies involved in intersubband absorption.[14]

Previous work on the optical properties of bismuth nanowire arrays inside the alumina template report a sharp absorption feature at about $1000\,cm^{-1}$.[3] The shape of the absorption peak, the frequency of observed absorption, the qualitative dependence on wire diameter, and the polarization of this absorption all indicate that intersubband transitions are the likely cause of the observed absorption.[3] In addition, a model for intersubband transitions in bismuth nanowires predicts peaks in the absorption spectra at energies that are in agreement with those observed in optical measurements of free standing wire arrays.[14] However, some aspects of the absorption curve remain unexplained. Firstly, although the energy of the absorption peak at around $1000\,cm^{-1}$ increases with decreasing wire diameter, as expected for intersubband transitions, it does not increase as rapidly as expected. Secondly, the relative intensities of the absorption peaks in the free standing wires are different from those predicted by theory.[14] In particular, the absorption peak at around $1000\,cm^{-1}$ is much more intense in the experimentally measured absorption than in the simulated intersubband absorption.[14] In this paper we report optical absorption measurements on ~ 30 nm wires as well as the fabrication of ordered 200 nm pores in alumina. With the newly developed ability to fabricate well ordered 200 nm wire arrays and with the currently developing ability to electro-chemically fill the 20 nm arrays, we hope to develop a better understanding of the observed absorption mechanisms in bismuth nanowire arrays.

Acknowledgements

The authors gratefully acknowledge the valuable discussions with of Prof. Jackie Ying and Dr. Gene Dresselhaus. The authors also gratefully acknowledge MURI subcontract 0205-G-BB953, NSF grant DMR-0116042, and US Navy contract N00167-98-0024 for support. This work made use of MRSEC Shared Facilities supported by the National Science Foundation contract DMR-9400334.

References

[1] Z. Zhang, X. Sun, M. S. Dresselhaus, J. Y. Ying, and J. Heremans, Phys. Rev. B **61**, 4850–4861 (2000).

[2] X. Sun, Z. Zhang, and M. S. Dresselhaus, Appl. Phys. Lett. **74**, 4005–4007 (1999).

[3] M. R. Black, M. Padi, S. B. Cronin, Y.-M. Lin, O. Rabin, T. McClure, G. Dresselhaus, P. L. Hagelstein, and M. S. Dresselhaus, Appl. Phys. Lett. **77**, 4142–4144 (2000).

[4] O. Rabin, Y. Lin, S. Cronin, and M. S. Dresselhaus. to be published.

[5] J. P. Sullivan and G. C. Wood, Proc. Roy Soc. Lond. A. **317**, 511–543 (1970).

[6] Feiyue Li, Lan Zhang, and Robert M. Metzger, Chem. Mater. **10**, 2470–2480 (1998).

[7] A. P. Li, F. Muller, A. Birner, K. Nielsch, and U. Gosele, Journal of Applied Physics **84**, 6023–6026 (1998).

[8] A. Jagminas, D. Bigelience, I. Mikulskas, and R. Tomasiunas, Journal of Crystal Growth **223**, 591–598 (2001).

[9] Hideki Masuda and Kenji Fukuda, Science **268**, 1466–1468 (1995).

[10] Y. Li, E. R. Holland, and P. R. Wilshaw, J. Vac. Sci. Technol. B. **18**, 994–996 (2000).

[11] T. E. Huber, M. J. Graf, C. A. Foss Jr., and P. Constant, J. Mater Res. **15**, 1816–1821 (2000).

[12] Z. Zhang, J. Ying, and M. Dressehaus, J. Mater. Res. **13**, 1745–1748 (1998).

[13] S. B. Cronin, Y.-M. Lin, P. L. Gai, O. Rabin, M. R. Black, G. Dresselhaus, and M. S. Dresselhaus. In *Anisotropic Nanoparticles: Synthesis, Characterization and Applications: MRS Symposium Proceedings, Boston, December 2000*, edited by S. Stranick, P. C. Searson, L. A. Lyon, and C. Keating, pages C571–C576, Materials Research Society Press, Pittsburgh, PA, 2001. pdf.

[14] M. R. Black, Y. M. Lin, S. B. Cronin, O. Rabin, and M. S. Dresselhaus, unpublished.

Optical Materials

Mat. Res. Soc. Symp. Proc. Vol. 707 © 2002 Materials Research Society

Silver-Polyimide Nanocomposite Films Yielding Highly Reflective Surfaces

Robin E. Southward,[1] C. J. Dean,[2] J. L. Scott,[2] S. T. Broadwater,[2] and D. W. Thompson[2]
[1]Structure and Materials Competency, NASA, Langley Research Center, Hampton, VA 23681
[2]Department of Chemistry, College of William and Mary, Williamsburg, VA 23187

ABSTRACT

Highly reflective surface-metallized flexible polyimide films have been prepared by the incorporation of the soluble silver ion complex (1,1,1-trifluoroacetylacetonato)silver(I) into dimethylacetamide solutions of the poly(amic acid) prepared from 2,2-bis(3,4-dicarboxyphenyl)-hexafluoropropane dianhydride (6FDA) and 2,2-bis[4-(4-aminophenoxy)phenyl]hexafluoro-propane (4-BDAF). Thermal curing of solution cast silver(I)-poly(amic acid) films leads to cycloimidization of the amic acid with concommitant silver(I) reduction and formation of a reflective surface-silvered film at 8 and 13 weight percent silver. The metallized films are thermally stable and flexible with mechanical properties similar to those of the parent polyimide. TEM reveals that the bulk (interior) of the polyimide composite films have 5-20 nanometer-sized silver particles with a surface layer of silver metal ca. 80 nm thick. Neither the bulk nor the surface of the films is electrically conductive. Adhesion of the surface metal to polyimide is excellent.

INTRODUCTION

The fabrication of specularly reflective and electrically conductive surface metallized polyimide films is of enormous interest as reviewed by Matienzo and Unertl [1]. Applications are numerous including: anti-infective coatings, contacts and circuit lines in microelectronics, enhancement of thermal conductivity, flexible surface conductive tapes, patternable conductive surfaces on dielectric bases, the terrestrial concentration of solar radiation for power generation, and gas permeability barriers. Of particular interest to us are space applications of metallized polyimides. These include highly reflective thin film reflectors and concentrators in space environments for solar thermal propulsion [2] and solar dynamic power generation [3,4], reflectors for flat panel solar power arrays for satellites [5], large scale radiofrequency antennas for the management of EM signals [6], solar sails [7,8], and sunshields to control device temperatures for projects such as the Next Generation Space Telescope. Polymeric supports offer advantages in weight, flexibility, elasticity, fragility, and deployability relative to inorganic supports such as glass and ceramics.

Southward et al. [9] and Taylor et al. [10] have have been successful in preparing surface-metallized, in particular silver-metallized, polyimide films by a novel single-stage, internal metallization technique which leads to flexible films with excellent specular reflectivity and/or electrical conductivity. "Single-stage" denotes the fabrication in one step of metallized films from a homogeneous solution of a positive valent metal precursor and a poly(amic acid). In contrast to traditional metallized film preparation protocols, in the present work the polyimide film is not prepared in a first stage and subsequently coated with metal (vapor deposition, sputtering, etc.) in a discrete second stage. "Internal metallization" refers to a film that is cast as a homogeneous silver ion-doped poly(amic acid) solution and then thermally treated to induce metal ion reduction to give the metallized surface with concomitant cycloimidization of the amic acid to the final polyimide. During the thermal cycle a portion of metal atoms and small clusters formed in the polyimide film aggregate at the surface to give a 50-200 nm metallic layer. Silver is the metal of interest since it has exceptional reflectivity and conductivity [11]. Polyimides were chosen as substrates owing to their excellent thermal-oxidative stability and film-forming properties [12,13].

The synthetic protocol with respect to the system reported herein is illustrated in Figure 1. Silver(I) acetate and trifluoroacetylacetone (TFAH) are allowed to react in the solvent dimethylacetamide (DMAc) to give an *in situ* solution of the (trifluoroacetylacetonato)silver(I) complex, AgTFA. A DMAc solution of the poly(amic acid) form of 6FDA/4-BDAF is added to the DMAc solution of AgTFA. A colorless homogeneous solution results. A film is then cast. Thermal curing of the Ag(I)-poly(amic acid) film effects reduction of Ag(I) to native metal and ring closure to the imide. During the cure silver atoms/clusters aggregate in part at the surface to give reflective films, usually with excellent adhesion of metal to polyimide.

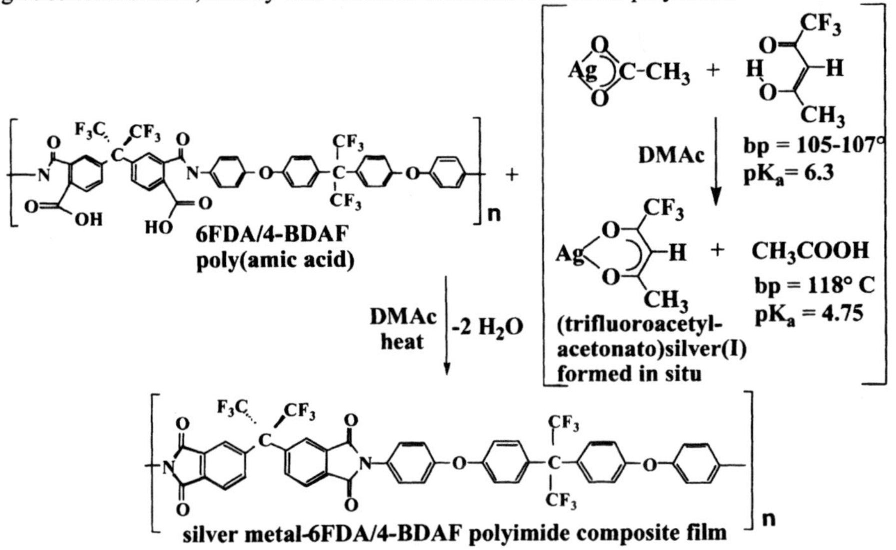

Figure 1. Synthetic route to a polyimide metallized film.

For present study described herein we chose the 6FDA/4-BDAF polymer because it has been reported to have high thermal stability, radiation resistance, and low absorption in the visible [14]. The low color of this polyimide is of importance since in previous studies of metallized polyimides exhibit a thin polyimide overlayer or significant polyimide at the surface. Since tradition polyimides such as PMDA/ODA, BTDA/ODA, BPDA/ODA, etc. absorb strongly in the visible, some of the loss of reflectivity has been due to surface polymer absorption.

EXPERIMENTAL DETAILS

Materials. All chemicals were obtained from commercial sources. 6FDA/4-BDAF poly(amic acid) solution was prepared with a 0.5% offset of dianhydride at 15% solids (w/w) in

DMAc. The resin was stirred for 5 h. The inherent viscosity was 1.2 dL/g at 35° C.

Preparation of BTDA/4,4′-ODA metallized films. Ag(I)-containing solutions were prepared by first dissolving silver(I) acetate in DMAc containing trifluoroacetylacetone. The 15% poly(amic acid) solution was then added to give the desired Ag to polymer ratio. Doped poly(amic acid) solutions were cast as films onto glass plates using a doctor blade set at 500-650 μm to obtain cured films 20-25 μm thick. After remaining in an atmosphere of slowly flowing dry air for 18 h, the films were cured in a forced air oven. The cure cycle involved heating over 20 min to 135 °C and holding for 1 h, heating to 300 °C over 4 h, and holding at 300 °C varying times.

RESULTS AND DISCUSSION

As seen in Figure 2, poly(amic acid)-DMAc or diglyme (2-methoxyethyl ether) films retain a substantial portion of solvent which cannot is not lost by evaporation at 25 °C. Thus, thermal cure of Ag ion-doped films occurs in a solvent rich state; the solvent then may play a role in metal ion reduction and also serve as a plasticizing agent. Figure 3 shows the development of reflectivity as a function of time/temperature for 8 and 13 wt% silver-6FDA/4-BDAF polyimide films. For the 8% film maximum reflectivity is observed after 2 h at 300 °C. The 13% film achieves maximum reflectivity after only 1 h at 300 after which the reflectivity diminishes dramatically due to metal promoted oxidative degradation of the polyimide. The observed reflectivities (relative to an optical Al mirror) are high and do not show a strong concentration dependence (8 versus 13%). The metal is firmly adhered to the polyimide and cannot be removed by adhesive tape test protocols.

Figure 4 displays TEM data for the 13% film. The surface silver is ca. 80 nm thick and composed of particles of globular shape. The bulk of the film contains silver particle with sizes in the 5-20 nm range. Thus, only a limited quantity of the original silver appears at the surface. While the 8 and 13% films have excellent reflectivity, neither is electrically conductive. The SEM for the 13% film (Figure 5) shows the globular form of the silver particles and reveals that the particles do not form a continuous network. Intervening polyimide keeps the nanoparticles effectively isolated.

While significant reflectivity for the two films is not observed until 300 °C, X-ray data shown in Figure 6 for the 13% films make clear that reduction to silver metal is occurring at temperatures as low as 175 °C. Thus, sufficient silver aggregation at the surface to form a mirror is much slower than silver(I) reduction. Interestingly, the early low temperature silver reflections are significantly broadened which is consistent with Scherrer broadening due to very small nanometer-sized particles. As curing temperature and time increase the reflections narrow with the larger particle sizes seen in the TEM and SEM at 300 °C for 1 h.

The metallized exhibited tensile moduli and strength which are not significantly different from those of the parent polyimide. This is consistent with the relatively low concentration of silver metal and with the fact that silver is a passive metal which does not interact strongly with polyimide functional groups. Thus, the metallic silver particles appear to behave as a inert nano-filler in the polymer. The glass transition temperature in the metallized films is unchanged from that of the parent. However, while thermal stability remains high, the temperature at which there is 10% weight loss in air is *ca.* 125 °C lower than for the parent polyimide. In nitrogen the thermal stability is the same as the parent indicating that silver metal, as expected, catalyzes polyimide degradation at higher temperatures.

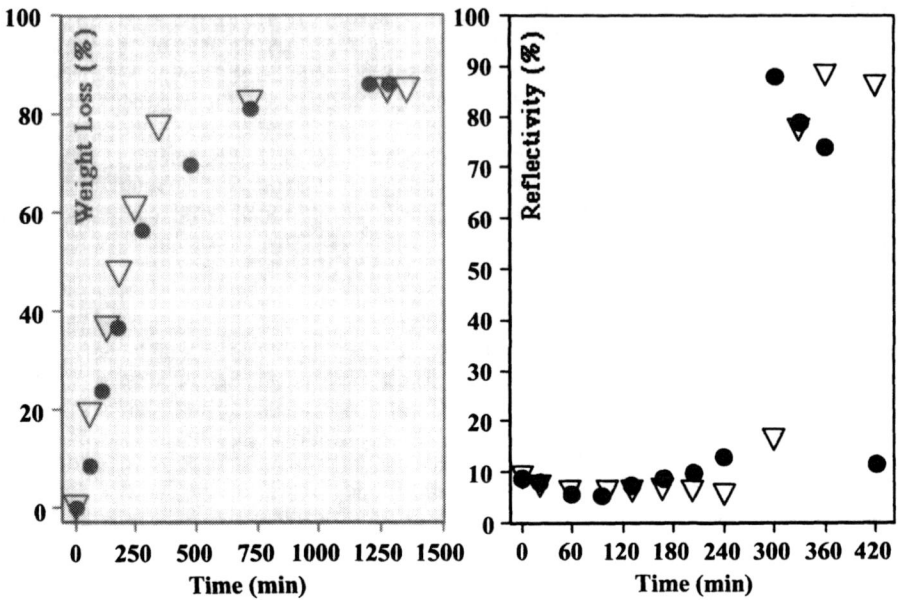

Figure 2. (Left) Evaporative solvent loss (DMAc-circle and diglyme-triangle) as a function of time under slowly flowing dry air.

Figure 3. (Right) Specular reflectivity as a function of time and temperature for 8% (triangle) and 13% (circle) silver-6FFDA/4-BDAF films

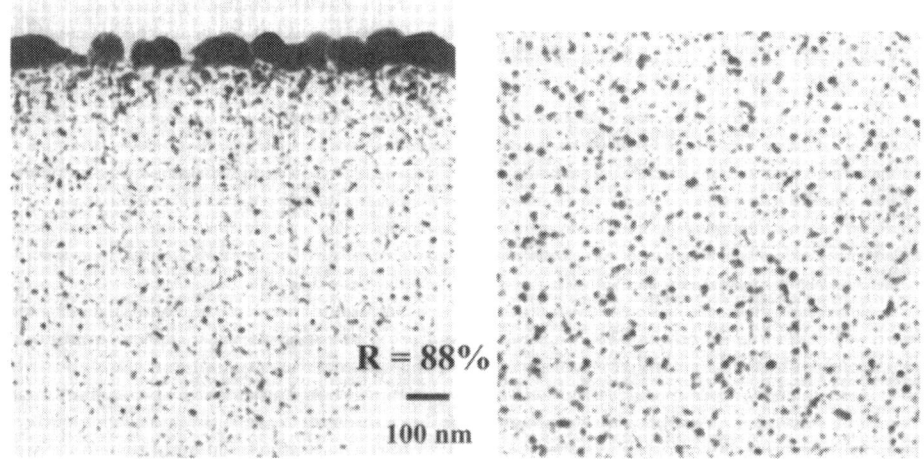

R = 88%

100 nm

Figure 4. TEM micrographs for the 13% silver-6FDA/4-BDAF film cured to 300 °C for 1 h. Left - surface view; Right - bulk view.

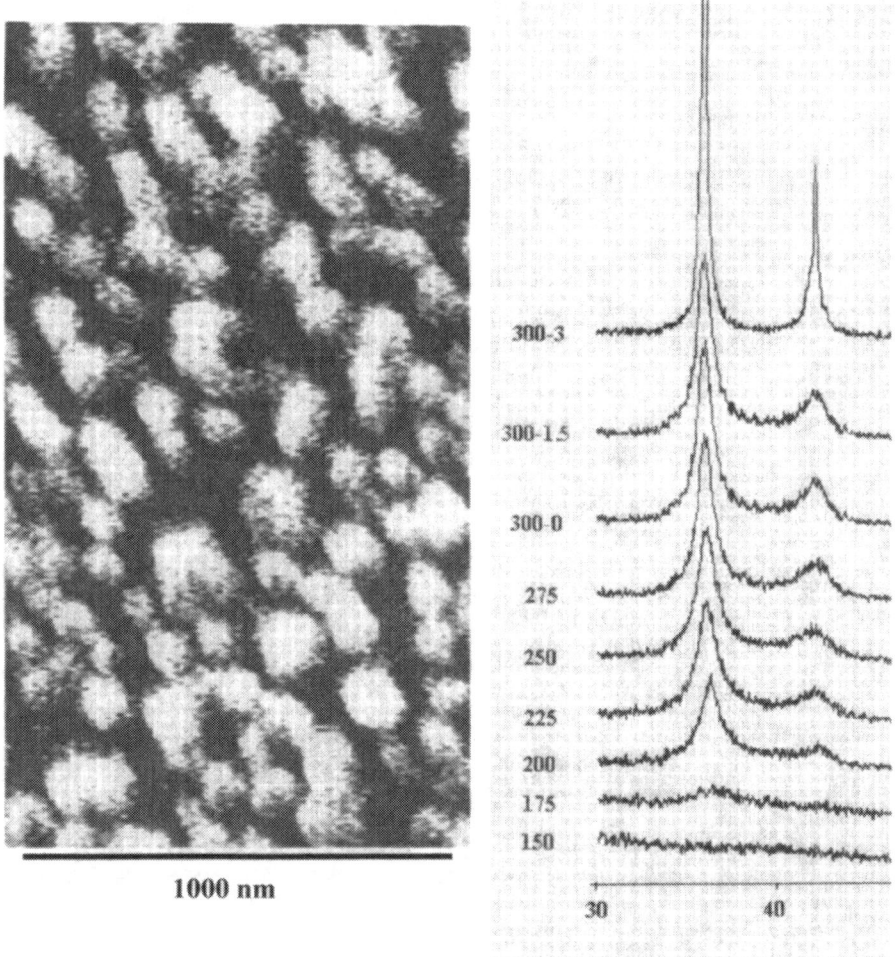

Figure 5. (Left) SEM micrograph for the 13% silver-6FDA/4-BDAF metallized film cured to 300 °C for 1 h.

Figure 6. (Right) X-ray reflections for a series of 13% silver-6FDA/4-BDAF metallized films as a function of cure temperature and time. The abscissa is in units of two theta; the ordinate is the temperature in °C at which the sample was withdrawn from the oven; at 300 °C samples where withdrawn after 0, 1.5, and 3 h. (See experimental section for cure cycle details.)

CONCLUSIONS

Silver surface-metallized films with high reflectivities can be prepared from AgTFA with the fluorinated polyimide 6FDA/4-BDAF in a thermally promoted single-stage process. These

305

metallized films are not electrically conductive. Metal-polyimide adhesion is excellent, and thermal and mechanical properties of the composite films remain near those of the parent polymide.

ACKNOWLEDGMENT

The authors thank the Petroleum Research Fund administered by the American Chemical Society for partial support of this work.

REFERENCES

1. Matienzo, L. J.; Unertl, W. N. "Adhesion of Metal Films to Polyimides,"in *Polyimides: Fundamental and Applications,* ed. Ghosh, M. K.; Mittal, K. L., Marcel Dekker, New York, 1996. pp. 629-696.
2. Gierow, P. A. in *Proceedings of the ASME-JSME-JSES Solar Energy Conference*: Reno, NV, 1991; pp 1-7.
3. Ehricke, K. in *Meeting of the American Rocket Society*: Cleveland, OH, June 18-20,1956; pp ARS paper 310-56.
4. Gulino, D. A.; Egger, R. A.; Bauholzer, W. F. *NASA Technical Memorandum 88865,* **1986**.
5. Naval Research Laboratory, "Solarcon - Concentrator Reflector System," described online: code8200.nrl.navy.mil/solarcon.html.
6. Freeland, R. E.; Bilyou, G. in *43rd Congress of the International Astronautical Federation, IAF-92-0301*: Washington, D.C., 1992
7. McInnes, C. R. *Solar Sailing Technology: Dynamics and Mission Applications*, Springer-Verlag, London, 1999.
8. Garner, C.; Diedeich, B.; Leipold, M. A Summary of Solar Sail Technology Developments, *AIAA/ASME/SAE/ASEE, 35th Joint Propulsion Conference and Exhibit*, Los Angeles, CA, June 21-23, 1999. Available online: techreports.jpl.nasa.gov
9. Southward, R. E.; Stoakley, D. M. *Progress in Organic Coatings*, **2001**, *41*, 99-119 and references therein.
10. Rubira, A. F.; Rancourt, J. D.; Taylor, L. T.; Stoakley, D. M.; St. Clair, A. K. *J. Macromolecular Sci., Pure and Applied Chemistry* **1998**, *A35*, 621-636 and references therein.
11. Jorgensen, G.; Schissel, P. in *Metallized Plastics*; Mittal, K. L.; Susko, J. R., eds.; Plenum: New York, 1989; Vol. 2, pp 79-92.
12. Bower, G. M.; Frost, L.W. *J. Polym. Sci., A* **1963**, *1*, 3135.
13. Sroog, C. E.; Endrey, A. L.; Abramo, S. V.; Berr, C. E.' Edwards, W. M.; Olivier, K. L. *J. Polym. Sci., A*, **1965**, *3*, 1373.
14. Clair, A. K. S.; Clair, T. L. S.; Slemp, W. S. "Optically Transparent/Colorless Polyimides:" in *Recent Advances in Polyimide Science and Technology*; Weber, W. D., Gupta, M. R., eds.; Society of Plastic Engineers, Mid-Hudson Section: Poughkeepsie, 1987, pp.16-34.

Mat. Res. Soc. Symp. Proc. Vol. 707 © 2002 Materials Research Society

The role of the interfaces in the optical effects of large-sized SiC$_x$O$_{1-x}$N nanocrystallites

K.J. Plucinski, H. Kaddouri[1], I.V. Kityk[2]
Military University of Technology, Dept of Electronics, Warsaw, POLAND
[1] Universite du Perpignan, Lab. LP2A, Perpignan, FRANCE
[2] Institute of Physics WSP, Częstochowa, POLAND

ABSTRACT

The band energy structure of large-sized (10-25) nm nanocrystallites (NC) of SiC$_x$O$_{1-x}$N (0.96<x<1.06) was investigated using different band energy approaches, as well as modified Car Parinello molecular dynamics simulations of interfaces. A thin carbon sheet (of about 1 nm) appears, covering the crystallites. This sheet leads to substantial reconstruction of the near-the-interface SiC$_x$O$_{1-x}$N crystalline layers. Numerical modeling shows that these NC may be treated as quantum dot-like SiC$_x$O$_{1-x}$N reconstructed crystalline surfaces, covering the appropriate crystallites. All band energy calculation approaches (semi-empirical pseudopotential, fully augmented plane waves and norm-conserving self-consistent pseudopotential approaches) predicted the experimental spectroscopic data. In particular, it was shown that the near-the-surface carbon sheet plays a dominant role in the behavior of the reconstructed band energy structure. Independent evidence for the important role of the dot-like crystalline layers are the excitonic-like states, which are not dependent on the particular structure of the SiC, but are sensitive to the thickness of the carbon layer.

INTRODUCTION

There has recently been increased interest in the possibility of using nano-technologies as materials for optics and electronics, the main reason being the size-dependent electronic and optical properties of these materials [1-4]. The properties of large-size nanocrystallites (larger than 10 nm) were treated as the so-called superposition of the NC and other structures [5-8]. One can predict that the latter will be directly dependent on the relative sizes of particular crystallites and the surrounding interface environment. Depending on these parameters, we will obtain different excitonic states which define the behavior of the interface band energy gradients. The coexistence of bulk-like and quantum confinement states presents the possibility of working with electron energy dispersion within the same crystallites contrary to the thin SiC$_x$O$_{1-x}$N films deposited on the SiC$_x$O$_{1-x}$N crystalline surfaces.

Among the many possible NC materials, SiC$_x$O$_{1-x}$N crystallites were chosen because the technology for their manufacture with the sizes needed is well developed, the energy gap of SiC$_x$O$_{1-x}$N may be manipulated within the large spectral ranges (2 – 4.5 eV), depending on polytype kinds, and because SiC$_x$O$_{1-x}$N is substantially more stable than other SiON materials, when external mechanical and thermo-treatments are applied.

The main goal of the present work was to study the optical properties of SiC$_x$O$_{1-x}$N NC, both experimentally, as well as theoretically. In particular, we investigated the influence of nanocrystallite size and carbon excess on the optical absorption of the SiC$_x$O$_{1-x}$N nanocrystallites, the contribution of the reconstructed near-interface states to the visible absorption of the NC and the contribution of the carbon sheet interfaces to the absorption spectra observed.

The nanopowders were synthesized by CO$_2$ laser pyrolysis of silane and acetylene gaseous mixtures. The reactant fluxes monitored the ratio C/O and induced a carbon or silicon-rich network in the outer-most SiC nanoparticle surfaces.

Size-dependent effects are usually studied for NC with sizes below 8 - 10 nm, where clear size-confined effects are observed. In such cases, k- space bulk-like dispersion disappears and discrete excitonic-like levels in the energy gaps appear. However, interface (reconstructed) dot-like states corresponding to an intermediate level, with both bulk-like, as well as dot-like quantised excitonic, properties can be promising in detecting experimentally obtained optical phenomena. These interface sheets are formed by reconstructed $SiC_xO_{1-x}N$ crystalline films separating the bulk-like and thin carbon layers.

On the basis of the relative presence of quantum dot states, one can estimate the relative contributions of reconstructed $SiC_xO_{1-x}N$ near-the–interface structural fragments to the band energy (BE) dispersion, as well as that of the excitonic state to the optical absorption [9-12] (being directly connected to the imaginary part of the dielectric susceptibilities).

RESULTS AND DISCUSSION

The SiCON nanocrystallite specimens were monitored using TEM, NMR, Raman and IR methods. Optical absorption caused both by band energy reconstruction, as well as by excitonic effects, is analyzed. The role of carbon excess was demonstrated experimentally by increasing the C/O ration over the C-rich particle surfaces.

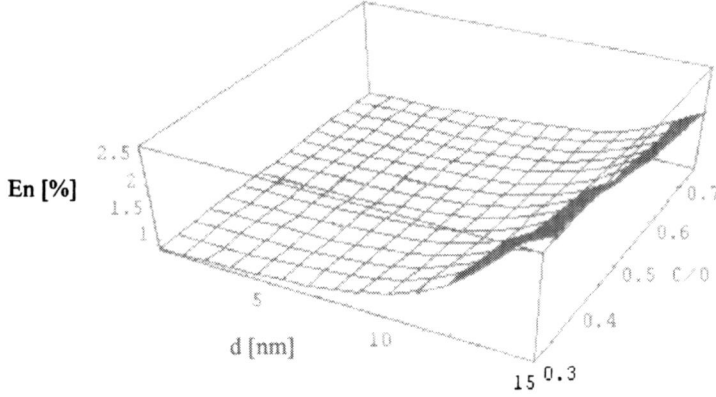

Figure 1. The relative contribution of the interfaces to the energy gap is presented dependent on the crystallite sizes d and ratio of the C/O.

In Fig. 2 the band energy dispersion for the interfaces in the case of the x=1.02 are presented. One can see substantial reconstruction of the band energy dispersion near the interfaces. Moreover the energy dispersion derivative is varied. This fact is caused by the reconstruction of the near-the surface states [13-16].

Figure 2. Band structure of the $SiC_xO_{1-x}N$ for the x=1.02. The points indicate the bands corresponding to the reconstructed surfaces.

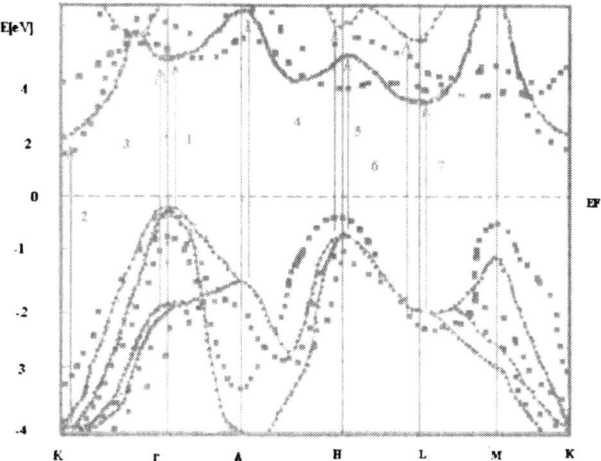

Figure 3. Band structure of the $SiC_xO_{1-x}N$ assuming an interface thickness of about 0.6 nm.

Fig. 3 shows additional band structure reconstruction when we undergo to the bands more close to the interfaces. This correlates well with the charge density distributions near the interfaces (see Fig. 4). One can see that a shift toward the deep interface levels stimulates additional charge density re-distribution.

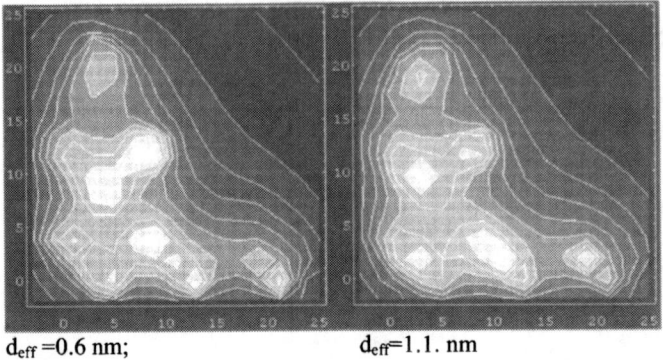

d_{eff} =0.6 nm;   d_{eff}=1.1. nm

Figure 4. Reconstruction of the near-the-interface charge density distribution.

This fact confirms the layered dot-like structure of the investigated nanocrystallites. Our investigations have shown that the dominant role in the optical spectra observed is played by the dot-like near-the–surface states which form strong sharp-like absorption excitonic

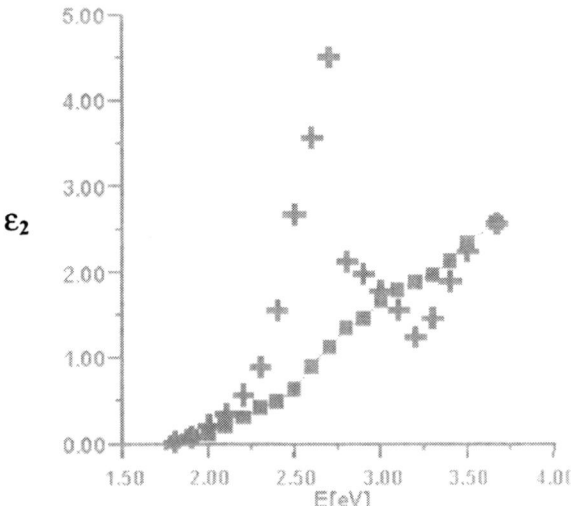

Figure 5. Excitonic spectra at LHeT - + and RT - ■.

spectra (Fig. 5). From Fig. 5 one can see an occurrence of the dot-like excitons playing a key role in the observed spectra.

The connected with the measured behaviours the dependencies of the transparency T versus the film thickness and O/C ratio at LHeT are presented in the Fig. 6.

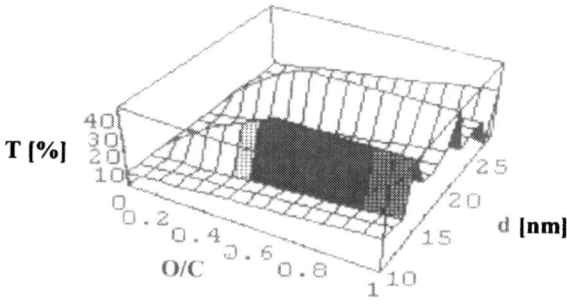

Figure 6. Typical dependencies of the transparency versus the O/C ratio and film thickness.

One can see that T possess modulated-like dependence versus the thickness d. This reflects an increase of competition between the long-range ordering and short-range disordering.

CONCLUSIONS

1. Our investigations have shown that the materials investigated have interface thin layers with thickness of about 1.5 nm. This layer may be considered as a reconstructed surface of the Si by the covered films.
2. The obtained results present new opportunities in semiconductor electronics, making it possible, for example, to receive 3D memory for computers on the trapping levels.
3. The calculations we made using the *ab initio* Car-Parinello method are an additional confirmation of the new type of the interfaces that will shortly find application in the electronic technique.
4. The simulations carried out predict the modulated-like features of the transparency versus the film thickness and O/C ratio.

ACKNOWLEDGMENTS

The study was supported by the Polish State Committee for Scientific research through grant No KBN-7-T11B 011 20.

REFERENCES

1. *Microcrystalline and Nanocrystalline Semiconductor*s, edited by R. W. Collins, C. C. Tsai, M. Hiros, F. Koch, and L. E. Brus, MRS Symposium Proceedings No. 358, MaterialsResearch Society, Boston, 1994.
2. *Light Emission from Novel Silicon Materials*, edited by Y. Kane-mitsu, M. Kondo, and K. Takeda, The Physical Society of Japan, Tokyo, 1994.
3. L. W. Wang and A. Zunger, in *Nanocrystalline Semiconductor Materials*, edited by P. V. Kamat and D. Meisel, Elsevier Science, Amsterdam, (1996); A. D. Yoffe, *Adv. Phys.* **42**, 173 (1993); K.D.Hirschman, L.Tsybeskov, S.P.Duttagupta, P.M.Fauchet, Nature (London) **384** (1996) 338; M.C.Schlamp, X.Peng, and A.P.Alivisatos, *J.Appl.Phys.* **82** (1997) 5837.
4. F. Buda, J. Kohanoff, and M. Parrinello, *Phys. Rev. Lett.* **69**, 1272 (1992).

5. Y.S.Park, SiC *Materials and Devices, Semiconductors and Semimetals*, Academic Press, London, UK, 1998; M.B.Yu, S.F.Rusli, S.F.Yoon, S.J.Xu, K.Chew, J.Cui, J.Ahn, and Q.Zhang, *Thin Solid Films* **177**, 377-378 (2000).
6. I.V.Kityk, M.Makowska-Janusik, A.Kassiba, and K.Plucinski, *Optical Materials* **13**, 449 (2000).
7. J.A.Stroscio and D.M.Eigler, *Science* **254**, 1319 (1991).
8. V.Derycke, *Phys. Rev. Lett.* **81**, 5868 (1998).
9. V.M.Bermudez, *Phys. Stat. Solidi* **B202**, 447 (1997).
10. V.I.Gavrilenko, S.I.Frolov, and N.I.Klyui, *Physica* **B185**, 394 (1993).
11. S.Albercht, L.Reining, R.Del Sole, and G.Onida, *Phys. Rev.* **B37**, 7486 (1999).
12. G.Galli, R.M.Martin, R.Car, and M.Parinello, *Phys. Rev.* **B42**, 7470(1990).
13. B.M.Bylander, and L.Kleinman, *Phys. Rev.* **B41**, 7868 (1990).
14. P.-A.Glans and L.I.Johansson, *Surf. Sci.* **465**, L759 (2000); P.-A.Glans, T. Balasubramanian, M.Syvajori, B.Yakimava, L.I.Johansson, *Surf. Science* **470**, 284 (2001).
15. N.Troullier, and J.L.Martins, *Phys. Rev.* **B43**, 8861 (1991).
16. C.Persson and U.Lindefelt, *J.Appl.Phys.* **82**, 5496 (1997).

Mat. Res. Soc. Symp. Proc. Vol. 707 © 2002 Materials Research Society   DD12.5

In Situ Measurement of the Second Harmonic Signal of Adsorbing Nonlinear Optical Ionically Self-assembled Monolayers.

C. Brands,[1] P.J. Neyman,[2] M.T. Guzy,[3] S. Shah,[3] K.E Van Cott,[3] R.M. Davis,[3] H. Wang,[4] H.W. Gibson,[4] and J.R. Heflin.[1]

[1]Department of Physics, Virginia Tech
Blacksburg, VA 24061
[2]Department of Materials Science and Engineering, Virginia Tech
Blacksburg, VA 24061
[3]Department of Chemical Engineering, Virginia Tech
Blacksburg, VA 24061
[4]Department of Chemistry, Virginia Tech
Blacksburg, VA 24061

ABSTRACT

Ionically self-assembled monolayers (ISAMs) have recently been shown to spontaneously exhibit a polar ordering that gives rise to substantial second order nonlinear optical response. The deposition of ISAMs has been studied *in situ* via second harmonic generation. This is a particularly sensitive probe of the growth of nanometer-thick films since the centrosymmetry of the immersion solutions, the substrate, and the container yields no SHG contribution from these bulk components. Upon immersion in the NLO-active polyelectrolyte solution, the SHG rises sharply over the first minute. When a film is immersed into salt water, the SHG decreases significantly only to be restored when the salt solution is replaced with deionized water.

INTRODUCTION

Ionically self-assembled monolayer (ISAM) films are grown, one monolayer at a time, by immersing a charged substrate alternately in anionic and cationic solutions. This has been shown to be an easy, economic, and fast method for creating laterally homogeneous, nanostructured thin films. [1,2] These films can be used to provide nanoscale control of thickness, composition and orientation in devices such as light-emitting diodes, photovoltaics, and electrochromics. One application we have been focusing on is the fabrication of films with a second order nonlinear optical response. These films show substantial $\chi^{(2)}$ values with outstanding thermal and time stability. [3-7] To increase the understanding of the formation of the ISAM films it is beneficial to be able to measure the growth of the layer *in situ*. In this paper, we describe studies of the deposition process using second harmonic generation (SHG) as a probe. We also compare the deposition process of a polymer with that of a monomer.

EXPERIMENTAL DETAILS

The measurements were done with a standard SHG setup employing a 30 picosecond pulsewidth modelocked Nd:YAG laser. The fundamental wavelength is 1064 nm and is p-polarized. Typical values for beam radius and pulse energy values were 100 μm and 1 mJ/pulse respectively. The SHG data are averaged over 10 shots and 100 shots per data point for *in situ* and *ex situ* measurements, respectively. The sample holder was

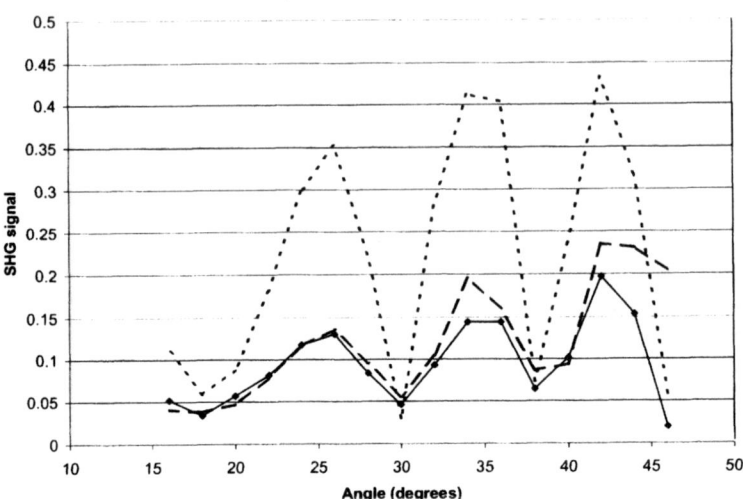

Figure 1. The second harmonic signal as a function of incidence angle. The solid line represents the glass slide, which has some SHG signal due to the fact that the interface is non-centrosymetric. The dashed and the dotted lines represent the same slide after depositing PAH and PCBS layers, respectively.

constructed so that the sample could be immersed in solution without moving the sample with respect to the incoming laser beam. This enabled us to measure the SHG signal while the layer is growing and the SHG signal between the deposition steps, always monitoring the same spot on the sample. [8] The optically active material used in these experiments was poly{1-4-(3-carboxy -4–hydroxyphenylazo)-benzenesulfoamido-1,2-ethanediyl, sodium salt} (PCBS). The optically inactive material used as counterion was poly (allylamine hydrochloride) (PAH). Both materials were purchased from Aldrich.

The SHG signals generated by the film on either side of the sample interfere with one other. The phase of the interference is determined by the thickness of the glass slide, the angle of incidence of the laser beam, and the refractive indices of the glass slide and the surrounding media. The thickness of the layers is negligible compared to the thickness of the glass slide, therefore one would expect that interference pattern of SHG versus angle of incidence is independent of layers deposited when the measurements are made *ex situ*. Figure 1 shows this to be true for the bare substrate, a PAH layer, and a PAH/PCBS bilayer. It is clear that signal strength is much stronger after deposition of PCBS, but the angular position of the maxima and minima is unaltered by depositing additional layers. The SHG signal observed from a bare glass slide is due to the fact that the air-glass interface, like any other interface, is intrinsically non-centrosymmetric.

The interference between the SHG signals generated on opposite sides of the substrate is determined by the propagation distance through the glass, which is $h = 1$ mm $/ \cos \theta'$ where 1 mm is the thickness of the slide and θ' is the refracted angle in the glass. The interference pattern is thus dependent on the refractive index of the surrounding

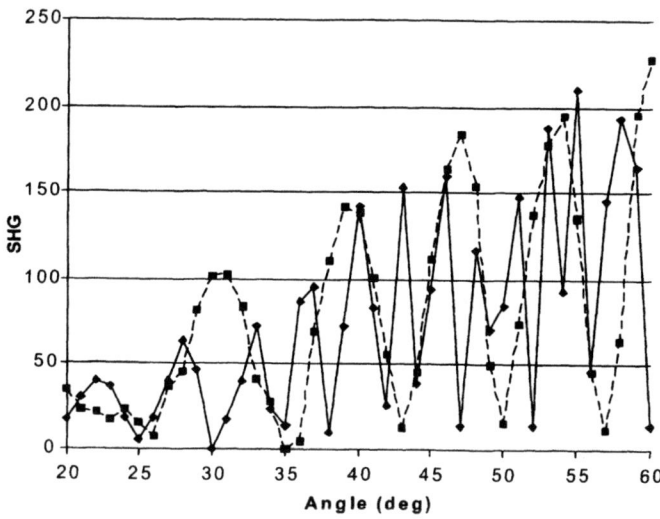

Figure 2. The SHG as a function of incident angle for a PAH/PCBS sample. The solid and dashed lines represent the sample in water and in air, respectively.

medium since that determines θ'. Figure 2 compares the interference patterns of a sample in air with that of the same sample in water. It is clear that the peaks are shifted closer together when the sample is in aqueous solution. This is due to the fact that the refractive index difference between water and glass is smaller than the refractive index difference between air and glass.

Figure 3 shows the calculated SHG interference pattern as a function of external incident angle for a glass slide (n = 1.5, 1 mm thickness) in air (n = 1.0) and water (n = 1.33). The calculation confirms the experimental finding that the peaks are closer together as the refractive index of the surrounding medium increases. In the remainder of this paper, the angle of maximum interference was determined for each glass slide in air and in aqueous solution. These angles were then used for the remainder of the experiment.

RESULTS

The deposition rate of PCBS on PAH was measured for three concentrations from 0.5 mMol down to 0.0025 mMol as shown in Figure 4. . The concentration of PAH was kept at 10 mMol. A decrease in the equilibrium SHG intensity was observed with decreasing concentration, indicating a decreased amount of adsorbed PCBS. The rate of adsorption is similar for each concentration.

In polyelectrolyte ISAM films, it is known that pH and salt concentration have a strong impact on thickness and SHG of the films. [7] Figure 5 shows that immersion of a

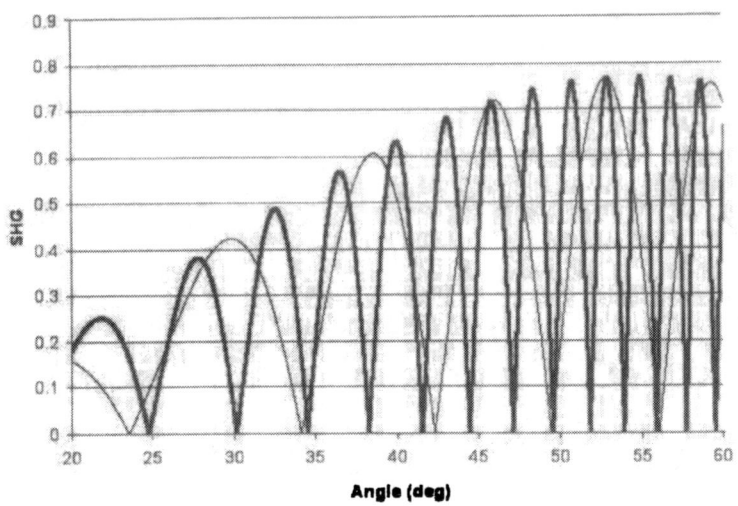

Figure 3. Calculated angular dependence of SHG interference fringes for thin films on opposite sides of the substrate immersed in water (thick curve) and in air (thin curve).

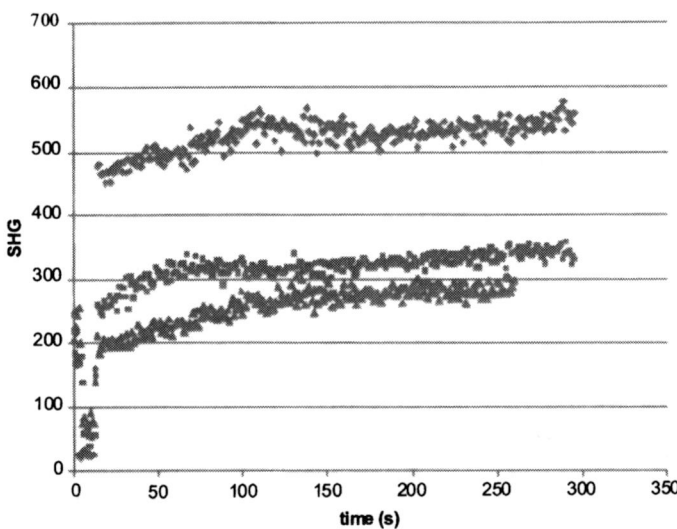

Figure 4. SHG signal as function of time during PCBS deposition on PAH for low concentrations of PCBS. The concentrations, top to bottom, are 0.5 mM, 0.05 mM, and 0.0025 mM.

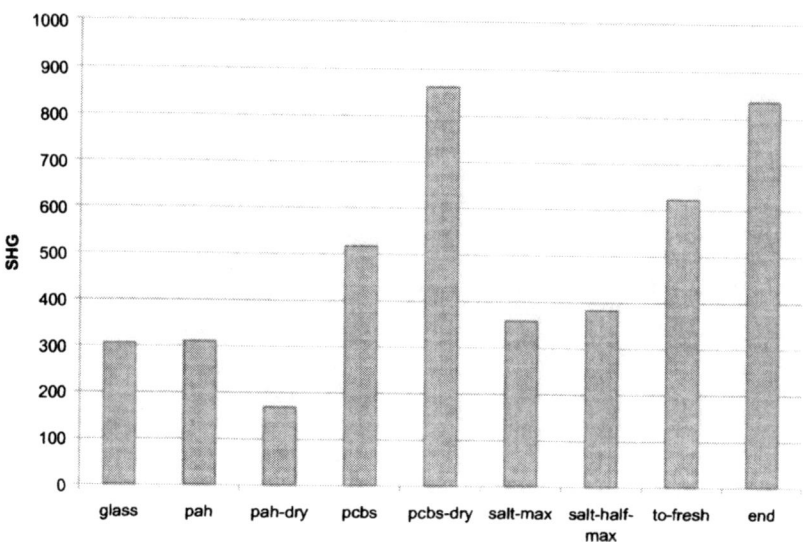

Figure 5. Final values of the *ex situ* SHG signal for different deposition steps. Immersion in salt solution results in a rapid decrease in signal, which is restored when the sample is immersed in deionized water.

PAH/PCBS bilayer in a saturated salt solution leads to decreased SHG intensity. This is believed to be due to the incorporation of salt ions into the monolayer, resulting in decreased polar order of the chromophores. Reducing by half the salt concentration of the ambient solution has hardly any effect on the SHG. However, when the sample is immersed in deionized water, the salt ions diffuse out of the film and the film conformation and therefore the SHG is fully restored. In situ measurements show that most of the restoration takes place in the first minute.

We have recently developed a novel, hybrid covalent/ionic approach to fabricating second order NLO ISAM-like films based on monomeric chromophores. [9] Procion Red (PR, Aldrich) was chosen as a prototype system, possessing a triazine ring which can covalently couple into the preceding PAH layer and two sulfonate groups which can promote electrostatic adsorption of the successive PAH layer. Figure 6 shows the *in situ* SHG as a function of time during the deposition of Procion Red on PAH. This deposition is essentially complete in two minutes.

SUMMARY

In this study, we showed that both Procion Red and PCBS can be deposited on PAH in roughly minute. Also, PAH can be deposited on Procion Red in less than a minute. The concentration of the PCBS solution can be lowered to 0.5 mMol without loss of signal or deposition speed. Lower concentrations reduce the SHG signal but significant deposition has been recorded at concentrations as low as 0.0025 mMol. Immersion in salt solution

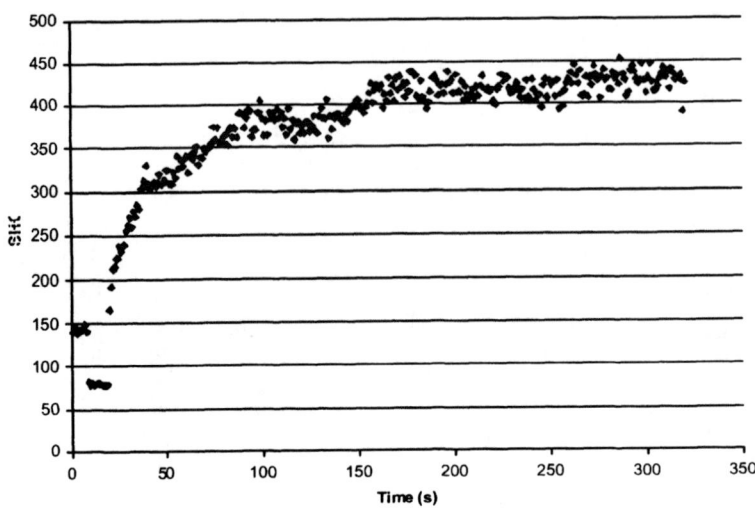

Figure 6. The SHG signal as function of time during deposition of procion red on PAH.

immediately decreases the SHG signal due to inclusion of salt ions in the film. Subsequent immersion into deionized water causes the salt ions to diffuse out of the layer causing near complete restoration of the SHG signal

ACKNOWLEDGEMENTS
This research was supported by National Science Foundation grant ECS-9907747.

REFERENCES
1. G. Decher, J.D. Hong, and J. Schmitt, Thin Solid Films **210**, 831 (1992).
2. G. Decher, Science **277**, 1232 (1997).
3. J.R. Heflin, C. Figura, D. Marciu, Y. Liu, and R.O. Claus, SPIE Proc. **3147**, 10(1997); Appl. Phys. Lett. 74, 595 (1999).
4. Y. Lvov, S. Yamada, and T. Kunitake, Thin Solid Films **300**, 107(1997).
5. X. Wang, S.Balasubramanian, L.Li, X. Jiang, D. Sandman. M.F. Rubner, J. Kumar, and S.K. Tripathy, Macromol. Rapid Commun. **18**, 451 (1997).
6. M.J.Roberts, G.A. Lindsay, W.N.Herman, and K.J. Whynne, J. Am. Chem. Soc. **120**, 11202(1998).
7. C. Figura, P.J. Neyman, D. Marciu, C. Brands, M.A. Murray, S. Hair, M.B. Miller, R.M. Davis, and J.R. Heflin. MRS Proc. vol. **598**, BB4.9.1-6(2000).
8. C. Brands, J.R. Heflin, P.J. Neyman, M. T. Guzy, S. Shah, H. W. Gibson, K. E. Van Cott, R. M. Davis, SPIE Proc. **4461**, 311 (2001).
9. P.J. Neyman, M.T. Guzy, S. Shah, H. Wang, H.W. Gibson, K. E. Van Cott, R.M. Davis, C. Brands, J.R. Heflin, (this proceedings).

Photonic Crystals

Mat. Res. Soc. Symp. Proc. Vol. 707 © 2002 Materials Research Society AA1.3/K1.3

Light pulse propagation in three-dimensional photonic crystals

H.Kitano, F.Minami, T.Sawada[1], S.Yamaguchi[2] and K.Ohtaka[2]
Department of Physics, Tokyo Institute of Technology, Meguro-ku, Tokyo 152-8551, Japan
[1]National Institute for Materials Science, Tsukuba, Ibaraki 305-0044, Japan
[2]Center for Frontier Science, Chiba University, Chiba 263-8522, Japan

ABSTRACT

The phase characteristics of transmitted optical pulses in three-dimensional photonic crystals were investigated in the frequency and time domain by using the spectrally resolved cross-correlation technique. The temporal evolution of femtosecond pulses passing through polystyrene colloidal crystals exhibits a large phase distortion near the stop bands. The phase discontinuity around the band gap was observed in the frequency-domain. The phase of the transmitted pulses is found to change by π across the band gap. The dispersion curve estimated from the phase shift shows good correspondence with those calculated from a photonic band calculation. The group velocity significantly slows down near the stop bands. A large change of the group velocity dispersion is also observed near the band edges. These results are in good agreement with the band theory.

INTRODUCTION

Photonic band gap (PBG) or photonic crystals are artificially manufactured periodic dielectric structures with lattice constants of the order of the wavelength of light [1]. Their periodicity gives rise to photonic band structures, analogous to electronic band structures, and does not allow propagation of electromagnetic waves for a certain range of frequencies, forming stop bands. The propagation of electromagnetic waves is strongly modified when the frequency of the waves is close to the stop bands. In particular, information on the phase distortion of transmitted electromagnetic waves is important because the dispersion relation $\omega(\mathbf{k})$ between frequency ω and wave vector $\mathbf{k}$ of propagating wave is directly connected to the frequency dependence of the phase shift [2-5].

In this paper, we present the intensity and the phase of light propagating through a three-dimensional optical photonic crystal fabricated from a colloidal suspension of polystyrene microspheres. The colloidal crystal, which has a lattice spacing comparable to the wavelength of light, does not exhibit a complete photonic band gap, because the concentration and the index of refraction of the polystyrene spheres relative to water are not sufficiently high. However, this crystal is very useful in studying PBG effects seen only in particular directions [6].

The experiments were based on a new phase-sensitive spectroscopic method: the frequency-resolved optical gating (FROG) which was recently developed to characterize phase properties of the laser pulses [7]. In the FROG method, a nonlinear auto-correlation signal is spectrally resolved, and from this datum both of the phase and the intensity of the light field are reconstructed. We applied this technique to a spectrally resolved second harmonic cross-correlation (SHG-cross FROG). We determined the phase of the weak transmitted pulse through the photonic crystal as a function of frequency and time with the SHG-cross FROG method [8].

EXPERIMENTAL PROCEDURES

The photonic crystals were composed of deionized suspensions of polystyrene spheres with a diameter of 194 nm dispersed in water. Colloidal suspensions of a volume fraction of 11.2 % were put in flat capillary flow cells of fused quartz with a pass length of 0.1 mm. The polystyrene spheres ordered into a face-centered cubic lattice with the (111) planes parallel to the capillary face. A centimeter-sized single crystal domain was created in the capillary and it was sufficiently stable against external disturbance [9].

For phase-sensitive time-resolved measurements, we used a femtosecond mode-locked Ti-Sapphire laser as an excitation source. The power incident upon the sample is about 80 mW. In the experiment, we first measured the spectrally resolved auto-correlation signals of the incident pulse. With use of the FROG algorithm, namely, the iterative Fourier transform algorithm with generalized projection [7], we determined the phase of the incident pulse, together with the intensity, as a function of frequency and time. We then measured the spectrally resolved cross-correlation between the transmitted pulse passing through the sample and the incident pulse, which had been characterized in advance. The phase characteristics of the weak transmitted pulse were reconstructed using a modified FROG algorithm developed in [8]. For the correlation measurements, we utilized a 100-μm thick KDP crystal.

RESULTS AND DISCUSSION

Figure 1 shows the transmission spectra of the polystyrene colloidal crystal. The spectra exhibit two sharp dips corresponding to optical stop bands. The dramatic variation of the stop bands is observed when changing the crystal angle. The zero degree corresponds to the [111] direction of the crystal. In moving the [111] direction (L point) to the [110] direction (W point), the position of the lower (higher) stop band shifts to a shorter (longer) wavelength. The transmission spectrum of the crystal along [111] direction exhibits a stop band centered around 2.34×10^{15} rad/sec with a bandwidth of 0.3×10^{14} rad/sec. For the time-resolved measurements,

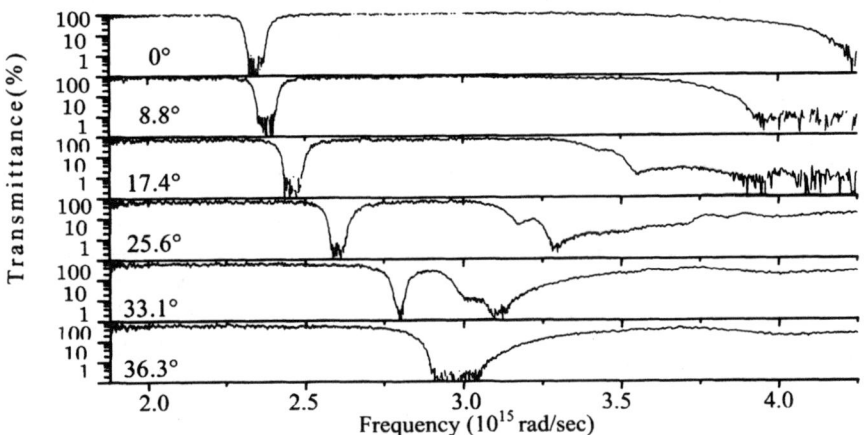

Figure 1. Transmission spectra taken along L-W directions for the polystyrene colloidal crystal.

we used the pulses propagating along the [111] direction and studied the intensity and phase distortions of the pulses near the stop band at 2.34×10^{15} rad/sec.

The intensity of the incident pulse and its phase as a function of frequency are shown in Figure 2(a). The same graph as a function of time is represented in Figure 2 (b). The pulse width is 50 fs and the spectral width is 0.6×10^{14} rad/sec. It is found that the present laser pulse is almost transform-limited. The phase is observed to be flat both in frequency domain and in time domain.

Figure 2(c),(d) illustrate the intensity and the phase of the transmitted pulse close to the stop band. The center frequency was chosen within the stop band. The changes become less pronounced when the frequency of the pulse is tuned away from the stop band. As compared to the data in Figure 2, the intensity and phase of the pulse are significantly modified. The spectrum exhibits two peaks because the light with frequencies within the stop band is strongly attenuated. The phase discontinuity around the band gap is clearly observed in the frequency-domain data. The phase of the transmitted pulses is found to change by π across the band gap.

At the low-frequency side of the stop band, the phase shows a parabola pointing down. As will be seen later, this parabolic shape reflects the fact that light with higher frequencies propagates slowly through the sample compared to that with lower frequencies, i.e., normal dispersion.• On the other hand, at the high-frequency side, the phase exhibits a parabola pointing up instead of down, reflecting anomalous dispersion.

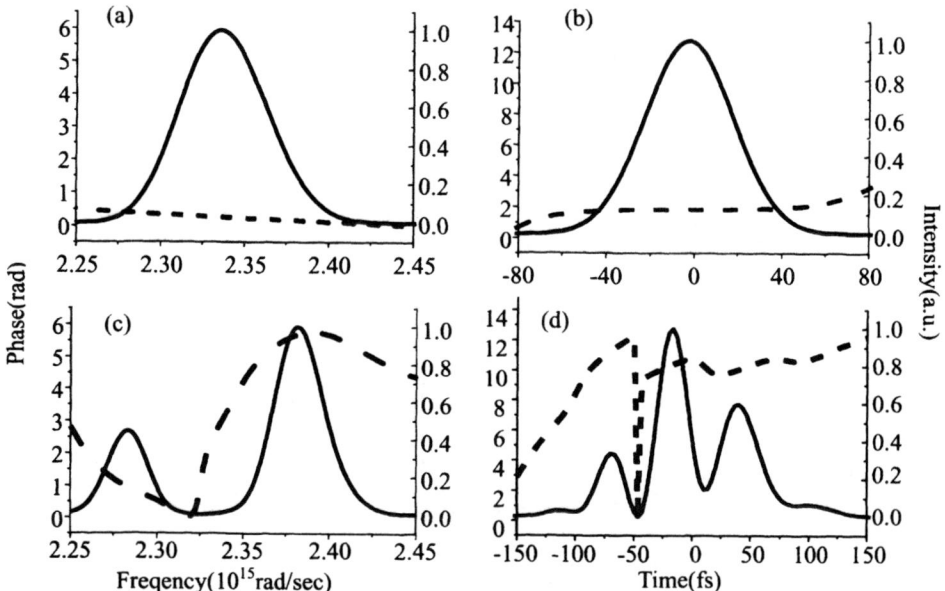

Figure 2. Change of the phase of the incident pulse together with the intensity in frequency domain (a) and time domain (b). And change of the phase of the transmitted pulse together with the intensity in frequency domain (c), and time domain (d). The solid line refers to the intensity and the dashed to the phase.

The wave vector k is proportional to the phase shift ϕ, i.e., $\phi(\omega)=\kappa(\omega)L$, where L is the thickness of the sample. Therefore, the dispersion relation (ω versus k) is directly connected to the phase shift of the transmitted pulses. The group delay is given by $d\phi/d\omega$, and the group delay dispersion (GDD) is written as $d^2\phi/d\omega^2$. The group delay and the GDD are proportional to the inverse group velocity and the group velocity dispersion (GVD) parameter, respectively.

The dispersion curve, the group delay and the GDD obtained from the data are shown in Figure 3-5. For comparison, the theoretical results calculated from the KKR photonic band calculation are also presented in the figures.

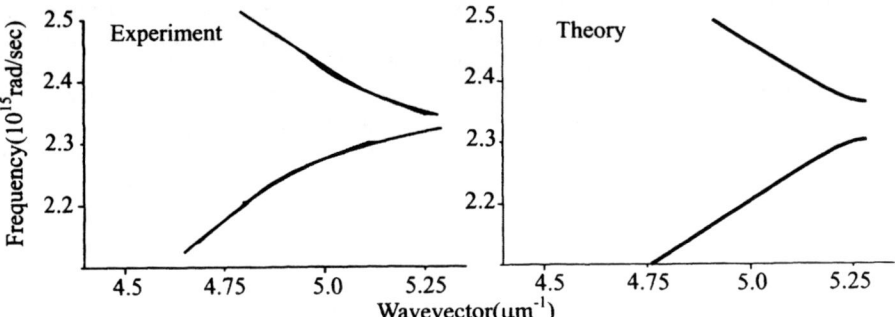

Figure 3. Dispersion relation of light along Γ -L direction in the fcc colloidal crystal.

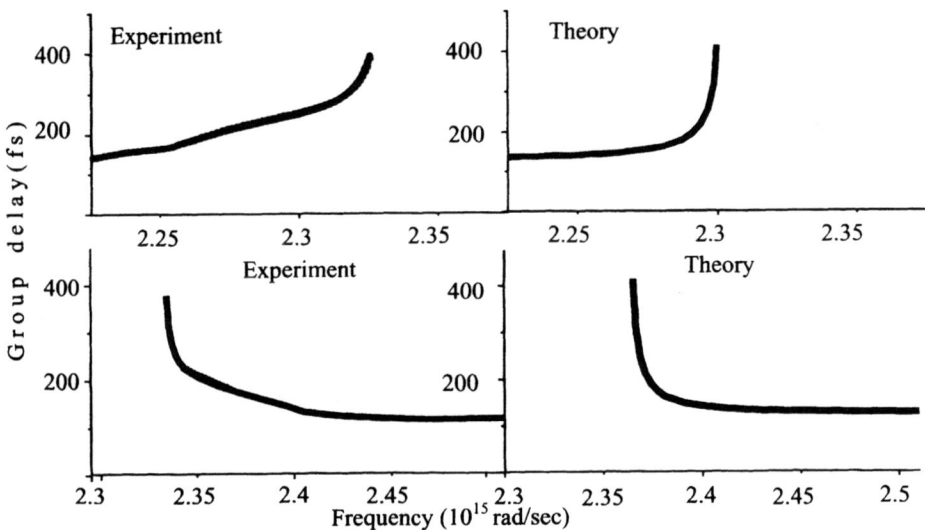

Figure 4.   The frequency dependence of the group delay near the stop band. Upper and bottom graphs represent the group delay of the higher and lower energy band, respectively.

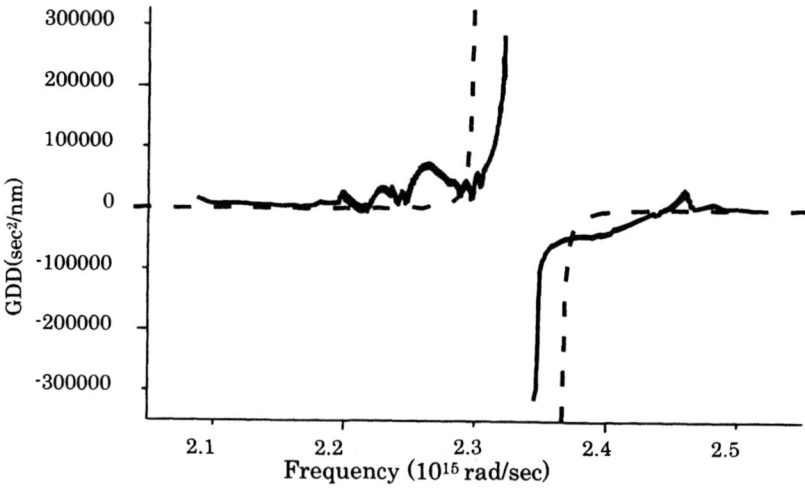

Figure 5. The frequency dependence of the group delay dispersion (GDD) near the stop band. Solid and dashed lines show the experimental and the theoretical results, respectively.

The frequency dependence of the phase shift shows a good correspondence with the theoretical dispersion curves along Γ-L direction. The group delay calculated from the phase shift by numerically differentiation shows a significant increase at frequencies near the stop band, reflecting a strong suppression of the group velocity. The frequency dependence corresponds well to the theoretical dependence. The bandwidth of the stop band is observed to be somewhat narrower than the calculated gap. This discrepancy might be originated from the oscillation of each colloidal sphere in the crystal. The GDD obtained experimentally diverges at the band edges and shows a branch of normal (abnormal) dispersion at low (high) frequency side of the gap. The measured GDD curves again agree comparatively well with the theoretical curves.

CONCLUSIONS

With use of the SHG-crossFROG technique, we determined the intensity and the phase of the pulse passing through the photonic crystal fabricated from polystyrene microspheres in water. In the energy domain, we have observed the phase distortion around the photonic band gap. The spectral phase distortion is in agreement with the theoretical photonic band. Furthermore, we have derived the group velocity and group delay dispersion from the phase of the transmitted pulses. The result shows that the optical pulse passing through the photonic crystal is strongly modified at the photonic band edges.

ACKNOWLEDGEMENT

This work was supported by a Grand-in-Aid for Scientific Research.

REFERENCES

1. Photonic Band Gap Materials, edited by C.M. Soukoulis, NATO ASI Ser. E, Vol.315 (Kluwer, Dordrecht, 1996).
2. A.I. Willem, L. Vos, R. Sprik and A. Largendijk, Phys. Rev. Lett. **83**, 2492 (1999).
3. T. Aoki, M.W. Takeda, J.W. Haus, Z. Yuan, M. Tani, K. Sakai, N. Kawai and K. Inoue, Phys. Rev. B **64**, 45106 (2001).
4. I.I. Tarhan, M.P. Zinkin and M.H. Watson, Opt. Lett. **20**, 1571 (1995).
5. B.T. Rosner, G.J. Schneider and G.H. Watson, J. Opt. Soc. Am. B**15**, 2654 (1998).
6. I.I. Tarhan and M.H.Watson, Phys.Rev. Lett. **76**, 315 (1996).
7. R. Trebino, K. W. DeLong, D. N. Fittinghoff, J. N. Sweetser, M. A. Krumbëgel, and B. A. Richman, Rev. Sci. Instrum. **68**, 3277 (1997), and references therein.
8. S. Linden, H. Giessen, and J. Kuhl, Phys. Stat. Sol. (b) **206**, 119 (1998).
9. T. Sawada, Y. Suzuki, A.Toyotama and N. Iyi, J. J. Appl. Phys. **40**, L1226 (2001).

Mat. Res. Soc. Symp. Proc. Vol. 707 © 2002 Materials Research Society   AA7.5/K7.5

Photonic Crystals at Near-Infrared and Optical Wavelengths

Alexander Moroz
Debye Institute, Utrecht University,
P.O. Box 80000, 3508 TA Utrecht, The Netherlands
http://www.amolf.nl/research/photonic materials theory/moroz/moroz.html

ABSTRACT

As demonstrated for the example of a diamond and zinc blende structure of dielectric spheres, small inclusions of a low absorbing metal with the volume fraction f_m can have a dramatic effect on a complete photonic band gap (CPBG) between the 2nd-3rd bands. For example, in the case of silica coated silver spheres, the CPBG opens for $f_m \approx 1.1\%$ and exceeds 5% for $f_m \approx 2.5\%$. Consequently, any dielectric material can be used to fabricate a photonic crystal with a sizeable and robust CPBG in three dimensions. Absorption in the CPBG of 5% remains very small ($\leq 2.6\%$ for $\lambda \geq 750$ nm). The structure enjoys almost perfect scaling, enabling one to scale the CPBG from microwaves down to ultraviolet wavelengths.

INTRODUCTION

Photonic crystals are structures with a periodically modulated dielectric constant [1]. In analogy to the case of an electron moving in a periodic potential, certain photon frequencies in a photonic crystal can become forbidden, independent of photon polarization and the direction of propagation - a complete photonic bandgap (CPBG) [1, 2, 3]. In the last decade, photonic crystals enjoyed a lot of interest in connection with their possibilities to guide light and to become a platform for the fabrication of photonic integrated circuits [4, 5]. Despite the research activities of a large number of experimental groups, achievement of a CPBG below infrared wavelengths for both two- and three-dimensional photonic structures is still elusive, mainly because the required dielectric contrast δ to open a CPBG is rather high. Even for the best geometries $\delta \approx 5$ is required [2, 6]. Already this threshold value of δ excludes the majority of semiconductors and other compounds and materials, such as (conducting) polymers, from many useful photonic crystal applications. However, the required δ is even higher. For applications one needs a sufficiently large CPBG to leave a margin for gap-edge distortions due to omnipresent defects. Let us define the relative gap width g_w as the gap width-to-midgap frequency ratio, $\Delta\omega/\omega_c$. Then in order to achieve g_w larger than 5%, $\delta \geq 9.8$ and $\delta \geq 12$ is required for a diamond [6] and face-centered-cubic (fcc) structure [7], respectively. This leaves only a couple of materials for photonic crystals applications at near infrared and optical wavelengths [8]. Surprisingly enough, there is a way to create a sizeable and robust CPBG with just any dielectric material, be it silica glass or a polymer. A price to pay is to accept a small volume fraction f_m of a low absorbing metal, the actual amount of which depends on an available material dielectric constant ε. Obviously, small

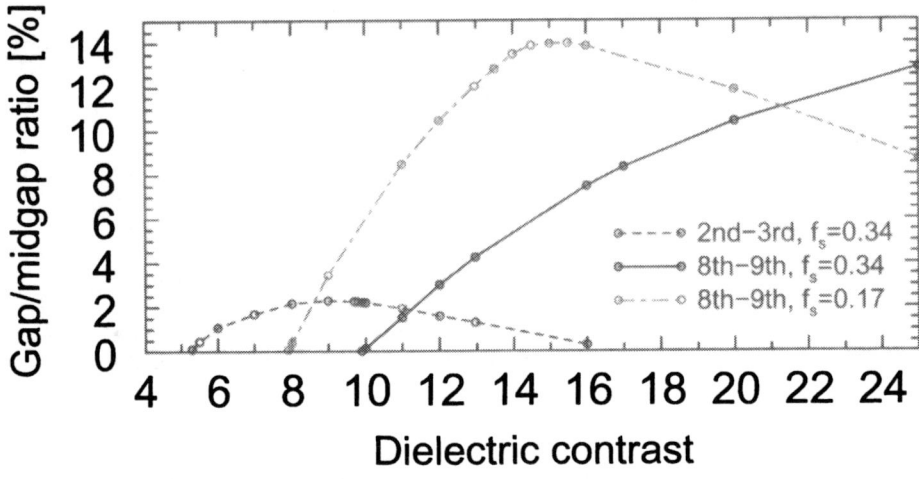

Figure 1: Gap width to midgap frequency ratio for a diamond lattice of dielectric spheres as a function of the dielectric contrast.

metal inclusions do not open a CPBG in every dielectric structure. For example, a simple face-centered-cubic (fcc) lattice of spheres with a metal core requires $f_m \approx 50\%$ to open a CPBG [9, 10, 11, 12]. For a fcc lattice of metal-coated dielectric spheres the required metal filling fraction f_m is slightly lower but still very high ($\approx 40\%$ [12]). Therefore, not surprisingly, when going further to shorter and shorter wavelengths, one is facing an increasing absorption: at $\lambda \approx 600$ nm the absorption exceeds 10% even within a CPBG [12]. Although such a metallo-dielectric fcc structure could provide a CPBG [9, 10, 11, 12] at near infrared, the extension to visible is difficult. We show that a zinc-blende and diamond structures of metallo-dielectric spheres [2, 13] can display much better properties. Photonic band structures are calculated using the photonic Korringa-Kohn-Rostocker (KKR) method [7, 14]. The KKR method can be used for scatterers of arbitrary shape [15] but is optimized for lattices of spheres. In our case convergence of bands was achieved well below 1%.

PURELY DIELECTRIC ZINC BLENDE PHOTONIC STRUCTURES

It turned out to be necessary to recalculate the earlier results of Ho, Chan, and Soukoulis [2] on the photonic band structure of a diamond lattice of nonoverlapping spheres with di-

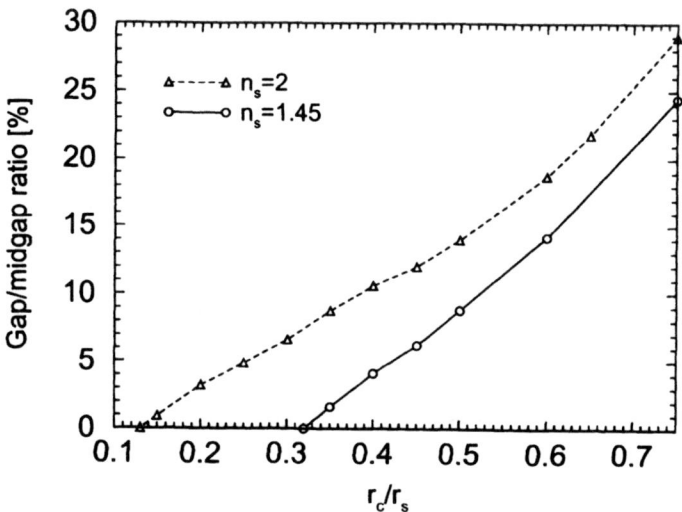

Figure 2: Gap to midgap frequency ratio g_w of the 2nd-3rd CPBG for a close-packed diamond lattice of dielectric $n_s = 1.45$ (silica) and $n_s = 2$ (ZnS) coated silver spheres of radius $r_s = 80$ nm in air. g_w is plotted as a function of the metal core radial filling fraction r_c/r_s. Metal volume fraction is then $f_m = 0.34 \times (r_c/r_s)^3$.

electric constant ε_s in air, which were not converged (see [6] for more details). According to Figure 1, for a sphere filling fraction f_s varying from 0 till the close-packed case $f_{cp} = 0.34$, two CPBGs can occur simultaneously, between the 2nd-3rd bands, and, as in an inverted fcc case, between the 8th-9th bands [6]. Contrary to previous calculations [2], the lower CPBG is not the dominant one (for its optimal $f_s = f_{cp}$ it does not exceed 2.3% (for $\varepsilon_s = 9$), only persists for $\varepsilon_s \in [5.2, 16.3]$, and closes already for $f_s = 32\%$) [6]. The dominant CPBG is the upper one. For $\varepsilon_s = 12.96$, the upper CPBG persists down to $f_s = 4\%$. For $f_s = 17\%$ and $\varepsilon_s = 12.96$ it can reach 12%, however, the threshold value of ε_s for its opening is 7.9, comparable to that for an inverted fcc lattice [7]. Unlike the case of a simple lattice (one scatterer per lattice primitive cell) [7], for the case of a diamond lattice of dielectric spheres, even when using the plane-wave method based MIT ab-initio program [16], one has to take a much higher number of plane waves than expected to reach a convergence comparable with the photonic KKR method. To reach convergence of the photonic band structure of a diamond lattice of dielectric spheres within 1% the number of plane waves N_c has to exceed 32768 (cf. Ref. [2]) and still an extrapolation $N \to \infty$ [17] has to be performed [18]. Relatively smaller differences were found [6] when compared to the results of Simeonov, Bass, and McGurn for zinc blende structures [13].

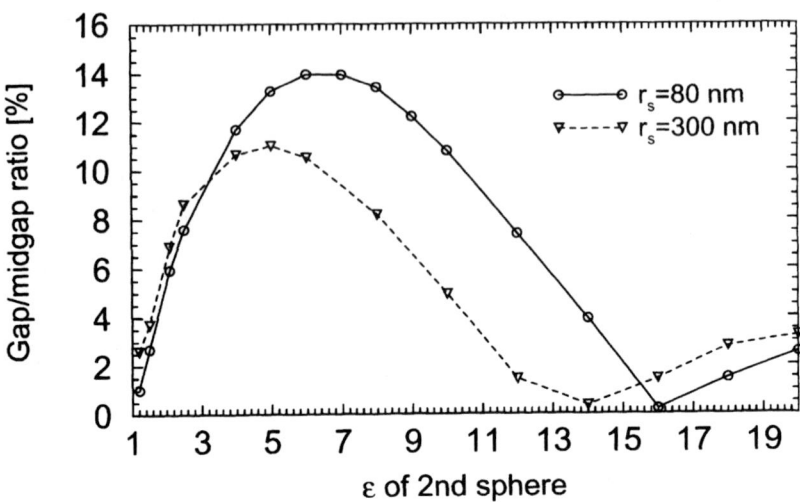

Figure 3: Gap to midgap frequency ratio g_w of the 2nd-3rd CPBG for a close-packed metallo-dielectric zinc blende lattice of spheres in air with identical radii r_s. There are a silver core - silica ($n_s = 1.45$) shell sphere with $r_c/r_s = 0.75$ and a homogeneous dielectric sphere in the unit cell. g_w is plotted as a function of the dielectric constant ε_s of the 2nd sphere for the cases $r_s = 80$ nm and $r_s = 300$ nm.

ZINC BLENDE PHOTONIC STRUCTURES WITH METALLIC INCLUSIONS

On purely experimental grounds, only the case of metal cores is investigated here. Indeed, a metal shell around dielectric core is formed by an aggregation of small metallic nanoparticles. The shell has to be around 20 nm thick before it becomes complete [19]. With emphasis on photonic structures in the visible and near infrared, the 20 nm shell thickness then would mean rather high threshold value of the metal filling fraction f_m (of the order of 5%). On the other hand, it is much easier to tune the metal filling fraction f_m from zero to a few percent by coating small metal nanoparticles with a dielectric in a controlled way [20]. We have considered the close-packed diamond lattice (both spheres in the primitive cell are identical metal core-dielectric shell spheres) and its various close-packed zinc-blende deformations (one sphere in the primitive cell is a metal core-dielectric shell sphere and the second sphere is purely dielectric, both spheres having the same radius).

Photonic band structure calculations revealed two remarkable features of the metallo-dielectric structure. First, a strong increase of the CPBG between the 2nd and 3rd bands with f_m (see Figure 2). For example, for a close-packed diamond lattice of silver spheres

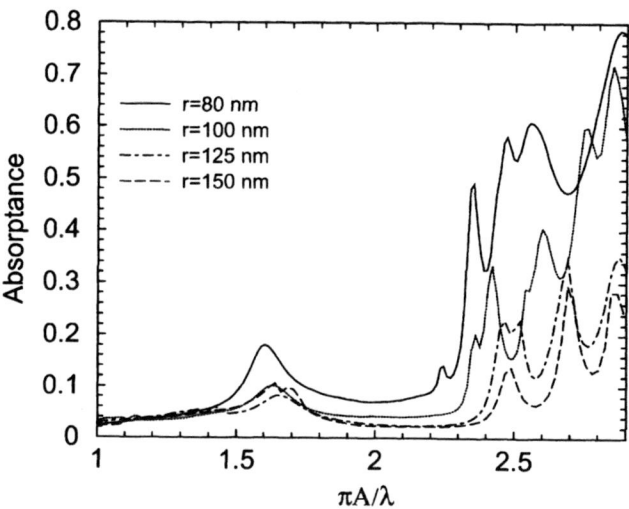

Figure 4: Absorptance of light incident normally on a two unit cells (12 planes) thick zinc blende lattice of spheres in air stacked in the (111) direction. One of the two spheres in the primitive cell is silica coated silver sphere ($n_s = 1.45$) with $r_c/r_s = 0.75$, whereas the other is a ZnS core-silica shell sphere with $r_c/r_s = 0.60$ of the same radius. Dimensionless frequency is used on the x-axis, where A is the unit cell size, to emphasize the scaling-like behavior - in all cases, the 2nd-3rd CPBG lies between ≈ 1.7 and 2.2.

coated with a dielectric with refractive index $n_s = 1.45$ (silica) and radius 80 nm, the CPBG between the 2nd and 3rd bands below 600 nm opens for $f_m \approx 1.1\%$ and reaches 5% already for $f_m \approx 2.5\%$ (see Figure 2). When the sphere refractive index n_s increases, one comes closer and closer to the threshold refractive index contrast of ≈ 2.3, for which the CPBG of the parent diamond structure of non-overlapping dielectric spheres begins to open [6]. Not surprisingly (Figure 2), the respective metal f_ms to open the 2nd-3rd CPBG and to have a CPBG of 5% rapidly decrease with increasing sphere refractive index n_s. However, quite counter-intuitively, the 2nd-3rd CPBG begins to narrow after the dielectric constant increases beyond a certain threshold (see Figures 1, 3).

For the diamond lattice of pure metallic spheres, only the CPBG between the 2nd-3rd bands opens and it can be huge. For silver spheres, depending on the sphere radius r_s, it can stretch from 60% ($r_s = 80$ nm) till 75% ($r_s \geq 300$ nm) [6, 11]. This is consistent with a previous estimate of $g_w \geq 60\%$ for the case of an ideal metal ($\varepsilon_x = -\infty$) [21]. A combination of the two limiting cases, i.e., purely dielectric (Figure 1) and purely metallic spheres, yields an indication of why only the lower CPBG (between the 2nd-3rd bands)

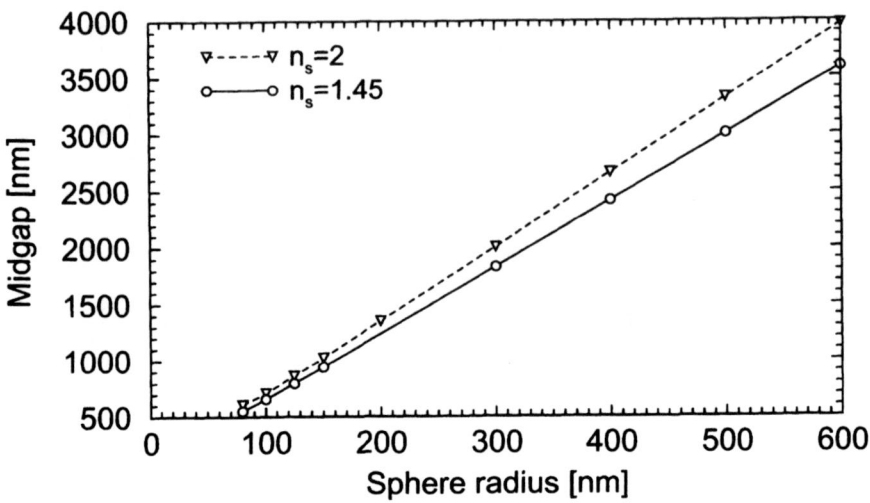

Figure 5: An example of scaling of the 2nd-3rd CPBG midgap wavelength for a close-packed diamond lattice of coated silver spheres in air with the sphere radius. Spheres are either $n_s = 1.45$ (silica) coated silver spheres with fixed $r_c/r_s = 0.6$, or, $n_s = 2$ (ZnS) coated silver spheres with fixed $r_c/r_s = 0.4$.

opens in the intermediate case of metallic inclusions for $n_s \leq 2.3$ and why the CPBG begins to contract after the dielectric constant increases beyond a certain threshold (Figure 3). The same reasoning also indicates why an order of magnitude higher metal volume fraction is required to open a CPBG in the case of a simple fcc structure [9, 10, 11]: in the purely dielectric case, an fcc lattice of spheres in air does not have any CPBG, irrespective of the sphere dielectric constant [7, 17]. Most crucially (see Figure 4), absorption within a CPBG of 5% can be kept below 2.6% for $\lambda \geq 750$ nm. (Absorption was calculated using the layer KKR method [22].) This should be tolerable in most practical applications.

A second remarkable feature of the metallo-dielectric structure is a surprising scaling-like behavior (see Figure 5), which is intrinsic only to ideal dispersionless structures. This scaling-like property is very useful from a practical point of view. It means that once a CPBG is found, with some midgap wavelength λ_c, the CPBG can be centered at any other wavelength by a simple scaling of all the sizes of a structure. Since metals are highly dispersive, the almost precise scaling behavior of the photonic structures is far from obvious, especially for a CPBG below 2 μm. It is true that a typical metal filling fraction f_m which is considered here is $\leq 5\%$, i.e., rather small. Yet, as a counterargument, even with such a small metallic content, the effect on photonic band gaps turns out to be very strong (see Figure 2). For sphere radii $r_s > 300$ nm the CPBG lies above 2 μm where the limit of perfect metal is approached. Here, the scaling becomes more and more precise, since it makes rather little difference if metal $\varepsilon = -200$ or $\varepsilon = -\infty$.

CONCLUSIONS

For the example of a zinc blende structure of dielectric spheres it has been demonstrated that small inclusions of a low absorbing metal with volume fraction f_m can have a dramatic effect on a CPBG between the 2nd-3rd bands. Surprisingly, the inclusions have the biggest effect for $\varepsilon \in [2, 12]$, which is a typical dielectric constant at near-infrared and in the visible for many semiconductors and polymers. For example, in the case of silica spheres, the 2nd-3rd CPBG opens for $f_m \approx 1.1\%$ of silver and exceeds 5% for $f_m \approx 2.5\%$. Absorption in the 2nd-3rd CPBG of 5% remains very small ($\leq 2.6\%$ for $\lambda \geq 750$ nm). The structure enjoys scaling-like behavior, enabling one to scale the 2nd-3rd CPBG from microwaves down to ultraviolet wavelengths. Our results imply that just any dielectric material can be used to fabricate a photonic crystal with a sizeable and robust CPBG in three dimensions. These findings (i) open a door for many other semiconductor and polymer materials to be used as genuine photonic crystal building blocks and (ii) significantly increase the possibilities for experimentalists to realize a CPBG in the visible. Moreover, due to a high sensitivity of a CPBG on f_m, one has the freedom to engineer g_w from zero to more than 60%.

ACKNOWLEDGMENTS

I like to thank my colleagues A. van Blaaderen, A. Imhof, M. Megens, A. Polman, A. Tip, and K. P. Velikov for careful reading of the manuscript and useful comments.

REFERENCES

1. V. P. Bykov, Sov. Phys. JETP **35**, 269 (1972); Sov. J. Quant. Electron. **4**, 861 (1975).
2. K. M. Ho, C. T. Chan, and C. M. Soukoulis, Phys. Rev. Lett. **65**, 3152 (1990).
3. E. Yablonovitch, T. J. Gmitter, and K. M. Leung, Phys. Rev. Lett. **67**, 2295 (1991).
4. E. Yablonovitch, Phys. Rev. Lett. **58**, 2059 (1987).
5. Proceedings of the NATO ASI School "Photonic Crystals and Localization in the 21st Century", ed. by C. M. Soukoulis (Kluwer, Amsterdam, 2001)
6. A. Moroz, unpublished.
7. A. Moroz and C. Sommers, J. Phys.: Condens. Matter **11**, 997 (1999).
8. B. G. Levi, Phys. Today, January 1999, p. 17.
9. A. Moroz, Phys. Rev. Lett. **83**, 5274 (1999).
10. A. Moroz, Europhys. Lett. **50**, 466 (2000).

11. W. Y. Zhang, X. Y. Lei, Z. L. Wang, D. G. Zheng, W. Y. Tam, C. T. Chan, and P. Sheng, Phys. Rev. Lett. **84**, 2853 (2000).
12. Z. Wang, C. T. Chan, W. Zhang, N. Ming, and P. Sheng, Phys. Rev. B **64**, 113108 (2001).
13. S. Simeonov, U. Bass, and A. R. McGurn, Physica B **228**, 245 (1996).
14. A. Moroz, Phys. Rev. B **51**, 2068 (1995).
15. A. R. Williams and J. van W. Morgan, J. Phys. C: Solid State **7**, 37 (1974).
16. S. G. Johnson and J. D. Joannopoulos, Opt. Express **8**, 173 (2001).
17. H. S. Sözüer, J. W. Haus, and R. Inguva, Phys. Rev. B **45**, 13962 (1992).
18. M. Megens, private communication.
19. C. Graf and A. van Blaaderen, Langmuir **18**, 524 (2002).
20. L. M. Liz-Marzán, M. Giersig, and P. Mulvaney, Langmuir **12**, 4329 (1996).
21. S. Fan, P. R. Villeneuve, and J. D. Joannopoulos, Phys. Rev. B **54**, 11245 (1996).
22. V. Yannopapas, N. Stefanou, and A. Modinos, Comp. Phys. Commun. **113**, 49 (1998).

Mat. Res. Soc. Symp. Proc. Vol. 707 © 2002 Materials Research Society   AA7.8/K7.8

Optical Characterization of Cadmium Telluride Doped Heterostructured Opaline Photonic Crystal

V. G. Solovyev, S.G. Romanov, C.M. Sotomayor Torres,
N. Gaponik*, A. Eychmüller* & A. L. Rogach*
Institute of Materials Science and Dept. of Electrical and Information Engineering, University of Wuppertal, Gauss-str. 20, 42097 Wuppertal, Germany
*Institute of Physical Chemistry, University of Hamburg, Bundesstr. 45, 20146, Hamburg, Germany

ABSTRACT

A double-layer photonic crystal (PhC) possessing incomplete photonic band gap (PBG) has been prepared by successive formation of one opaline film on top of another and subsequent impregnation with CdTe colloidal nanocrystals as the light source. Transmission, reflection and photoluminescence (PL) spectra of this nanocomposite have been measured from 1.8 to 2.5 eV at different angles of the light incidence and detection. The suppression of emission intensity has been found at both stop bands of the PhC heterostructure. Depending on the excitation power, either suppression or enhancement of the emission rate at the stop band has been observed.

INTRODUCTION

The development of PBG materials towards functional devices for waveguiding and light emission requires their intentional internal structuring. In the case of opaline materials the complexity of this task is related to the necessity of combining the macroscale structuring with the self-organized growth of the PhC. The feasibility of this approach was first demonstrated with a PhC based on bulk opal [1] and then confirmed with thin film opals [2]. In spite of the still unidentified role of the heterointerface in optical properties of the PhC, the overall optical transmission/reflectance of multilayered opals can be represented as the linear superposition of stop bands of individual layers. With respect to light emission, the suppression of the emission from opaline PhC is known as the less robust characteristic of the PhC [3], which is very sensitive to the peculiarities of the PBG. Moreover, the most desirable feature of the PBG material, namely, the suppression/enhancement of the emission rate has not yet been unambiguously demonstrated in the case of opals. It seems reasonable to expect that co-existence of different PBG structures in a single piece of material will enrich the variety of emission phenomena.

PhC heterostructures were designed to incorporate an artificial defect in the body of the opal and this double-layered opaline film can be considered as a pre-requisite to realize such structures. So far, light diffraction in PhC heterostructures has been studied. In this work we examine the effect of multiple stop bands upon the emission properties of the light source embedded in heterostructure. Correspondingly, we compare the propagation of an external light inside the crystals with the emission generated internally.

MATERIAL PREPARATION AND EXPERIMENTAL TECHNIQIES

The samples under study were double-layer opaline films made of latex beads with sphere diameters ~ 240nm and ~270 nm for the "top" and "bottom" layers, respectively.

Polystyrene (PS) beads (Duke Scientific Inc.) of 270 nm diameter were deposited on glass slides by slow sedimentation from 1% aqueous colloidal solution. Glass slides (1×1 cm^2) were washed under sonication in 1% Extran solution during 30 min. A moderate air blowing was used as a vibration source in order to assist the self-assembly of beads during the evaporation of the solvent and to accelerate the drying process. After drying ~50 μm thick films were obtained and then sintered for 2 hours at 100°C. CdTe nanocrystals (~3 nm size) were prepared as aqueous colloidal solutions following the method described elsewhere [4] in the presence of 2-mercaptoethylamine as size-regulating and stabilising agent. Infiltration of CdTe nanocrystals in opaline films was performed by dipping the latter in 0.02 M CdTe colloidal solution for 1 min and subsequent drying at 30°C in air. Successively, the second film of 240 nm beads (~20nm thick) was deposited on top of the bottom layer.

Reflectance and transmission spectra of opaline heterostructures were collected by illuminating a 0.03cm^2 sample area with the white light from a tungsten halogen lamp and the subsequent detecting of the diffracted light with the spectral resolution of 2 nm and the angular resolution of, approximately, 2° in the specular configuration, where the angle of incidence θ is equal to the angle of collection. To trace the angular dependence, the angle of incidence was varied from 0° to 70°.

PL spectra were excited by the cw 457.9 nm line of an Ar-laser focused on a 0.1 mm diameter spot, the excitation power has been varied between 0.15 and 15 mW. PL spectra were collected within a 5° fraction of the solid angle from the film face opposite to that exposed to laser beam, thus allowing the emission to propagate through the structure. Emission spectra were measured with the spectral resolution of 1.5 nm at several angles θ between the [111] axis of the face centered cubic (FCC) lattice of opal and the detection direction in order to trace the dependence of emission characteristics upon the energy dispersion of the stop band. The explored excitation power range covers four orders of magnitude from 10^{-5} to 10^{-1} W. To observe the changes in the spectrum of the source embedded in the opaline photonic crystal the emission was collected within a small fraction of the solid angle (about 2°) along the direction of the stop-band. In order to keep the consistency of the discussion, all figures relate to the spectra collected at 20°, but the studied effects appears at other angles as well.

RESULTS AND DISCUSSION

The reflectance spectra from the "front" and "back" sides of the heteroPhC display pronounced Bragg resonances (figure 1). For small values of the angle of incidence, θ, (θ<45°) the positions of the first order diffraction resonance wavelength λ can be described by the Bragg and Snell laws $\lambda^2 = 4a^2(n^2 - \sin^2 \theta)$, where $a = 0.816D$ is the interplane distance for (111) planes of the FCC lattice, D is the bead diameter and n represents the effective refraction index (RI) of the corresponding layer. This dependence is shown in figure 2.

A small deviation of the Bragg estimate of the bead diameters from the initial values ($D_2 \approx$ 240 nm, $D_1 \approx$ 270 nm) is probably due to the shrinkage of the lattice parameters during the sintering process. The gentler slope of the "back" stop band dispersion (figure 2) corresponds to both the larger index of refraction (RI) due to the heavier doping with CdTe in accord with the preparation procedure and the larger bead diameter.

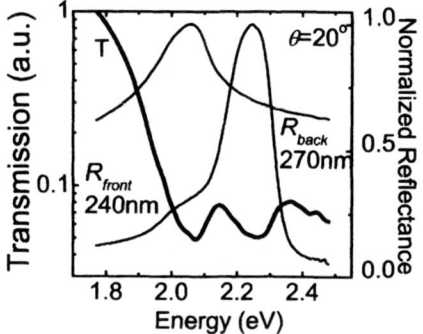

Figure 1. "Front" and "back" reflectance spectra of the heteroPhC composed from two opaline films at the $\theta = 20°$. The diameter of beads is 270 nm for the back film and 240 nm for the front film. Transmission and reflectance spectra of the sample at the $\theta = 20°$.

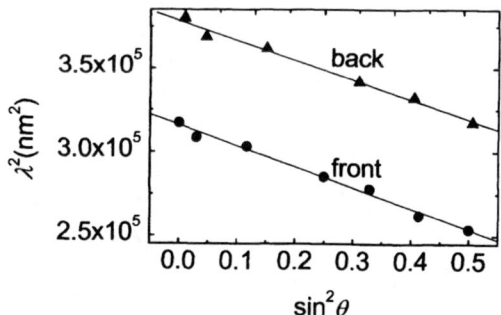

Figure 2. Experimental angular dispersions $\lambda(\theta)$ of the Bragg resonances and their linear approximations according to the Bragg and Snell laws.

Both Bragg resonances can be observed in the transmission spectra of the heteroPhC. Figure 1 shows that both transmission minima correlate well with maxima in "front" and "back" reflectance. The less pronounced reflectance peak as well as the weaker transmission dip of the "back" PhC is, likely, due to the larger disorder, which results from its heavier doping as compared with the "front" PhC. Taking into account the growth conditions, it is reasonable to conclude that the ordering of the "front" opal lattice is not destroyed by the incommensurable landscape provided by the underlying "back" opal.

To obtain the transmission function from the PL spectra, the PL spectrum of the PhC was divided by that of a reference sample, which does not show any gap in the emission spectrum,

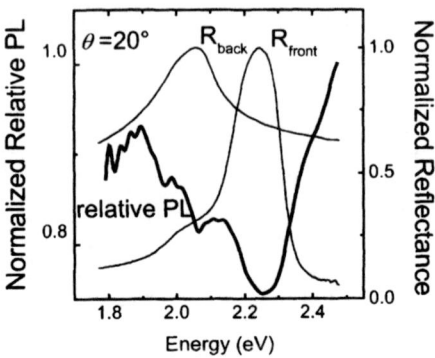

Figure 3. Normalized relative PL spectrum and reflection spectra of the sample at $\theta = 20°$.

and the spectrum obtained in this way is referred as the relative PL. The relative PL spectrum (figure 3) clearly demonstrates the emission intensity reduction at stop band frequencies. The angular dependence of these dips correlate well with the dispersion of corresponding Bragg resonances. The explicit difference between the "front" and "back" PL spectra can be seen in figure 4. To obtain this plot, the PL spectrum from the "front" side of the sample (excited through the glass substrate) was divided by that from the "back" side of the heterostructure (excited by illumination of the "top" layer). Each curve of "front" to "back" divided PL spectra distinctly displays the angular-dependent minimum and maximum. Their positions and half-widths agree well with corresponding values obtained from the reflectance spectrum.

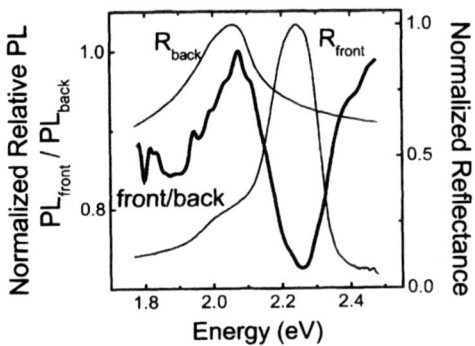

Figure 4. Relative intensity of "front" to "back" luminescence and reflectance spectra of the heteroPhC at $\theta = 20°$.

In our opinion, the deep minimum is due to the stop band-suppressed emission propagating along the normal to the "top" layer, whereas the maximum is due to the suppression of the emission intensity outgoing along the normal to the "back" layer. The angular-dependence of this maximum-minimum set of the "front/back" relative PL spectra correlates well with similar dispersions of the "front" and "back" reflectance peaks.

The ratio PL spectra of the heteroPhC were constructed by dividing PL spectra, which are excited at two different powers. Figure 5 (left) shows the ratio spectrum obtained at the weak excitation regime. It contains two minima centered at the Bragg resonance frequencies for the corresponding angle θ. With the increase of the excitation power the ratio maxima replace the ratio minima. Qualitatively, this evolution can be understood in terms of the interplay between

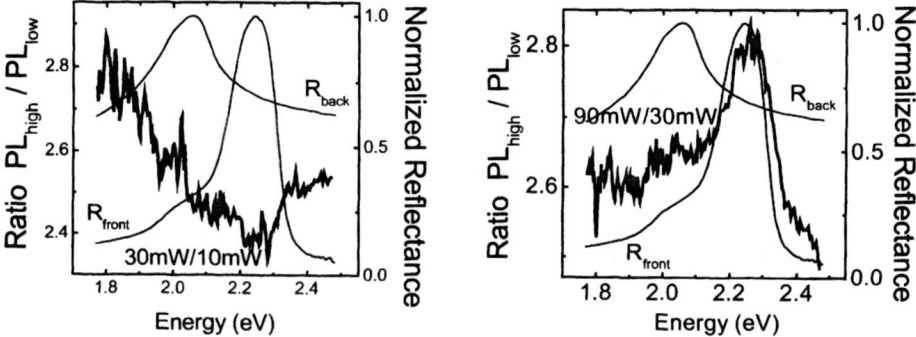

Figure 5. The ratio of PL spectra at $\theta = 20°$ for different laser pump powers, left panel – weak pumping and right panel – heavy pumping regimes. Ratios at the curves indicate the excitation powers, for which the low and the high pumping PL spectra were collected. Reflectance spectra are shown for comparison.

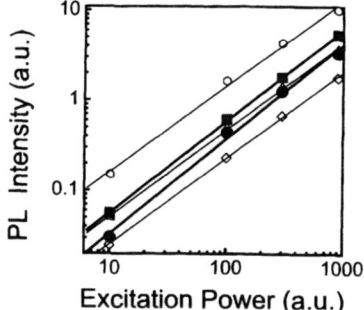

Figure 6. PL intensity vs excitation power for the $\theta = 20°$ at energies outside the stop bands - 1.90 eV (empty rhombs), 2.10 eV (empty triangles) and 2.47 eV (hollow hexagons) and inside the stop bands – at 2.07 eV (solid squares) and 2.25 eV (solid circles). Exponents m in the expression of the PL intensity upon the power $I \sim P^m$ are 0.95 for energies outside the stop band and 1.05 for the "back" PhC stop band and 1.01 for the "front" PhC stop band.

the suppression of the emission, which occurs due to the reduction in the number of optical modes available for coupling at the stop band, and the enhancement of the emission, which takes place in accord to the Purcell effect due to the modification of optical modes in the stop band as compared with modes of frequencies outside the stop band. Analysis of this observation, which is given elsewhere [5], shows that the emission rate in the stop band exceeds that outside by about 10%.

Thus, the experimental results of figure 5 provide an indirect evidence of the amplification of a spontaneous emission within the Bragg stop band in opaline PhCs. Another observation of the emission enhancement is shown in figure 6, where the PL intensity in the middle of the stop bands (for E=2.07 and 2.25 eV at θ =20°) and outside stop bands is plotted against the laser pump power. Power law approximation of these curves gives the exponent values, which are higher in the stop bands by the same 10%. The latter number agrees with the enhancement trend obtained from ratio spectra.

CONCLUSIONS

Double-layer opaline PhCs infilled with CdTe nanoparticles have been prepared and their optical properties have been examined. The intensity of the light, which is reflected, transmitted or emitted by this heterostructure, depends strongly on the direction of light propagation according to the anisotropy of PBG structure of this PhC. The bands of amplified spontaneous emission have been observed, moreover, the amplification occurs at the stop band frequencies. Two spectral ranges of emission enhancement were observed, following the designed structure of the PhC.

ACKNOWLEDGEMENTS

The authors acknowledge the support from the EU project IST 1999-19009 PHOBOS and the DFG Schwerpunktprogramm "Photonic Crystals".

REFERENCES

1. S.G.Romanov, H.M.Yates, M.E.Pemble, R.M. De La Rue, *J. Phys.: Condens. Matter* **12**, 8221 (2000).
2. P.Jiang, G.N.Ostojic, R.Narat, D.M.Mittelman, V.L.Colvin, *Adv. Materials* **13**, 389 (2001).
3. S. G. Romanov, T. Maka, C.M. Sotomayor Torres, M. Müller R. Zentel, in *"Photonic Crystals and Light Localisation in the 21st Century"*, NATO ASI *series* ed. C.M. Soukoulius, NATO Science Series C, v. 563, Kluwer, Dordrecht, 2001, pp.253-262.
4. A. L. Rogach, L. Katsikas, A. Kornowski, D. Su, A. Eychmüller, H. Weller, Synthesis and characterization of thiol-stabilized CdTe nanocrystals, *Ber. Bunsenges. Phys. Chem.,* **100**, 1772 (1996).
5. S. G. Romanov, C. M. Sotomayor Torres, N. Gaponik, A. Eychmüller, A. L. Rogach, submitted to *IEEE Quant. Electronics*.

Modeling/Simulation

Mat. Res. Soc. Symp. Proc. Vol. 707 © 2002 Materials Research Society   AA3.3.1/N2.3

Self organized array of quantum nanostructures via a strain induced morphological instability

David Montiel[1], Judith Müller[2] and Eugenia Corvera Poiré[1].
[1] Departamento de Física y Química Teórica, Facultad de Química, UNAM. Ciudad Universitaria. México, D.F. 04510, MEXICO
[2] Instituut-Lorentz, Universiteit Leiden. Postbus 9506, 2300 RA Leiden, THE NETHERLANDS

Abstract

Motivated by the work of Li et al. [1], we have studied the strain induced morphological instability at the submonolayer coverage stage of heteroepitaxial growth on a vicinal substrate with regularly spaced steps. We have performed a linear stability analysis and determined for which conditions of coverage a flat front is unstable and for which conditions it is stable. For low coverages the instability will cause the front to break in an array of islands. Assuming that the fastest growing mode of the instability determines the properties of the array, we make an estimation of the islands sizes and aspect ratios as well as an estimation of the separation length between islands of the array formed when the dominant mechanism for transport of matter is diffusion of particles along the growing front. These estimations are given as functions of the terrace width and coverage. Since these ones are experimentally controllable parameters, our results could be used to tailor the spontaneous formation of quantum nanostructures.

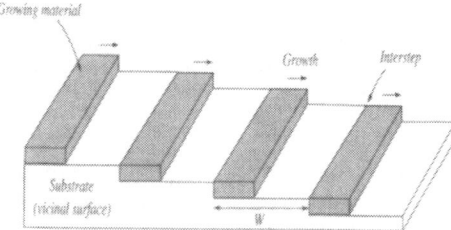

FIG. 1: Step flow on a vicinal surface.

In a recent letter [2] we have considered the stability of a flat front during heteroepitaxial growth on a vicinal surface. Since misfit strain is present, the deposited material causes a force distribution on the substrate. By using theory of continuum media and by assuming a relaxational mechanism consisting on diffusion of particles along the growing fronts we found a dispersion relation of the form

$$\omega = \frac{2E_s(1 - \sigma)D\pi}{W} \cot(\pi\theta)k^2 - D'\gamma k^4 \tag{1}$$

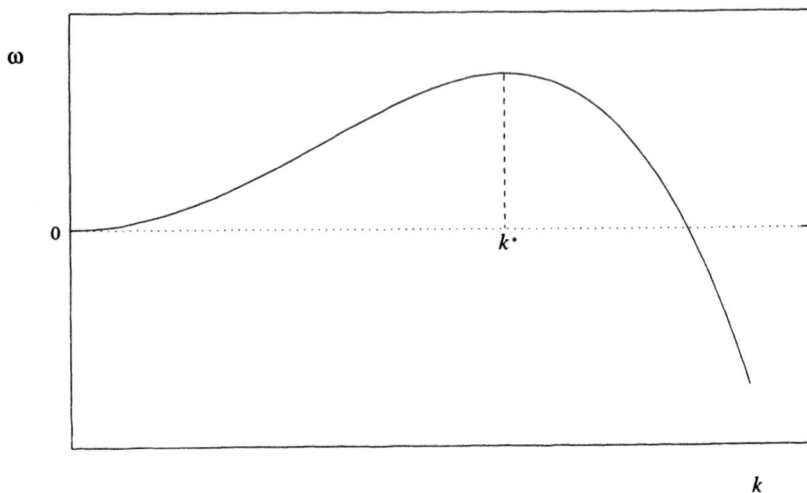

FIG. 2: Dispersion relation for coverages $\theta < 0.5$.

Here D and D' are proportional to the diffusion constant, $E_s = \frac{1+\sigma}{2\pi E} F_0^2$ is a unit strain energy introduced in reference [3]. E and σ are the Young's modulus and Poisson's ratio of the substrate respectively. F_0 is the magnitude of the force per unit length. The terrace width W is shown in figure 1 and the coverage θ, is the fraction of the terrace width W covered by the growing film. This dispersion relation gives the growth rate of a perturbation of the form $y(x,t) = y_0 + \delta e^{\omega t} cos(kx)$, where y_0 is the position of a flat unperturbed growing front and $\delta e^{\omega t}$ is the amplitude of the Fourier mode with wavenumber k. t stands for time and x and y for spatial coordinates. It is worth to emphasize that a general perturbation can be
expressed as a superposition of Fourier modes and that for a linear stability analysis all modes decouple in such a way that one can treat each of them individually.

Our results indicate that a straight stripe is unstable for coverages $\theta < 0.5$ and stable for coverages $\theta > 0.5$. Also, in the absence of step energy, there is no wavelength selection, that is, for the case $\theta < 0.5$, there is not a finite mode that grows faster than the other ones. In order to have wavelength selection, the effect of the step energy (or line tension) γ must be taken into account. In general the line tension γ is anisotropic. Here we have taken an isotropic γ to account for the overall effect of line tension. The effect of step energy is stabilizing. For the instability region ($\theta < 0.5$) a wavelength is selected [see figure 2].

The wavenumber that dominates the pattern at short times is given by

$$k^* = \left(\frac{(1-\sigma)E_s D\pi \cot(\pi\theta)}{W\gamma D'} \right)^{1/2} \tag{2}$$

At short times the amplitude of the perturbation grows exponentially fast, and therefore, the growing front will touch the interstep (defined in figure 1) causing the breaking of the stripe into

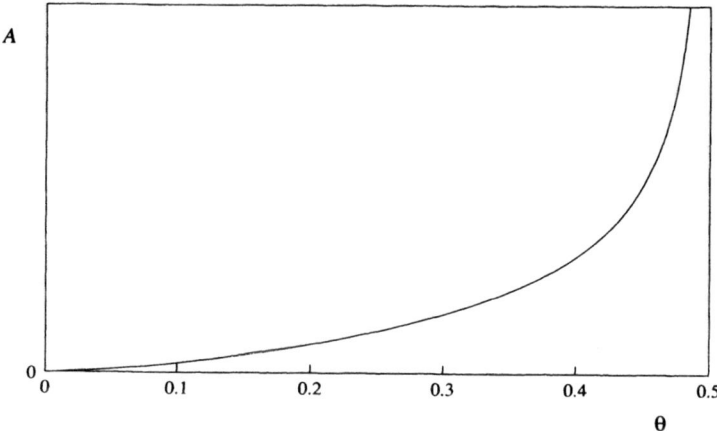

FIG. 3: Island size as a function of coverage.

an array of islands. We assume that the fastest growing mode of the instability determines the properties of the array.

We obtain the islands size, spacing, and aspect ratio as determined by the wavenumber k^*.

The island size (area) is given by

$$A = 2\theta W^{3/2} \left(\frac{\gamma D' \pi}{(1 - \sigma) E_s D \cot(\pi\theta)} \right)^{1/2} \tag{3}$$

Figure 3 shows how the area of the islands increases with coverage. For the figure we have chosen arbitrary units. For given material parameters and terrace width, the area of the islands is a monotonically increasing function of coverage.

The separation length between islands is given by

$$d = 2W^{1/2} \left(\frac{\gamma D' \pi}{(1 - \sigma) E_s D \cot(\pi\theta)} \right)^{1/2} \tag{4}$$

Figure 4 shows how the distance between islands is also a monotonically increasing function of coverage. For the figure we have chosen arbitrary units.

The aspect ratio $c \equiv \tan \chi$ where χ is shown in figure 5, is given by

$$c = \frac{1}{2}\theta W^{1/2} \left(\frac{\gamma D' \pi}{(1 - \sigma) E_s D \cot(\pi\theta)} \right)^{-1/2} \tag{5}$$

Figure 6 shows the angle χ as a function of coverage. According to which, for coverages close to zero and coverages very close to one half of the terrace width, the islands would be elongated in the direction parallel to the steps, while for intermediate coverages the islands will be elongated in the direction perpendicular to the steps.

345

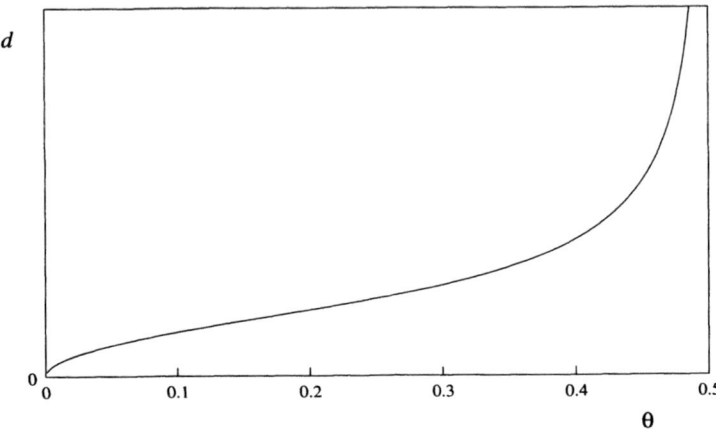

FIG. 4: Distance between islands as predicted by the fastest growing mode of the stability analysis.

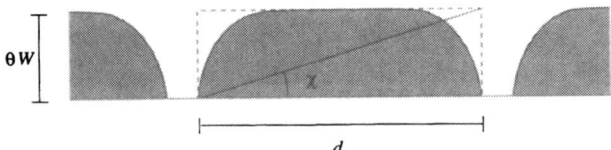

FIG. 5: The aspect ratio is given by $c \equiv \tan \chi$.

Expressions (3), (4) and (5) provide a way to control the domains sizes and spacing. Experimentally it is possible to control the terrace width through the cutting angle of the vicinal surface in which the material grows. It is also possible to control the coverage. Our results predict that for small coverages the separation between islands goes like $W^{1/2}$, the area of the islands as $W^{3/2}$ and the aspect ratio like $W^{1/2}$.

Even though, for our work, the island-array properties where computed for very low coverages, our predictions for the island size and aspect ratio as a function of coverage are surprisingly similar to predictions of reference [1] in the whole range of coverages. For the separation length though, our predictions are different from the ones predicted in reference [1], but a detail comparison of the two approaches is not our present goal.

We expect our predictions to be valid when the experimental conditions are such that the external flux is slow and diffusion on terraces is very fast compared to other time scales in the system.

Our results can be used to induce the spontaneous formation of an array of quantum structures with the desired size and spacing by controlling the cutting angle of the vicinal surface and the fraction of the surface covered by the growing film.

This work was supported by Conacyt under grant 33920-E and fellowship 145167 and by the TMR network *Pattern, Noise and Chaos* ERBFMRXCT960085.

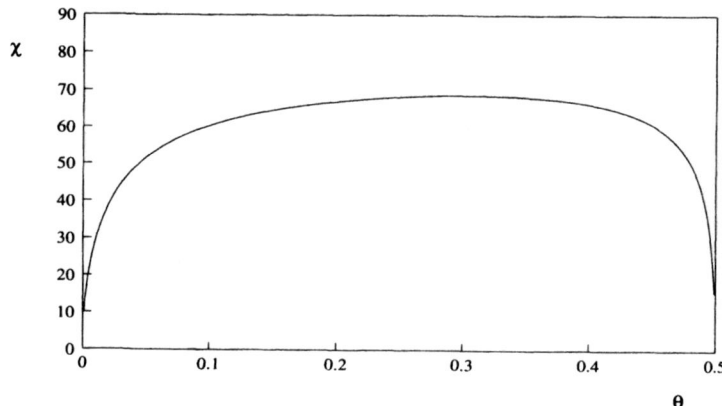

FIG. 6: χ in degrees defined by the aspect ratio as a function of coverage. For the figure arbitrary units have been chosen for c.

[1] Adam Li, Feng Liu, D. Y. Petrovykh, J.-L. Lin, J. Viernow, F. J. Himpsel, and M. G. Lagally , Phys. Rev. Lett. 85, 5380 (2000).

[2] David Montiel, Judith Müller and Eugenia Corvera Poiré. J. Phys.: Condens. Matter 14 L49-L55 (2002).

[3] A. Liu, F. Liu, and M.G. Lagally. Phys. Rev. Lett. 85, 1922 (2000).

AUTHOR INDEX

SUBJECT INDEX

CPSIA information can be obtained at www.ICGtesting.com
Printed in the USA
W06s1016220514

LV00011B/434/P

9 781107 411982